D1085503

5 81
Stl

Plant-Microbe Interactions

Volume 4

Edited by

Gary Stacey
University of Tennessee, Knoxville

Noel T. Keen
University of California, Riverside

APS PRESS
The American Phytopathological Society
St. Paul, Minnesota

98-544

This book was formatted from computer files submitted
to APS Press by the editors of the volume. No editing
or proofreading has been done by the Press.

Reference in this publication to a trademark, pro-
prietary product, or company name by personnel of
the U.S. Department of Agriculture or anyone else is
intended for explicit description only and does not imply
approval or recommendation to the exclusion of others
that may be suitable.

Library of Congress Catalog Card Number: 95-10088
International Standard Book Number: 0-89054-228-7

Printed in the United States of America on acid-free paper

The American Phytopathological Society
3340 Pilot Knob Road
St. Paul, Minnesota 55121-2097 USA

Series Editors

Gary Stacey
Departments of Microbiology and
  Ecology and Evolutionary Biology
Center for Legume Research
University of Tennessee, Knoxville

Noel T. Keen
Department of Plant Pathology
University of California, Riverside

Advisory Board

Fred M. Ausubel
Massachusetts General
  Hospital
Boston, MA

George Bruening
University of California,
  Davis

Jeff Dangl
University of North Carolina,
  Chapel Hill

Michael Daniels
Sainsbury Laboratory
Norwich, United Kingdom

Pierre de Wit
Wageningen Agricultural
  University
Wageningen, The Netherlands

Klaus Hahlbrock
Max Planck Institut für
  Züchtungforschung
Köln, Germany

Luis Herrera-Estrella
CINVESTAV, IPN
U-Irapuato, Mexico

Ben Lugtenberg
Institute of Molecular Plant
  Sciences
Leiden University
Leiden, The Netherlands

Eugene Nester
University of Washington, Seattle

John Ryals
Paradigm Genetics
Research Triangle Park, NC

Preface to the Plant-Microbe
Interactions Series

If one were to plot the number of research publications per year dealing with various aspects of plant-microbe interactions, it would be clear that information in this area is increasing exponentially. This work is of obvious importance, since plant-microbe interactions, in the form of pathogenicity, beneficial symbioses, biocontrol, etc., greatly impact agriculture. The recent rapid increase in knowledge can be largely correlated with the application of modern molecular methods to the understanding of plant-microbe interactions. Indeed, researchers interested in how plants and microbes interact were among the first to apply such molecular methods to biology.

For example, one of the great scientific discoveries of the twentieth century was the elucidation of interkingdom gene transfer through analysis of crown gall disease. In the years to come, we will see the full fruition of this work through its practical impact on agriculture. Likewise, the discovery of the lipo-chitin nodulation signals produced by rhizobia and their effect on the legume host serves as a model for host-symbiont interactions. These initial studies on *Agrobacterium* and *Rhizobium* led to the development of technologies that are widely applied not only to studies of plant-microbe interactions but in all areas of biology. Considering that plants and microbes have been interacting for eons on this planet, can we have any doubt that similar amazing discoveries await the prepared investigator of these interactions?

Indeed, the recent discovery of the first plant disease resistance genes heralds the beginning of a new age of discovery. It seems that many avenues are open to uncover the signal pathways involved in plant resistance, the nature of bacterial and fungal virulence mechanisms, and the interplay of regulatory signals between plant and pathogen. We are in the developing era of genomics with several important plants/pathogens/symbionts now under active investigation. The knowledge of the DNA sequence of these organisms will clearly lead to new routes to investigate and manipulate for human benefit important plant-microbe interactions.

These investigations will add to the long list of contributions that the study of plant-microbe interactions has made to plant biology and biology in general. Indeed, the long-debated role of auxin/cytokinin ratios in controlling plant morphogenesis received its strongest support from the study of the function of T-DNA genes of *Agrobacterium*. Likewise, recent studies have provided the first solid evidence that secondary plant metabolites are important defense agents, an idea that has been forcibly argued for many years. The ability of microorganisms to perturb the normal growth and development of plants has long been used as a method to study plant processes. This has led to the discovery of important plant growth regulators (e.g., gibberellins).

The goal of this book series is to chronicle the future research on plant-microbe interactions. Moreover, this series will hopefully prepare the new generations of scientists that will make these future breakthroughs. Without a doubt, remarkable discoveries will be made. It is safe to predict that new plant-growth regulators will be discovered. It will be a pleasure to see plant cell biology reach and then surpass the level of understanding that now exists for animal cells. The ability to easily manipulate the genetics of the plant and microbial genome will give investigators advantages not seen in many animal systems. Besides the excitement to be expected from such basic studies, practical applications of this work are now appearing and the pace of such advances will accelerate. It is the goal of this series to contribute to the advancement of the science of plant-microbe interactions.

Gary Stacey
Noel T. Keen

Preface to Plant-Microbe Interactions, Volume 4

This volume explores examples of microorganisms and their sometimes unexpected effects on plants.

Chapter 1 by Moez Hanin et al. and Chapter 6 by Peter Gresshoff explore these effects in *Rhizobium*-plant relationships and graphically demonstrate their reciprocal nature. Thus, Rhizobia and plants carry on their own ping pong game—signal perception and response by one partner followed by signal recognition and response by the other. It is likely that similar complex dialogues will eventually be recognized in many other plant-microbe interactions, including pathogens.

One such system, the *Xanthomonas avr/pth* gene family is described in Chapter 2 by Dean Gabriel. These bacterial proteins appear to behave as virulence factors but are most studied because they are the products of avirulence genes and as such recognized by particular plants to elicit active defense. It will be most interesting to elucidate the recognitional mechanisms whereby the *avr/pth* proteins cause the quite diametric effects of promoting virulence on one hand and active defense on the other.

Dennis Fulbright discusses the classic chestnut blight disease in Chapter 3 and recent work that is defining the basis of the use of hypovirulent fungal strains as biocontrol agents.

Chapter 4 by Jacyn Baker and Elizabeth Orlandi investigates the role of active oxygen, particularly in initiation of active defense. Similar to the case in vertebrates, the various active oxygen species generated during the hypersensitive response in plants are closely related to subsequent signaling mechanisms that lead to actual resistance.

In Chapter 5, Tomonori Shiraishi et al. discuss examples whereby pathogens produce suppressor substances that block active plant defense responses and accordingly are required for full pathogen virulence. Several recent genetic experiments have indicated that the extent of suppressor production by plant pathogens is probably much greater than we now appreciate.

In Chapter 7, Peter Weisbeek and Han Gerrits investigate the much fought over element iron and its essential role in the strategy of many biological disease-control agents. The great competition for iron in biological systems makes one wonder whether other trace elements (copper, zinc, cobalt, boron?) might also be argued over in certain environments and accordingly might be rationales for biological disease-control schemes.

Finally, in Chapter 8, Paul Rushton and Imre Somssich describe our current understanding of how pathogen signals (e.g., elicitors) activate plant gene expression. Such gene activation is of clear importance to the development of plant resistance, and recent research as outlined in this chapter is bringing new insight to this area.

The editors gratefully acknowledge the assistance of Rosemary Loria in the preparation of this volume.

Gary Stacey
Noel T. Keen

Contributors

C. Jacyn Baker
USDA-ARS
Beltsville, MD

William J. Broughton
L.B.M.P.S.
Universite Geneve
Geneva, Switzerland

Rémy Fellay
L.B.M.P.S.
Universite Geneve
Geneva, Switzerland

Dennis W. Fulbright
Department of Botany
  and Plant Pathology
Michigan State University,
  East Lansing

Dean W. Gabriel
Department of Plant
  Pathology
University of Florida,
  Gainesville

Han Gerrits
Rijksuniversiteit Utrecht
Utrecht, The Netherlands

Peter M. Gresshoff
Plant Molecular Genetics
University of Tennessee,
  Knoxville

Moez Hanin
L.B.M.P.S.
Universite Geneve
Geneva, Switzerland

Yuki Ichinose
Faculty of Agriculture
Okayama University
Okayama City, Japan

Saïd Jabbouri
L.B.M.P.S.
Universite Geneve
Geneva, Switzerland

Toshiaki Kato
Faculty of Agriculture
Okayama University
Okayama City, Japan

Akinori Kiba
Faculty of Agriculture
Okayama University
Okayama City, Japan

Yasuhiro Murakami
Faculty of Agriculture
Okayama University
Okayama City, Japan

Elizabeth W. Orlandi
Department of Microbiology
University of Maryland,
  College Park

Dolores Quesada-Vincens
L.B.M.P.S.
Universite Geneve
Geneva, Switzerland

Paul J. Rushton
Department of Biochemistry
Max-Planck-Institut für
  Züchtungforschung
Köln, Germany

Hikaru Seki
Faculty of Agriculture
Okayama University
Okayama City, Japan

Tomonori Shiraishi
Faculty of Agriculture
Okayama University
Okayama City, Japan

Imre E. Somssich
Department of Biochemistry
Max-Planck-Institut für
  Züchtungforschung
Köln, Germany

Kazuhiro Toyoda
Faculty of Agriculture
Okayama University
Okayama City, Japan

Peter J. Weisbeek
Rijksuniversiteit Utrecht
Utrecht, The Netherlands

Tetsuji Yamada
Faculty of Agriculture
Okayama University
Okayama City, Japan

Contents

CHAPTER 1

Molecular Aspects of Host-Specific Nodulation

Moez Hanin, Saïd Jabbouri, William J. Broughton, Rémy Fellay, and Dolores Quesada-Vincens

Under conditions of nitrogen limitation, soil bacteria of the genera *Azorhizobium, Bradyrhizobium,* and *Rhizobium* (collectively referred to as rhizobia) may induce formation of highly specialized organs on the roots or stems of their leguminous hosts. Within these structures, called nodules, rhizobia convert to an endosymbiotic form, the bacteroids, in which dinitrogen is reduced to ammonia. Specificity in symbiotic associations varies greatly amongst the symbionts. *Azorhizobium caulinodans* nodulates only *Sesbania rostrata, Rhizobium meliloti* can initiate nodule formation on few host plants (*Medicago, Melilotus* and *Trigonella*), whereas *Rhizobium* sp. NGR234, nodulates more than 110 genera of legumes, as well as the non-legume *Parasponia adersonii* (S. G. Pueppke and W. J. Broughton, *unpublished*).

Symbiotic interactions are controlled by signal exchange between the two partners. Plants secrete flavonoids, phenolic compounds that, in conjunction with the bacterial activator protein NodD, induce the expression of rhizobial nodulation (*nod, nol* and *noe*) genes (collectively *nod*-genes). As a result, rhizobia produce lipo-chito-oligosaccharidic (LCO's) signal molecules known as Nod-factors. Nod-factors induce various plant responses, including root-hair deformation, cortical cell-division, pseudo-nodule and nodule formation (1–4). Nodulation occurs via a cascade of developmental steps, which begin with bacterial colonization of the rhizosphere and attachment to root-hairs. Rhizobia are then entrapped in the folds of curled root hairs, which they eventually penetrate forming infection threads. Concomitantly, certain cortical cells divide to form nodule primordia, and it is toward these primordia that the infection thread grows. Further development gives rise to nodules, which differ from tumors in having defined anatomical structures. Bacteria multiply within the infection threads, which grow centripetally towards the root. Rhizobia begin to enlarge within the infection threads before being released into the cortical cells where they differentiate into bacteroids and begin to fix nitrogen (for reviews see 5, 6).

In this chapter we will deal only with the early events of nodulation, particularly those involving production of Nod-factors and their role in host-specificity with special emphasis on *Rhizobium* sp. NGR234.

Rhizobial Nodulation Genes
Are Host-Range Determinants

Nodulation and nitrogen fixation genes (*nif*- and *fix*-genes) are clustered on large transmissible symbiotic plasmids (pSym) in many rhizobia (7). In *Azorhizobium, Bradyrhizobium,* and *Rhizobium loti,* symbiotic genes are located on the chromosome, however (8–11). Mutation of nod-genes may result in the inability to form nodules (Nod⁻), delay nodulation (Nod^d, Fix⁺) or cause a change in host range.

By selectively reacting with various flavonoids, regulatory nod-genes are major determinants of host range. In contrast, there are two kinds of structural nod-genes: those that are found in all rhizobia and others that are limited to certain species (12). Enzymes encoded by the common *nodABC* genes direct the synthesis of the lipo-oligosaccharide backbone of the of Nod-factors, while host-specific *nod*-genes (*hsn*) modify the basic Nod-factor structure (Table 1-1), so conferring upon Nod-factors their host specificity (13).

"Common" Nod-Genes

The *nodABC* genes have been found in all *Azo(Brady)Rhizobium* strains studied. (12,14). Usually, they form a single operon, but in *R. etli, nodA* is separated from the *nodBC* genes by approximately 20 kb (15). Mutations in *nodA*, *nodB* or *nodC*, abolish all symbiotic reactions with the host plant, including root-hair curling (Hac⁻), and nodule formation (Nod⁻). Active *nodABC*, are necessary and sufficient to cause expression of early nodulins (16). *nodIJ* are present in most *Rhizobium* species (17-19). In some strains such as in *R. meliloti, R. leguminosarum,* and NGR234, *nodIJ* are downstream of *nodC* and their inactivation results in a delay of nodulation (18,20). Their role in LCO secretion will be discussed later.

Host-Specificity Genes

In addition to the common *nodABC* genes, numerous host specificity (*hsn*) genes have been identified. *hsn* genes are not present in all rhizobia (12), and are defined by one more of the following criteria:

- Mutations in *hsn* genes cannot or are only partially complementable by similar loci from other rhizobia (20,21).
- *hsn* genes either extend or restrict the host range. (e.g., *nodH, nodSU*) (22-24).
- Proteins encoded by *hsn* genes modify the basic structure of LCO's by adding various substituents.

The symbiotic plasmid of NGR234 (pNGR234*a*) contains at least three Hsn loci: HsnI, HsnII, and HsnIII (Figure 1-1) (21). Sequencing the 536Kb

Table 1-1. Nodulation genes (*nod*, *nol*, and *noe*) and probable functions

Genes	Functions or homologies	Strains[a]	References
Common *nod*-genes			
nodM	D-glucosamine synthase	Rm, Rlv, Rlt	95,194
nodA	Acyltransferase	All	107,108
nodB	Deacetylase	All	105,107
nodC	Chitin synthase	All	101,103, 104
Genes involved in modifying the non-reducing terminus			
nodE	β-Ketoacylsynthase	Rm, Rlv, Rlt	101, 104
nodF	Acyl carrier protein	Rm, Rlv, Rlt	111, 113, 114
nodS	Methyltransferase	RN, Ac, Bj, Rf, Rt	23, 117, 120
nodL	Acetyltransferase	Rm, Rlv, Rlt	124, 195, 196
nodU	Carbamoyltransferase	RN, Ac, Bj, Rf, Rt	23, 117
nolO	Carbamoyltransferase	RN, Bj	197
Genes involved in modifying the reducing terminus			
nodP	ATP-sulfurylase	Rm, Rt, RN	123, 131
nodQ	ATP-sulfurylase and APS kinase	Rm, Rt, RN	123, 131
nodH	Sulfotransferase	Rm, Rt	123, 129, 130
nodX	Acetyltransferase	RlvTOM	136, 137
nolL	Acetyltransferase	RN, Rl	141
noeK	Phosphomannomutase	RN	198
noeJ	Mannose-1-phosphate guanyltransferase	RN	198
noe L	Dehydratase	RN	198
nolK	Sugar epimerase	Ac, RN	128 198
nodZ	Fucosyltransferase	Bj, Ac, RN	29, 128, 138, 139
noeE	Sulfotransferase	RN	142
noeI	Methyltransferase	RN	197
Genes encoding transporters			
nodI	ATP-binding protein for transmembrane transport	All	197
nodJ	Integral membrane protein	All	197
nodT	Outer membrane protein transport?	Rlv, Rlt	199, 200
nolFGHI	Transport?	Rm	152
nodO	Secreted pore- forming protein	Rlv	177, 178
Genes encoding regulators			
nodD	LysR regulator protein family	All	9, 54, 56, 68, 74, 89, 182
nolR	DNA binding-protein, repressor	RN, Rm	88, 201
syrM	LysR regulator protein family	RN, Rm	73, 77

[a] Rm = *Rhizobium meliloti*, Rlv = *R. leguminosarum* bv. *viciae*, RlvTOM = *R. leguminosarum* bv. *viciae* strain TOM, Rlt = *R. leguminosarum* bv. *trifolii*, RN = NGR234, Ac = *Azorhizobium caulinodans*, Bj = *Bradyrhizobium japonicum*, Rf = *R. fredii*, Rt = *R. tropici*, Rl = *R. loti*.

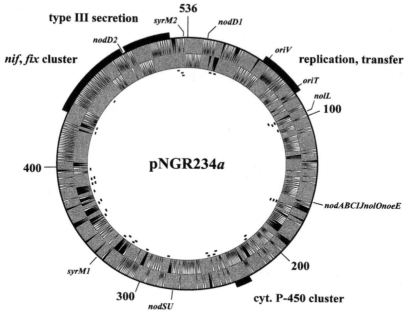

Fig. 1-1. Circular representation of the DNA sequence of the symbiotic plasmid (pNGR234*a*) of *Rhizobium* sp. NGR234. The outer and inner circles represent coding regions identified on the plus and minus DNA strands, respectively. Black arrows pointing inward mark the positions of insertion and/or mosaic sequences and white arrows all other open reading frames. *oriV* and *oriT* = origins of replication and transfer, respectively. *hsn*I to III = host-specificity of nodulation loci. Small dashes inside the circles show the positions of gene fragments. Numbers ($\times 10^{-3}$) represent nucleotide positions. Other abbreviations are listed in the text. (Courtesy X. Perret)

symbiotic plasmid showed that these Hsn regions contain many genes required for Nod-factor synthesis and host-specific nodulation (25).

Transcriptional Regulation of Nodulation Genes

Rhizobial *nod* gene expression requires (i) flavonoids excreted from the host root, (ii) the transcriptional activator NodD, and (iii) the NodD-binding *cis*-regulatory element, the *nod*-box. In the presence of plant signals, NodD activates transcription of *nod*-genes by binding to *nod*-box promoters (26, 27). Originally, these promoters were identified as conserved sequences of about 50 pb in length lying 100 to 250 pb upstream of translational start sites of inducible *nod* genes (28).

Activation of *nod*-gene expression via *nod*-box promoters is not the only regulatory system used in rhizobia (29, 30). *nolX* of *R. fredii* USDA257 lacks a *nod*-box, yet it is regulated in the conventional flavonoid and *nodD*-dependent manner (31). *nodZ* of *B. japonicum* also lacks a *nod*-box, and its expression is not dependent on NodD. *nodZ* is constitutively expressed under aerobic and anaerobic conditions. A puzzling characteristic of *nod*-gene regulation in *B. japonicum* is the high levels of *nodZ* expression observed in isolated bacteroids. Apparently, this expression is independent of the NifA regulator, and may suggest that another, unknown symbiotic regulator is involved (29).

Plant Signals

Symbiotic bacteria in the rhizosphere first come into contact with small molecules exuded by the legume. Plant exudates have been shown to activate or repress transcription of inducible *nod*-genes (32–36). Amongst these plant inducers, flavonoids that are produced via the phenylpropanoid biosynthetic pathway are the strongest inducers. Significant differences in the flavonoid compositions of seeds, roots and root-exudates of *Glycine max* (37–39), *Phaseolus vulgaris* (40–42), *Medicago sativa* (43,44) *Trifolium repens* (34) and *Vicia sativa* (45,46) exist. Many flavonoids are stored and released as glycosides or related conjugates (43). These conjugates, which are more soluble in water than aglycones, are usually less active in inducing *nod*-gene, but they are diffusible and can be hydrolyzed to more active substances by exoenzymes (44). As a further complication, inoculation of *Vicia sativa* with *R. leguminosarum* bv. *viciae* results in an increase in *nod*-gene inducing activity of root exudates (47). This change might be due to *de novo* synthesis of flavonoids, since accumulation of flavonoids have been observed in soybean root exudate after inoculation with *B. japonicum* and *R.* sp. NGR234 Nod-factors (48). An alternative explanation is based on the finding that rhizobia degrade a wide range of flavonoids (49,50). Degradation leads first to the appearance of chalcones, which can be (depending on the strain) more efficient inducers of *nod*-genes than other flavonoids (51). This is possibly due to their open C-ring structure which may provide additional conformational flexibility for binding to NodD proteins (50).

Interestingly, some, though not all *nod* inducing flavonoids also serve as chemo-attractants for rhizobia (52). However, chemotaxis and *nod* gene induction do not necessarily correlate. Amongst the flavonoid inducers of *R. leguminosarum* bv. *phaseoli nod*-genes, apigenin and luteolin evoke similar chemotactic responses as the inhibitors of *nod*-gene induction, umbelliferone and acetosyringone. On the other hand, the inducer naringenin has almost no effect on chemotaxis (52). Nevertheless, chemotaxis, *nod*-gene induction and symbiotic infection all appear to be focused on same part of the root (53).

NodD is a Member of the lysR Family of Transcriptional Activators

The *nodD*-gene product binds to *nod*-box promoters and activates transcription of *nod* genes in the presence of flavonoids (26,27). On the basis of homology, NodD is a member of the *lysR* family of transcriptional regulators (54). All known NodD of different rhizobia share the following: (for review see 55): (i) an N-terminal helix-turn-helix DNA binding motif; (ii) the absence of sequence homology in the C-terminal domain, and; (iii) the ability to bind to *nod*-boxes *in vitro* which is independent of the presence or absence of inducers (28,56). An inducer is necessary for NodD-dependent gene expression however (57-59).

NodD-genes control the first level of host specificity (60–65). Thus, mutations on *nodD* cannot always be complemented by *nodD* from other species (61,62), while point mutations may cause extension of the host range (66). Furthermore, transfer of *nodD1* of NGR234 into *R. meliloti* results in host range extension to *Macroptilium atropurpureum* whereas *R. meliloti nodD1* is incapable of restoring the ability of an NGR234 *nodD1* mutant to nodulate *Macroptilium*. Similarly, a *nodD*1 mutant of *R. trifoli* does not regain the ability to nodulate *Trifolium repens* when it is complemented with *nodD1* of *R. meliloti* (62). Mobilization of *nodD1* of NGR234 into *R. leguminosarum* bv. *trifolii* extends the host range to the non-legume *Parasponia andersonii* (64).

In other words, the products of the various *nodD* genes differ from one another in that they respond, in species-specific ways, to different sets of flavonoids and exudates (62). Correlations exist between spectrum of flavonoids able to interact with NodD proteins, and the broadness of host range. NodD of narrow host-range-rhizobia such as *R. meliloti*, *R. leguminosarum* bv. *viciae* and *R. leguminosarum* bv. *trifolii* respond to few flavonoids, while NodD1 of the broad host range species *Rhizobium.* sp. NGR234 reacts with a large spectrum of inducing compounds, including simple phenolic substances such as vanillin and iso-vanillin but also to compounds which have been identified as inhibitors in other rhizobia. (67,68).

Some rhizobia, e.g., *A. caulinodans* and *R. leguminosarum* bv. *viciae* possess, only one copy of *nodD* (54,69) and mutation of this gene leads to a Nod⁻ phenotype. In other species, the copy number may be higher, e.g. *R. tropici* has five copies (70), but there is no correlation between the copy number and host range. *R. meliloti* which possesses three copies of *nodD* nodulates only a few hosts (*Medicago*, *Melilotus* and *Trigonella*) (71), whereas the broad host range *Rhizobium* sp. NGR234, which has only two copies of *nodD* (72) forms symbiotic associations with 110 genera of legumes. In *R. meliloti*, *nodD*-dependent regulation is complex. In addition to the three copies of *nodD*, one structurally related gene (*syrM*) has also been characterized (73). Depending on the host plant, the symbiotic significance of the three *nodD* copies varies, and only the triple mutation *nodD1*, *nodD2*, *nodD3* completely abolishes nodulation (71). Only two copies, *nodD1* and *nodD2* function in the conventional manner by activating *nod*-gene expression in the presence of flavonoids (74). When present on multicopy plasmids, *nodD3* and *syrM* can induce expression of *nod*-genes in the absence of plant signals. Indeed, *syrM/nodD3* form a self-amplifying regulatory circuit, which may involve the molecular chaperons GroESL for activation (75).

Regulation *via nodD* is different in NGR234 however. Mutation of the constitutively expressed *nodD1* is sufficient to abolish nodulation on all plants tested (76). *nodD2* is only expressed late after induction with flavonoids, and plays a more discrete role in symbiosis, since its activation is *nodD*1 dependent. Other transcriptional activators are present on pNGR234*a*, (Figure 1-1) (25). Two copies of *syrM* (*syrM*1 and *syrM*2) have been identified and they are more than 60% similar. *syrM*1 plays a role in sulphation of NodNGR factors and is required for optimal nodulation of many hosts (77). It is also involved in the activation of the *nolXBTUVW* locus (77). Although *syrM*2 seems to be induced by flavonoids, nothing is known about its role in the symbiosis (25).

Other Regulatory *nod* Genes

The *nodVW* genes are unique to *B. japonicum* (which has also two *nodD* genes, *nodD1* and *nodD2*). Both *nodV* and *nodW* are required for nodulation of *M. atropurpureum*, *Vigna radiata*, and *V. unguiculata* which are alternative hosts for *B. japonicum* (78). Remarkably, neither mutations in *nodD1* nor *nodD2* abolish nodulation of any host. In fact, NodV and NodW are directly involved in isoflavone-mediated induction of the *nodD*1 and *nodYABC* operons (79). In the absence of NodD1, NodV and NodW are essential and sufficient for isoflavone-mediated *nod*-gene expression. How the NodVW regulatory system functions is not yet clear, however, especially at the level of interactions between NodW and *nod*-box promoters as well as the NodD-mediated transcription system. Nodulation of *G. max* is not affected by mutations in *nodV* and *nodW*, whereas *nodD1/nodW* or *nodD1/nodD2/nodW* mutants are unable to nodulate soybeans (79). Perhaps this is due to the production of isoflavones by soybeans that interact with either NodVW or NodD1. The requirement of NodV and NodW for nodulation of *Macroptilium* and *Vigna* species suggests that these plants may excrete inducers that interact with NodVW and not NodD1. The predicted amino acid sequence of NodVW predicted these gene products to be members of the family of two-component regulatory systems. NodV is most probably a sensor kinase responding to a yet-unidentified plant signal. In turn, NodV phosphorylates NodW which is a response regulator (80). Groβ and co-workers found homologues to *nodVW* (*nwsAB*) in *B. japonicum* which are expressed at very low levels. When overexpressed, *nws*B suppresses the Nod⁻ phenotype of a *nodW* mutant (81). Cross talk between members of the two component regulatory systems could explain this phenotype (82).

Negative Regulators of *nod* Gene Expression

Gene regulation frequently involves positive and negative control. *Nod*-gene expression is inhibited by phenolic compounds and is growth-phase dependent (36). After initial induction, transcription of the *nodABC* and *nodSU* operons of NGR234 decreases to undetectable levels 24 h after incubation with purified flavonoids (198). The *nodABC* genes of *R. leguminosarum* bv. *viciae* are not transcribed in bacteroids (83), and their introduction on a multi-copy plasmid into the wild-type strain inhibits nodulation (84).

The *nolA* gene of *B. japonicum,* which is required for genotype-specific nodulation of soybeans (85), was thought to be a repressor of *nod*-gene expression (86). In fact, NolA activates *nodD2* which then represses *nod*-gene expression (87).

In *R. meliloti*, NolR down-regulates transcription of *nodD1*, *nodD2* and the *nodABC*-genes by direct attachment to their overlapping promoters. Negative control by NolR might allow fine-tuning of *nod*-gene induction (via the *nod* regulon) by preventing over-production of Nod-factors (59,88). Even though NGR234 also possesses a copy of *nolR*, the *nodD2*-gene is involved in late repression of the *nodABC* operon (89). Inactivation of *nodD2* results in the formation of Fix⁻ nodules on *V. unguiculata* and *Tephrosia vogelii* perhaps because Nod-factors are still produced in the bacteroids (normally, the *nodABC*-genes are repressed in nodules, 89). Production of Nod-factors seems to be undesirable after the bacteria are released from the infection thread into the plant cytoplasm. Nevertheless, whether continued Nod-factor production affects nodulation or not greatly depends on the host plant, since nodulation of *Leuceana leucocephala* (indeterminate nodule-forming plant), and *Calopogonium caeruleum* (determinate nodule-forming plant) is not affected by mutation of *nodD2* NGR234 (89).

Nod-Factors: Structure and Biosynthesis

Structures of Nod-Factors

Nod-factors are the key signals in symbiotic recognition (90–92). Nod-factors of *R. meliloti* were the first characterized (2), and since then, the structures of Nod-factors from many different rhizobia have been determined (Table 1-2). All Nod-factors consist of a chitin oligomer backbone of β-1,4-linked *N*-acetyl-D-glucosamine (GlcNAc) in which the non-reducing terminus is acylated by various fatty acids. Major differences between the Nod-factors of various strains include: (i) length of the chitin oligomer backbone. The number of GlcNAc units varies from 3 to 5; (ii) the nature of the lipid chain. A variety of fatty acid substituents which include saturated, unsaturated, α/β unsaturated or (ω-1) hydroxylated fatty acids (93), and; (iii) a variety of specific substituents which include: acetyl-, arabinosyl-, carbamoyl-, fucosyl-, glyceryl-, hydroxyl-, mannosyl-, and sulphyl-groups (R1 to R3 and R5 to R7) (Table 1-2).

Biosynthesis of the Glucosamine Precursor

The *nodM* gene of *R. leguminosarum* bv. *viciae* and *R. meliloti,* which share homology with a glucosamine synthase (*glmS*) of *E. coli,* are involved in production of glucosamine (Figure 1-2) (94,95). *R. meliloti nodM* mutants excrete lower levels of Nod-factors and are Nod[d] on *Medicago* (95). Since glucosamine is also an important constituent of the bacterial envelope, a second, housekeeping copy of glucosamine synthase exists (94). Thus, the symbiotic copy of glucosamine synthase (NodM) simply increases the concentration of GlcNAc for Nod-factor synthesis. Uridyl diphosphate-GlcNAc (UDP-GlcNAc), is the common precursor of Nod-factors but the enzyme(s) required for its synthesis from β-D-glucosamine-6-P has not been described (Figure 1-2).

Biosynthesis of the Backbone

NodABC are responsible for synthesis of the core Nod-factor molecule. When introduced into *E. coli,* the *nodABC* genes of *R. meliloti* produce the basic Nod-factors (96,97). NodC shows significant homologies to chitin synthase and cellulose synthase (98,99). NodB is similar to a non-specific oligochitin deacetylase from the fungus *Mucor rouxii* (100). NodA completes assembly of the "core" Nod-factor by catalysing the transfer of the fatty acid chain to the non reducing GlcNAc.

NodC. By using radio-labeled UDP-GlcNAc and extracts from cells which over-produce NodC, *in vitro* assays showed that NodC is able to polymerize GlcNAc into chitin oligomers. NodC cannot use oligo-chitins as

Table 1-2. Structure of different Nod-factors[a]

Species	n	R1	R2	R3	R4	R5	R6	R7	References
NGR234	3	Cb or H	Cb or H	Cb or H	C18:1(11) C16:0 C16:1	Me	3-O-SO$_3$H, 2-O-MeFuc or 4-O-Ac 2-O-MeFuc	H	125
R. fredii (USDA257)	1,2,3	H	H	H	C18:1(11)	H	2-O-MeFuc or Fuc	H	167
A. caulinodans	2,3	Cb	H	H	C18:1(11) C18:0(11)	Me	Ara or Fuc	H	164
B. japonicum (USDA110)	3	Ac	H	H	C18:1(9) C18:1(9,Me) C16:1(9) C16:0	H	2-O-MeFuc	H	165, 166
R. loti	3	H	Cb	H	C18:1(11) C18:0	Me	4-O-Ac Fuc	H	202
R. etli	3	H	Cb or H	Cb or H	C18:1(11)	Me	4-O-Ac Fuc	H	160, 162
B. elkanii	2,3	Ac	Cb	H	C18:1(11)	Me , H	2-O-MeFuc or Fuc	H	122, 165
R. tropici	3	H	H	H	C18:1(11)	Me	SO$_3$H	Glyceryl	161
R. meliloti	1,2,3	Ac	H	H	C16:2(2,9) C16:3(2,4,9)	H	SO$_3$H	Mannosyl	2, 3, 123
R. leguminosarum bv. viciae	2,3	Ac	H	H	C18:1(11) C18:4(2,4,6,11)	H	H or Ac[b]	H	112 137[b]

[a] Cb = carbamaoyl, Me = methyl, Ac = acetyl, Fuc = fucose, and Ara = arabinose.
[b] R. leguminosarum bv. viciae strain TOM.

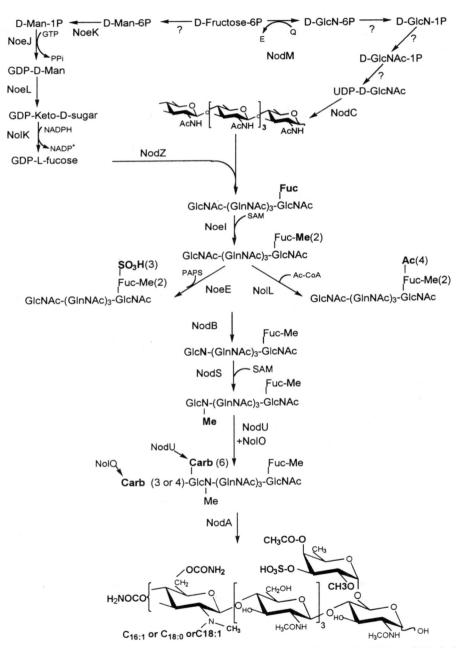

Fig. 1-2. Proposed biosynthetic pathway of NGR234 Nod-factors. Ac = acetate; C16:1 = palmitoleyl; C18:0 = stearyl; C18:1 = vaccenyl; Carb = carbamate; CoA = coenzyme A; GlcNAc = N-acetyl-D-glucosamine; Fuc = fucose; Man = mannose; Me = methyl; PAPS = 3'-phosphoadenosine 5'-phosphosulphate; and SAM = S-adenosylmethionine.

the substrate for initiation of elongation—GlcNAc monomers are the preferred acceptor molecules. Elongation proceeds towards the reducing end and degrees of polymerization of two to five units were obtained (101).

Interestingly, the number of GlcNAc residues within Nod-factors is well defined for each *Rhizobium*. NGR234 for example, produces only pentamers, whereas, the majority of Nod-factors excreted by *R. meliloti* contain four GlcNAc residues but how the length of the chitooligosaccharide backbone is controlled, is poorly understood. Genetic evidence suggests that *nodC* is involved in chain-length control, however, since introduction of *nodC* of *R. tropici* into *R. meliloti* results in the production of pentamers which are less active than tetramers on *Medicago* (3,102).

Iannino *et al* (103) proposed that the concentration of UDP-GlcNAc is an important factor in control of LCO chain length during biosynthesis of *R. fredii* Nod-factors. *in vitro* assays have shown, however, that the difference in length between LCO's of *R. meliloti* and *R. leguminosarum* bv. *viciae* is independent of the UDP-GlcNAc concentration and probably due to the NodC proteins (104). In this sense, *nodC* is a determinant of host specificity. Furthermore, NodS of NGR234 seems to be involved in chain-length determination *in vivo*. *nodS* mutants produce a mixture of LCO's of different lengths, while the wild-type NGR234 only produces pentameric Nod-factors. Similarly, conjugation of *nodS* of NGR234 into *R. fredii* abolishes the production of tri and tetrameric LCO's (23).

NodB. The *nodB* gene encodes a chitin deacetylase that is specific for the non-reducing GlcNAc. Purified NodB of *R. meliloti* deacetylates oligochitins with two to six GlcNAc's but not monomeric glucosamine (105, 106). NodB stoichiometrically releases acetate from chitin-oligomers so that the free amino group at the non-reducing terminus can be acylated with a fatty acyl chain.

NodA. Cell-free extracts of *R. meliloti* that over produce *nodA* convert deacetylated radiolabeled chitotetraoses to hydrophobic compounds (107, 108). Chromatographic and chemical analyses show that these are equivalent to acylated Nod-factors. NodA exhibits specificity for both donors and acceptors of the acyl groups. For example, when *nodA* of *R. tropici* is introduced into a *R. meliloti nodA⁻* mutant, Nod-factors acylated with vaccenic acid are produced instead of the C_{16} unsaturated or hydroxylated fatty acids (109). In this sense, *nodA* may also be considered as host-range gene.

Together, NodA, NodB, NodC are responsible for the synthesis of mono-*N*-acylated chitin-oligosaccharides. Synthesis of the core Nod-factors is a cascade in which NodC synthesizes the chitin oligomer, then NodB deacetylates the non-reducing GlcNAc, which is finally acylated by NodA

(Figure 1-2). This raises an interesting point. The *nodABC* genes were originally called "common" because they were believed to be functionally equivalent in various species. Yet, *nodA* and *nodC* are also host-range determinants (110).

Modifications to the Non-reducing Moiety

Rhizobia possess a spectrum of structural modifications to either or both the reducing- and non-reducing termini, each of which is the product of the host-specific nodulation genes. A number of these genes have been characterized.

N-acylation. The *nodE* and *nodF* genes of *R. meliloti* and *R. leguminosarum* specify the synthesis of a rare polyunsaturated fatty acid found on the non-reducing end of their respective Nod-factors (Table 1-2). In *R. meliloti*, *nodE* and *nodF* mutants are unable to produce C16:2 fatty acids which are replaced with vaccenic acid (C18:1) (111). *nodE* of *R. leguminosarum* bv. *viciae* is required for the synthesis of the C18:4 acyl moiety (112).

NodF, which is homologous to acyl-carrier proteins, contains a characteristic 4'-phosphopantetheine prosthetic group (113,114). NodE is homologous to various β-ketoacetyl synthases (115). NodE and NodF of *R. leguminosarum* and *R. meliloti*, are similar but produce different fatty acids (111). Indeed, introduction of *R. leguminosarum nodEF* genes into a *nodEF⁻* mutant of *R. meliloti*, leads to the production of polyunsaturated fatty acids (C18:2, C18:3, and C18:4). Perhaps the presence of C18:2 and C18:3 fatty acids is due to a house keeping enoyl reductase present in *R. meliloti* but absent in *R. leguminosarum*. These findings suggest that allelic variation in *nodEF* genes controls the length and the degree of unsaturation of the acyl chains.

N-methylation. Some rhizobial strains produce *N*-methylated Nod-factors. *nodS* was first identified in NGR234 and in *B. japonicum* (116,117). Since then, it has been found in many *Rhizobium* species (Table 1-1). NodS shares homology with S-adenosylmethionine-utilising enzymes (118). Mutation of *nodS* in NGR234 partially abolishes production of *N*-methylated Nod-factors (23), and, drastically reduces Nod-factor production (23). Even though the closely related strain, *R. fredii* USDA257 possesses the entire *nodS* gene, it is unable to produce *N*-methylated Nod-factors because a defective *nod*-box prevents the expression of the gene (119). Conjugation of *nodS* of NGR234 into USDA257 not only results in the synthesis of *N*-methylated Nod-factors, but the suppression of tri- and tetrameric LCO's (23). These findings suggest that *nodS* not only controls methylation but is also involved in control of both the degree of polymerization and the level of Nod-factors produced.

Using [^{3}H-methyl]-SAM as the methyl donor Geelen *et al* showed that NodS of either *A. caulinodans* or NGR234 methylate deacetylated chito-oligosaccharides, suggesting that NodS acts after deacetylation (120). Time course experiments in which the *nodABCS* genes of *A. caulinodans* (121) were introduced into *E. coli* confirmed that methylation occurs after deace-tylation and prior to acylation of Nod-factors (121).

O-acetylation. Acetate groups are present at C6 on the non-reducing terminus of Nod-factors produced by *B. japonicum* (122), *B. elkanii* (122), *R. meliloti* (123), *R. leguminosarum* bv. *trifolii* and *viciae* (112) (Table 1-2). 6-*O*- acetylation is controlled by NodL, which is homologous to several bacterial acetyl transferases, and is essential for acetylation on the C6 position. *In vitro* assays using purified NodL of *R. leguminosarum* bv. *viciae* and acetyl-CoA as the acetyl donor showed that the protein is able to acetylate various substrates, including LCO's, chitin fragments and GlcNAc (124).

O-carbamoylation. Among the large family of NodNGR-factors are non-carbamoylated, mono-carbamoylated or *bis*-carbamoylated species (125). Carbamoyl groups are restricted to positions C6, C3 or C4 of the non-reducing terminus (Table 1-2). Mutation of *nodU* of NGR234 leads to the disappearance of *bis*-carbamoylated species, while USDA257(*nodU*) trans-conjugants are able to produce 6-*O*-carbamoylated NodUSDA-factors (23). Carbamoylation on C3 or on C4 is dependent on the *nolO* gene (S. Jab-bouri, *unpublished*). *nolO* is homologous to *nodU* and to a protein of *Norcardia lactamdurans* which is involved in the biosynthesis of carba-moylated β-lactam antibiotics. Mono-carbamoylated Nod-factors have been also found in other rhizobia (Table 1-2). Nod-factors of *A. caulinodans* are carbamoylated at C6 of the non-reducing terminus. Introduction of *nodU* into *E. coli* containing *nodABCS* results in the production of 6-*O*-carba-moylated Nod-factors, confirming that that *nodU* controls carbamoylation at this position (126).

Modifications on the Reducing Moiety

Most modifications to the reducing terminus are at position C6, although in some species, there are adducts at other positions. For example, the Nod-factors of *B. elkanii* and *R. tropici* are substituted on C1 with glycerol and mannose, respectively, while Nod-factors of *A. caulinodans* have arabinose at C3 (122,127,128). Modifications on C6 are of two types: those in which an additional mono-saccharide (usually fucose) is added and those in which simple *O*-sulphate or *O*-acetate substitutions occur (Table 1-2).

O-Sulphation. 6-*O*-sulphated Nod-factors are only found in a few *Rhizo-bium* species including *Rhizobium* sp. GRH2, *R. meliloti*, and *R. tropici*,

(Table 1-2). Three *hsn*-genes (*nodPQH*) are involved in sulphation of NodRm-factors, and they ensure that all Nod-factors secreted by *R. meliloti*, are sulphated (2,24). Mutation of *nodH* abolishes the production of sulphated NodRm-factors, while *nodP* and *nodQ* mutants produce a mixture of sulphated and non-sulphated Nod-factors (24). NodH is a sulphotransferase and tetrameric Nod-factors are the preferred substrate for sulphate transfer (129,130). NodPQ are required for production of the activated form of sulphate, 3′-phosphoadenosine 5′-phosphosulphate (PAPS) which is the substrate for NodH. Together, NodPQ form a multi-functional protein complex that exhibits both ATP-sulfurylase and APS kinase activities (131,132). *nodPQH* genes have been also characterized in *R. tropici* (127,133) and *Rhizobium* sp. strain N33 (134). In NGR234, two *nodPQ* loci have been mapped on the chromosome (72).

In *R. tropici*, the relatively small PAPS pool limits production of 6-*O*-sulphated Nod-factors, since *R. tropici* transconjugants harbouring the *nodPQ* genes of *R. meliloti,* only produce sulphated NodRt factors instead of the mixture of sulphated and non-sulphated molecules produced by the wild-type strain (135).

O-Acetylation. *O*-acetylation of the reducing terminus is exclusive to *R. leguminosarum* bv. *viciae* strains (e.g., TOM) which form Fix$^+$ nodules on primitive cultivars of *Pisum sativum* (e.g., cv. Afghanistan). In this system, cultivar specificity is positively regulated by *nodX* (136). NodX mediates *O*-acetylation of the reducing GlcNAc of pentameric but not tetrameric Nod-factors (137). Strains like TOM thus have two acetyltransferases, NodL and NodX. The first acetylates C6 of the non-reducing GlcNAc while the latter acetylates C6 of the reducing GlcNAC. Perhaps not surprisingly, only weak homology exists between the two enzymes and they are not functionally interchangeable. The mechanism by which each acetyltrans-ferase distinguishes between reducing and non-reducing GlcNAc's remains unclear.

O-Fucosylation. C6-fucosylated Nod-factors are found in *A. caulino-dans, B. japonicum, B. elkanii,* NGR234, *R. etli, R. fredii* USDA257 and *R. loti* (Table 1-2). Different adjuncts are present on the fucose residue. Most combinations of acetylation, methylation and sulphation are found at positions C2, C3 and C4. *nodZ* encodes a fucosyltransferase (128,138,139) which is specific for C6 of the reducing GlcNAc. Pentameric GlcNAC is the preferred substrate for the NGR234 NodZ fucosyltransferase, but purified NodZ also fucosylates various chitin-oligomers (139). This sug-gests that fucosylation of Nod-factors occurs before acylation. Mutation of *nodZ* of *A. caulinodans* strongly reduces the production of fucosylated Nod-factors which are completely abolished by similar mutations in *B.*

japonicum and in NGR234. It thus seems likely that a second, constitutive fucosyltransferase might be present in *A. caulinodans* which is responsible for the residual fucosylation (128). Moreover, it cannot be excluded that NodZ of *A. caulinodans* possesses arabinosyl transferase activity, since the nodZ⁻ mutant is no longer capable of producing arabinosylated Nod-factors (128).

In NGR234, biosynthesis of GDP-L-fucose from GDP-D-mannose has been proposed (Figure 1-2). GDP-D-mannose-4,6-dehydratase is the first enzyme in this pathway. The product of dehydratase action is an unstable sugar, GDP-4-keto-6-deoxy-D-mannose that is converted to GDP-fucose by an epimerase (140). Genetic and molecular analysis of the NGR234 *nodZ* locus identified genes involved in biosynthesis of fucose (Figure 1-1). *noeL* and *nolK,* which are part of *nodZ* operon, share homologies with mannose-4,6-dehydratase and epimerase/reductase respectively (Figure 1-2). Mutation of any of these genes abolishes production of fucosylated NodNGR-factors. *noeK* and *noeJ* are not essential for fucosylation however since their mutation only leads to a reduction in the level of fucosylated Nod-factors, not their absence. These genes should then possess functional homologues (D. Quesada-Vincens, *unpublished*).

2-O-methylation of fucose. Little is known about methylation of the fucose residue. The *noeI* gene of NGR234 shares weak homology with a methyltransferase of *Streptomyces hygroscopicus* and its mutation abolishes production of the 2-*O*-methyl on the fucose of Nod-factors (S. Jabbouri, *unpublished*).

4-O-acetylation of fucose. The *nolL* gene of NGR234 and *R. etli* share homologies with the *O*-acetyltransferase of *R. leguminosarum* bv. *viciae* encoded by *nodX* (141). An NGR234 *nolL* mutant is unable to produce *O*-acetylated Nod-factors, and transfer of this gene to *R. fredii* USDA257 leads to the production of *O*-acetylated Nod-factors (S. Berck, *unpublished*).

3-O-sulphation of fucose. Nod-factors that are 3-*O*-sulphated on the methylfucose have been found in NGR234 (125) and in *Bradyrhizobium* species which nodulate *Acacia albida* (M. Ferro, *personal communication*). noe*E* of NGR234 encodes a sulphotransferase and its mutation results in the absence of sulphated NodNGR-factors. This sulphotransferase activity is specific for the fucose residue since mutation of *nodZ* produces non-fucosylated, non-sulphated Nod-factors. Similarly, *R. fredii* USDA257 (*noeE*) transconjugants acquire the ability to produce sulphated Nod-factors. NoeE has only weak homology to other sulphotransferases including NodH of *R. meliloti*. Biochemical analyses show that NodH is less specific than NoeE, since it is able to transfer sulphate to both, the fucose residue, and C6 of the reducing GlcNAc of NodNGR factors (142).

Secretion of Nod-Factors

Since most Nod-enzymes are located in the cytosol and inner membrane, Nod-factor biosynthesis probably also occurs there using the lipid carrier undecaprenyl phosphate as proposed (143). Inhibitors of carbohydrate backbone do not affect Nod-factors biosynthesis, however (121). Once synthesized, Nod-factors are usually excreted into the extra-cellular medium, although in some strains such as *R. leguminosarum* bv. *trifolii* they accumulate in bacterial membranes (144). Most probably, one or several transport systems are involved in Nod-factor secretion.

So far, only *nodIJ* are known to be involved in transport of Nod-factors (145,146). *nodIJ* is required for the secretion of Nod-metabolites synthesized by *E. coli* harbouring the *A. caulinodans nodABCSU* genes (146). As expected from these data, *nodIJ* mutants of *R. leguminosarum* bv. *trifolii*, are unable to excrete Nod-factors (147). In *R. etli* Nod-factors begin to appear in the extra-cellular medium about 1 h after induction of the *nod*-genes. Mutation of either *nodI* or *nodJ* decreases secretion and increases intra-cellular accumulation of Nod-factors (148). In other strains such as *R. leguminosarum* bv. *viciae,* mutations on *nodIJ* have only moderate effects on Nod-factor secretion however (145). On the basis of sequence homology, it has been proposed that the NodIJ complex is an ATP-binding cassette (ABC) transporter of the ABC-2 subfamily. This subfamily comprises capsular polysaccharide secretion systems of Gram-negative bacteria. ABC transporters consist of an integral membrane protein and a protein that binds ATP. NodI corresponds to the ATP-binding cytoplasmic part of the complex, and NodJ to the membrane-inserted hydrophobic protein (146, 149,150). Apparently, NodIJ are associated with the inner membrane (151), suggesting that they may function to translocate Nod-factors from the cytoplasm to the periplasm. In this scenario, a second protein may be needed to excrete the Nod-factors into the medium. In *R. leguminosarum*, it is possible that NodT plays this role, and that together with NodI and NodJ, forms a secretion system which exports the molecules directly into the medium (145). In *R. meliloti* it is possible that NolGHI which are homologous to members of families of functionally related bacterial membrane proteins are involved (152). The suggestion is that NolGHI are cytoplasmic membrane proteins which act together to transport substances (e.g., Nod-factors) across the inner membrane and into the periplasm.

"Duality" of Host Specificity and Nod-Factor Structures

Loss and gain of function experiments show that host specificity is linked to Nod-factor structure. NodH-dependent sulphation of the reducing terminus of Nod-factors is a major determinant of host range in *R. meliloti* (2,24). Mutation of *nodH* prevents nodulation of *M. sativa* but confers the ability to nodulate the non-host plant, *Vicia sativa* (13,153). Similarly, transfer of *R. meliloti nodEFGHPQ* to *R. leguminosarum* bv. *trifolii* or bv. *viciae*, confers on these strains the ability to nodulate *Medicago* but prevents nodulation of the normal hosts, *Trifolium repens* and *Vicia*, respectively (22,154). *nodE* of *R. leguminosarum* controls the synthesis of polyunsaturated fatty acids and plays a crucial role in determining host-range (112). NodX, of *R. leguminosarum* strain TOM is necessary for nodulation of primitive peas (136). *nodL* mutants of *R. meliloti* are significantly impaired in their ability to elicit infection thread formation on *Medicago*, while *nodL/nodF* double mutants are unable to penetrate their hosts (155). Inactivation of *nodS* of *A. caulinodans*, NGR234, and *R. tropici*, abolishes nodulation of *Leucaena* and *Phaseolus* (117,156). Introduction of either *nodS* or *nodU* into *R. fredii* extends its host-range to include *Leucaena* (23,119). The cross complementation between *nodS* of NGR234 and *nodO* of *Rhizobium* sp. BRB16 suggests that both determine specificity for nodulation of *L. leucocephala* and that their functions are interchangeable (157). A similar effect has also observed with *R. leguminosarum* bv. *viciae* in which *nodO* partially complements a deletion of *nodEF* for nodulation of *Pisium* and *Viciae* (158) and extends the host range of *nodE R. trifolii* mutant to *Viciae* (159).

The *nodZ* gene encodes a fucosyltransferase (138,139) which in *B. japonicum* is required for effective nodulation of *Macroptilium* (29). NodZ͞ mutants of NGR234 loose the capacity to nodulate *Pachyrhizus tuberosus* (139). In contrast, nodulation of many host plants is not significantly affected by mutation of *nodZ* despite the resulting changes in Nod-factor structure. Furthermore, although *R. leguminosarum* complemented with *nodZ* of *B. japonicum* acquires the capacity to nodulate *Macroptilium*, the nodules are unable to fix nitrogen (138).

Mutation of *noeE* of NGR234 also blocks nodulation of *Pachyrhizus* while its introduction into USDA257 extends the host range of *R. fredii* to include *Calopogonium caeruleum* (142). Yet, USDA257 (*noeE*) transconjugants do not acquire the ability to nodulate *P. tuberosus* (M. Hanin *unpublished*). In other words, the importance of different Nod-factor adjuncts depends on the host plant.

Other subtle influences on host or cultivar specificity seem to be independent of Nod-factor structure. Some strains (e.g. *R. etli* and *R. loti*) have different host ranges, even though the Nod-factors produced by both strains are identical (160). Perhaps this is due to a differential induction of the respective *nodD*-genes by varying spectra of flavonoids excreted by the various hosts. In contrast, two rhizobia that nodulate the same plant may secrete different Nod-factors. Thus, *R. tropici* and *R. etli* produce sulphated and acetylfucosylated Nod-factors respectively but both effectively nodulate *Phaseolus vulgaris* (161,162). Nod-factor levels are also important in determining host specificity. Introduction of *nodD1* of NGR234 into *R. meliloti* increases Nod-factor production by about two-fold and permits nodulation *Vigna unguiculata*, a non-host (163).

Biological Activity of Nod-Factors

The biological activity of Nod-factors is shown in (Table 1-3). In most ways, Nod-factors faithfully mimic free-living rhizobia.

Morphological Responses

After the first contact with epidermal cells of the root, rhizobia induce hair-deformation (Had), curling (Hac) and infection thread formation (Inf) (5,6). At nano to pico-molar concentrations, Nod-factors elicit Had and Hac

Table 1-3. Known biological and molecular responses of legume roots and root-hairs to application of Nod-factors

Responses	Nod factors	Concentration	Plants	References
Accumulation of flavonoids	Nod Rlv(Ac) NodNGR(OH or Ac) NodBj	10^{-9} M	Vicia Glycine	48, 184
Increase of $[Ca^{2+}]$	NodNGR(Ac,S), NodRmIV(S)	10^{-9} M	Vigna	175
Depolarization of root-hairs	NodRm(S)	10^{-9} M	Medicago	171-173
Calcium spiking in root-hairs	NodRmIV(Ac, S)		Medicago	176
Had	NodNGR(S)	10^{-7} to 10^{-12} M	Vigna	125
	NodNGR(Ac)	10^{-8} to 10^{-11} M	Macroptillium	182
	NodNGR(S)	10^{-7} to 10^{-10} M	Medicago	125
	NodRmIV(S)	10^{-9} to 10^{-12} M	Medicago and Melilotus	2 1
	NodRlv, NodRm	10^{-9} to 10^{-12} M	Vicia	112
Hac	NodNGR		Vigna	125
Penetration of bacteria	NodNGR(S)	10^{-6} to 10^{-7} M	Vigna, Glycine, Calopogonium	170
Induction of pre-infection threads (PIT)	NodRlv(Ac)		Vicia	168
Induction of nodule meristems	NodBe		Glycine	203
	NodRlv(Ac)		Vicia	112
	NodRm(S)		Medicago	1
Induction of Nodulin genes (ENOD5, 12)	NodRlv (Ac)	10^{-12} to 10^{-13} M	Pisum Medicago sativa and M. truncatula	204-206 207-209
Induction of other genes (LTP, CHS, Extensin)	NodNGR	10^{-6} to 10^{-7} M	Vigna	210-213

but not Inf activity (2–4,112,123,125,164–167). In the outer cortex of *Vicia*, mitogenic Nod-factors of *R. leguminosarum* bv. *viciae* induce formation of pre-infection threads and stimulate formation of root hair-like structures (168). At higher concentrations (>1 nM), Nod-factors elicit cortical cell division and the formation of meristems which differentiate into nodule-like structures. (24,112,168). At 10^{-11} to 10^{-9} M, NodNGR factors provoke Had, and "shepherd's crook" type curling of *Macroptilium artropureum* root-hairs. At higher concentrations (10^{-7} to 10^{-6} M), pseudo-nodules are formed (4). Direct application of cytokinins or inhibitors of auxin transport also induces formation of pseudo-nodules on *Macroptilium*, suggesting that Nod-factors may act by modifying the cytokinin-auxin balance.

In most plants tested, Nod-factors induced nodule primordia that do not develop into genuine nodules, although in *Medicago*, formation of full-size nodular structures has been observed (1). Perhaps this is related to spontaneous nodulation of *Medicago* (169).

Entry of Rhizobia

Nod-factors are also the signals required for entry of rhizobia into legumes. Co-inoculation of NodNGR factors with a *nodABC⁻* mutant of NGR234 or *B. japonicum* USDA110 permits these strains to nodulate and fix nitrogen on *V. unguiculata* and *G. max* respectively (170). NodNGR factors also allow entry of *R. fredii* USDA257 into the roots of *Calopogonium caeruleum,* a non-host (170).

Physiological Responses

Morphological responses are preceded by other physiological events. In *Medicago* root-hairs, depolarization of the cytoplasmic membrane potential occurs within two minutes of NodRm factor application. (171). Both the sulphate group as well as an unsaturated acyl moiety are necessary for this response (172,173). Similarly, an increase in proton efflux and changes in the flux of calcium have been observed in root hairs perfused with NodRm-IV(S) (174). Application of nano-molar concentrations of NodNGR- and NodRm-factors to *Vigna unguiculata* results, within seconds, in plateau-like increases in intracellular free calcium in root hairs and epidermal cells (175). This elevation may be associated with membrane depolarisation of the root hairs. Recently, Long and co-workers have observed periodic calcium spiking in *Medicago* roots exposed to *R. meliloti* Nod-factors. This root-hair Ca^{++} variation was not observed with chitin oligomers or Nod-factors from *R. leguminosarum*. Since calcium spiking initiates after a long delay (10 min), it is probably not responsible for depolarisation of the membranes (176).

nodO of *R. leguminosarum* bv. *leguminosarum* does not appear to be involved in synthesis of Nod-factors; rather, this gene encodes a secreted Ca^{++} binding protein which shares homology with the bacterial pore-forming hemolysins (177). Purified NodO can form a cation-selective ion channel when inserted into artificial bilayers (178). These are thought to be Ca^{++} permeable, thus mediating increases in cytosolic Ca^{++} and initiation of Ca^{++} dependent signal transduction.

Activation of Plant Gene Expression

Understandably, Nod-factors activate expression of genes involved in regulation of the cell cycle and nodulation (for reviews see 179–183). Unfortunately, the regulatory pathway and function of the early nodulation genes (ENOD's) genes are largely unexplored. Moreover, flavonoid accumulation, as well as the expression of the isoflavone reductase gene, are induced by Nod-factors (48,184).

Nod-Factor Recognition and Signal Transduction

The diverse responses of plants to Nod-factors suggest that many receptors are involved. Nod-factor binding proteins have been described by Bono *et al* (185). In an other study, Ardourel *et al* (155) proposed a model in which two distinct receptors are involved, one of which is able to interact with non host-specific Nod-factors, thereby triggering cell division. A second, a stringent entry receptor, would interact with host-specific Nod-factors.

The intriguing similarities between Nod-factors and chitins prompted many researchers to test whether legume chitinases can hydrolyse Nod-factors. Modifications to both sugar termini of Nod-factors confer resistance against chitinases from roots of hosts but not against those non-host plants (186,187). It is thus possible that host specificity of rhizobia resides not only in Nod-factor receptors but also in their resistance to inactivation by chitinases or other hydrolases (188).

Nod-Factor Independent Nodulation Genes

Several nodulation genes have no apparent role in Nod-factor biosynthesis, yet they are involved in ensuring successful symbioses. Unfortunately, their functions have not been fully elucidated. One locus that is present in *R. fredii* USDA257 contains six genes, *nolXWBTUV*, which together negatively regulate cultivar specificity. Wild-type USDA257 nodulates *G. max* in a cultivar-specific manner. USDA257 is also able to nodulate Asian soybean cultivars such as Peking, but not developed American lines such as McCall. Inactivation of any of *nolXWBTUV* allows formation of Fix⁺ nodules on cultivar McCall (189). Significant homologies between NolW, NolX, NolT and Hrp proteins of plant pathogenic bacteria and with proteins of the Yop secretion system of animal pathogenic *Yersinia* exist (189). Since both protein families are thought to mediate secretion of proteins involved in pathogenicity, *nolXWBTUV* might, by analogy play a role in excretion of proteins that inhibit interaction with McCall plants. Interestingly, this locus also exists in NGR234 (190).

Mutation of other genes simply delays nodulation. Amongst them, *nodN*, *nodR*, *nolP*, and *nolE* are flavonoid-inducible (191–193), and may play important roles in special environmental conditions.

Conclusions

Symbiotic interactions involve continued and reciprocal signal exchange. Either growth of legume roots towards pockets of rhizobia in the soil or, chemotaxis towards root-exudates provides the initial impetus for intimate contact. Then, flavonoids induce *nod*-genes to synthesize and

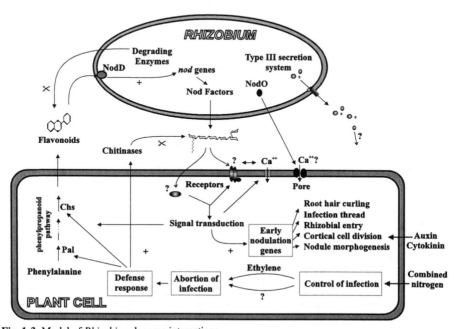

Fig. 1-3. Model of *Rhizobium*-legume interactions.

secrete Nod-factors that interact with roots, so stimulating the synthesis and release of flavonoids (Figure 1-3). More intimate contact amongst the symbionts sets a cascade of molecular events in motion which, within seconds, results in plateau-like increases in intracellular free calcium $\{[Ca^{2+}]_i\}$ in root-hairs and root epidermal cells. Most probably these elevated calcium levels trigger a signal transduction pathway which activates the expression of early nodulin genes. In turn, the enzymatic and structural functions encoded by these genes provoke curling of the root-hairs and allow rhizobia to enter the plant. Although it is exceedingly difficult to demonstrate, it seems likely that continued signal exchange occurs between rhizobia within the infection threads and the host permitting bacteroid development and nitrogen fixation.

By and large, the host-range of rhizobia is reflected in structural variations in Nod-factors. Yet rhizobial NodD proteins are not simple mediators of flavonoid messages. Rather, their inherent variation allows activation with some or many different phenolic compounds so conferring an additional degree of host-specificity. Finally, the response of different legumes to divergent Nod-factors varies. Nod-factor concentrations in the rhizosphere are an important component of specificity, but it is not yet possible to completely correlate Nod-factor structure with the ability to nodulate a specific legume.

References

1. Truchet, G., P. Roche, P. Lerouge, J. Vasse, S. Camut, F. De Billy, J-C. Promé, and J. Dénarié. 1991. Sulphated lipooligosaccharides signals of *Rhizobium meliloti* elicit root nodule organogenesis in alfalfa. Nature 351:670-673.
2. Lerouge, P., P. Roche, C. Faucher, F. Maillet, G. Truchet, J-C. Promé, and J. Dénarié. 1990. Symbiotic host-specificity of *Rhizobium meliloti* is determined by a sulphated and acylated glucosamine oligosaccharide signal. Nature 344:781-784.
3. Schultze, M., B. Quiclet-Sire, E. Kondorosi, H. Virelizier, J. N. Glushka, G. Endre, S. D. Géro, and A. Kondorosi. 1992. *Rhizobium meliloti* produces a family of sulfated lipooligosaccharides exhibiting different degrees of plant host specificity. Proc. Natl. Acad. Sci. USA 89:192-196.
4. Relic', B., F. Talmont, J. Kopcinska, W. Golinowski, J-C. Promé, and W. J. Broughton. 1993. Biological activity of *Rhizobium* sp. NGR234 Nod-factors on *Macroptilium atropurpureum*. Mol. Plant-Microbe Interact. 6:764-774.
5. Brewin, N. J. 1991. Development of the legume root nodule. Annu. Rev. Cell. Biol. 7:191-226.
6. Kijne, J. W. 1992. The *Rhizobium* infection process, p. 349-398. *In* G. Stacey, R. H. Burris, and H. J. Evans (ed.), Biological Nitrogen Fixation. Chapman & Hall, New York.
7. Martinez, E., D. Romero, and R. Palacios. 1990. The *Rhizobium* genome. Crit. Rev. Plant Sci. 9:59-93.

8. Pankhurst, C. E., W. J. Broughton, and U. Wieneke. 1983. Transfer of an indigenous plasmid of *Rhizobium loti* to other rhizobia and *Agrobacterium tumefaciens*. J. Gen. Microbiol. 129:2535-2543.

9. Appelbaum, E. R., D. V. Thompson, K. Idler, and N. Chartrain. 1988. *Rhizobium japonicum* USDA 191 has two *nodD* genes that differ in primary structure and function. J. Bacteriol. 170:12-20.

10. Armitage, J. P., A. Gallagher, and A. W. B. Johnston. 1988. Comparison of the chemotactic behaviour of *Rhizobium leguminosarum* with and without the nodulation plasmid. Mol. Microbiol. 2:743-748.

11. Arnold, W., A. Becker, M. Keller, A. Roxlau, and A. Pühler. 1994. The role of *Rhizobium meliloti* surface polysaccharides in the infection of *Medicago sativa* nodules. Endocytobiosis Cell Res. 10:17-28.

12. Kondorosi, E., Z. Banfalvi, and A. Kondorosi. 1984. Physical and genetic analysis of a symbiotic region of *Rhizobium meliloti*: Identification of nodulation genes. Mol. Gen. Genet. 193:445-452.

13. Faucher, C., F. Maillet, J. Vasse, C. Rosenberg, A. A. N. van Brussel, G. Truchet, and J. Dénarié. 1988. *Rhizobium meliloti* host range *nodH* gene determines production of an alfalfa-specific extracellular signal. J. Bacteriol. 170:5489-5499.

14. Goethals, K., M. Gao, K. Tomekpe, M. van Montagu, and M. Holsters. 1989. Common *nodABC* genes in *nod* locus 1 of *Azorhizobium caulinodans*: nucleotide sequence and plant inducible expression. Mol. Gen. Genet. 219:289-298.

15. Vàzquez, M., A. Davalos, A. De las Penas, F. Sanchez, and C. Quinto. 1991. Novel organisation of the common nodulation genes in *Rhizobium leguminosarum* bv. *phaseoli* strains. J. Bacteriol. 173:1250-1258.

16. Govers, F., M. Moerman, J. A. Downie, P. Hooykaas, H. J. Franssen, J. Louwerse, A. van Kammen, and T. Bisseling. 1986. *Rhizobium nod* genes are involved in inducing early nodulin genes. Nature 323:564-566.

17. Fellay, R., P. Rochepeau, B. Relic', W. J. Broughton. 1995. Signals to and emanating from *Rhizobium* largely control symbiotic specificity, p. 199-220. *In* K. Kohmoto, R. P. Singh, and U. S. Singh (ed.), Pathogenesis and host Specificity in Plant Diseases Vol. I. Prokaryotes. Elsevier Press, Oxford, U. K.

18. Evans, I. J., and J. A. Downie. 1986. The *nodI* gene product of *Rhizobium leguminosarum* is closely related to ATP-binding bacterial transport proteins; nucleotide sequence analysis of the *nodI* and *nodJ* genes. Gene 43:95-101.

19. Jacobs, T. W., T. T. Egelhoff, and S. R. Long. 1985. Physical and genetic map of a *Rhizobium meliloti* nodulation gene region and nucleotide sequence of *nodC*. J. Bacteriol. 162:469-476.

20. Djordjevic, M. A., P. R. Schofield, and B. G. Rolfe. 1985. Tn-5 mutagenesis of *Rhizobium trifolii* host specific nodulation genes results in mutants with altered host range ability. Mol. Gen. Genet. 200:463-471.

21. Lewin, A., C. Rosenberg, H. Meyer z. A., C. H. Wong, L. Nelson, J.-F. Manen, J. Stanley, D. N. Downing, J. Dénarié, and W. J. Broughton. 1987. Multiple host-specificity loci of the broad host-range *Rhizobium* sp. NGR234 selected using the widely compatible legume *Vigna unguiculata*. Plant. Mol. Biol. 8:447-459.

22. Faucher, C., S. Camut, J. Dénarié, and G. Truchet. 1989. The *nodH* and *nodQ* host range genes of *Rhizobium meliloti* behave as avirulence genes in *R. leguminosarum* bv. *viciae* and determine changes in the production of plant-specific extracellular signals. Mol. Plant-Microbe Interact. 2:291-300.

23. Jabbouri, S., R. Fellay, F. Talmont, P. Kamalaprija, U. Burger, B. Relic', J-C. Promé, and W. J. Broughton. 1995. Involvement of *nodS* in *N*-methylation and

nodU in 6-*O*-carbamoylation of *Rhizobium* sp. NGR234 Nod factors. J. Biol. Chem. 270:22968-22973.

24. Roche, P., F. Debellé, F. Maillet, P. Lerouge, C. Faucher, G. Truchet, J. Dénarié, and J-C. Promé. 1991. Molecular basis of symbiotic host specificity in *Rhizobium meliloti*: *nodH* and *nodPQ* genes encode the sulfation of lipo-oligosaccharide signals. Cell 67:1131-1143.

25. Freiberg, C., R. Fellay, A. Bairoch, W. J. Broughton, A. Rosenthal, and X. Perret. 1997. Molecular basis of symbiosis between *Rhizobium* and legumes. Nature 387: 397-401.

26. Wang, S. P., and G. Stacey. 1991. Studies of the *Bradyrhizobium japonicum nodD*1 promoter: a repeated structure for the *nod* box. J. Bacteriol. 173:3356-3365.

27. Goethals, K., M. van Montagu, and M. Holsters. 1992. Conserved motifs in a divergent *nod* box of *Azorhizobium caulinodans* ORS571 reveal a common structure in promoters regulated by LysR-type proteins. Proc. Natl. Acad. Sci. USA 89:1646-1650.

28. Rostas, K., E. Kondorosi, B. Horvath, A. Simoncsits, and A. Kondorosi. 1986. Conservation of extended promoter regions of nodulation genes in *Rhizobium*. Proc. Natl. Acad. Sci. USA 83:1757-1761.

29. Stacey, G., S. Luka, J. Sanjuan, Z. Banfalvi, A. J. Nieuwkoop, J. Y. Chun, L. S. Forsberg, and R. W. Carlson. 1994. *nodZ*, a unique host-specific nodulation gene, is involved in the fucosylation of the lipopolysaccharide nodulation signal of *Bradyrhizobium japonicum*. J. Bacteriol. 176:620-633.

30. Kovacs, L. G., P. A. Balatti, H. Krishnan, and S. G. Pueppke. 1995. Transcriptional organization and expression of *nolXWBTUV*, a locus that regulates cultivar-specific nodulation of soybean by *Rhizobium fredii* USDA257. Mol. Microbiol. 17:923-933.

31. Bellato, C. M., P. A. Balatti, S. G. Pueppke, and H. Krishnan. 1996. Proteins from cells of *Rhizobium fredii* bind to DNA sequences preceding *nolX*, a flavonoid-inducible *nod* gene that is not associated with a *nod* box. Mol. Plant-Microbe Interact. 9:457-463.

32. B. G. Rolfe. 1988. Flavones and isoflavones as inducing substances of legume nodulation. Biofactors. 1:3-10.

33. Peters, N. K., J. W. Frost, and S. R. Long. 1986. A plant flavone, luteolin, induces expression of *Rhizobium meliloti* nodulation genes. Science 233:977-980.

34. Redmond, J. W., M. Batley, M. A. Djordjevic, R. W. Innes, P. L. Kuempel, and B. G. Rolfe. 1986. Flavones induce expression of nodulation genes in Rhizobium. Nature 323:632-635.

35. Firmin, J. L., K. E. Wilson, L. Rossen, and A. W. B. Johnston. 1986. Flavonoid activation of nodulation genes in *Rhizobium* reversed by other compounds present in plants. Nature 324:90-92.

36. Djordjevic, M. A., J. W. Redmond, M. Batley, and B. G. Rolfe. 1987. Clovers secrete specific phenolic compounds which either stimulate or repress *nod* gene expression in *Rhizobium trifolii*. EMBO J. 6:1173-1179.

37. Sutherland, T. D., B. J. Bassam, L. J. Schuller, and P. M. Gresshoff. 1990. Early nodulation signals of the wild type and symbiotic mutants of soybean (*Glycine max*). Mol. Plant-Microbe Interact. 3:122-128.

38. Kape, R., M. Parniske, S. Brandt, and D. Werner. 1992. Isoliquiritigenin, a strong *nod* gene- and glyceollin resistance-inducing flavonoid from soybean root exudate. Appl. Environ. Microbiol. 58:1705-1710.

98.544

39. Kosslak, R. M., R. Bookland, J. Barkei, H. E. Paaren, and E. R. Appelbaum. 1987. Induction of *Bradyrhizobium japonicum* common *nod* genes by isoflavones isolated from *Glycine max.* Proc. Natl. Acad. Sci. USA 84:7428-7432.
40. Hungria, M., C. M. Joseph, and D. A. Philips. 1991. *Rhizobium nod* gene inducers exuded naturally from roots of common bean (*Phaseolus vulgaris* L) Plant Physiol. 97:759-764.
41. Hungria, M., C. M. Joseph, and D. A. Philips. 1991. Anthocyanidins and flavonols: major *nod* gene inducers from seeds of a black-seeded common bean (*Phaseolus vulagaris L*). Plant Physiol. 97:751-758.
42. Hungria, M., A. W. B. Johnston, and D. A. Philips. 1992. Effects of flavonoids released naturally from bean (*Phaseolus vulgaris*) on *nodD*-regulated gene transcription in *Rhizobium leguminosarum* bv. *viciae.* Mol. Plant-Microbe Interact. 5:199-203.
43. Maxwell, C. A., and D. A. Philips. 1990. Concurrent synthesis and release of *nod*-gene-inducing flavonoids from alfalfa roots. Plant Physiol. 93:1552-1558.
44. Hartwig, H. A., and D. A. Philips. 1991. Release and modification of *nod*-gene-inducing flabvonoids from alfalfa seeds. Plant Physiol. 95:804-807.
45. Recourt, K., M. Verkerke, J. Schripsema, A. A. N. van Brussel, B. J. J. Lugtenberg and J. W. Kijne. 1992. Major flavonoids in uninoculated and inoculated roots of *Viciae sativa* subsp. *nigra* are four conjugates of the nodulation gene inhibitor kaempferol. Plant. Mol. Biol. 18:505-513.
46. Zaat, S. A. J., J. Schripsema, C. A. Wijffelman, A. A. N. van Brussel, and B. J. J. Lugtenberg. 1989. Analysis of the major inducers of the *Rhizobium nodA* promoter from *Viciae sativa* root exudate and their activity with different *nodD* genes. Plant. Mol. Biol. 13:175-188.
47. Recourt, K., J. Schripsema, J. W. Kijne, A. A. N. van Brussel, and B. J. J. Lugtenberg. 1991. Inoculation of *Viciae sativa* subsp. *nigra* roots with *Rhizobium leguminosarum* bv. *viciae* results in release of *nod* gene activating flavanones and chalcones. Plant. Mol. Biol. 16:841-852.
48. Schmidt, P. E., W. J. Broughton, and D. Werner. 1994. Nod factors of *Bradyrhizobium japonicum* and *Rhizobium* sp. NGR234 induce flavonoid accumulation in soybean exudate. Mol. Plant-Microbe Interact. 7:384-390.
49. Rao, J. R., and J. E. Cooper. 1995. Soybean nodulating rhizobia modify *nod* gene inducers daidzein and genistein to yield aromatic products that can influence gene inducing activity. Plant-Microbe Interact. 8:855-862.
50. Rao, J. R., and J. E. Cooper. 1994. Rhizobia catabolize *nod* gene inducing flavonoids via C-ring fission mechanisms. J. Bacteriol. 176:5409-5413.
51. Hartwig, U. A., C. A. Maxwell, C. M. Joseph, and D. A. Philips. 1990. Effects of alfalfa *nod* gene-inducing flavonoids on *nodABC* transcription in *Rhizobium meliloti* strains containing different *nodD* genes. J. Bacteriol. 172:2769-2773.
52. Aguilar, J. M., A. M. Ashby, J. M. Richards, G. J. Loake, M. D. Watson, and C. H. Shaw. 1988. Chemotaxis of *Rhizobium leguminosarum* biovar *phaseoli* towards flavonoids inducers of the symbiotic nodulation genes. J. Gen. Microbiol. 134:2741-2746.
53. Dharmatilake, A. J., and W. D. Bauer. 1992. Chemotaxis of *Rhizobium meliloti* towards nodulation gene-inducing compounds from alfalfa roots. Appl. Environ. Microbiol. 58:1153-1158.
54. Goethals, K., G. van den Eede, M. van Montagu, and M. Holsters. 1990. Identification and characterization of a functional *nodD* gene in *Azorhizobium caulinodans.* J. Bacteriol. 172:2658-2666.

55. Schlaman, H. R. M., R. J. H. Okker, and B. J. J. Lugtenberg. 1992. Regulation of nodulation gene expression by NodD in rhizobia. J. Bacteriol. 174:5177-5182.
56. Fisher, R. F., and S. R. Long. 1993. Interactions of NodD at the *nod* box: NodD binds two distinct sites on the same face of the helix and induces a bend in the DNA. J. Mol. Biol. 233:336-348.
57. Hong, G.-F., J. E. Burn, and A. W. B. Johnston. 1987. Evidence that DNA involved in the expression of nodulation (*nod*) genes in *Rhizobium* binds to the product of the regulatory gene *nodD*. Nucleic Acids Res. 15:9677-9690.
58. Fisher, R. F., T. T. Egelhoff, J. T. Mulligan, and S. R. Long. 1988. Specific binding of proteins from *Rhizobium meliloti* cell-free extracts containing NodD to DNA sequences upstream of inducible nodulation genes. Genes Dev. 2:282 - 293.
59. Kondorosi, E., J. Gyuris, J. Schmidt, M. John, E. Duda, B. Hoffman, J. Schell, and A. Kondorosi. 1989. Positive and negative control of *nod* gene expression in *Rhizobium meliloti* is required for optimal nodulation. EMBO J. 8:1331-1340.
60. Dénarié, J., F. Debellé, and C. Rosenberg. 1992. Signaling and host range variation in nodulation. Annu. Rev. Microbiol. 46:497-531.
61. Horvath, B., C. W. Bachem, J. Schell, and A. Kondorosi. 1987. Host-specific regulation of nodulation genes in *Rhizobium* is mediated by a signal plant interacting with the *nodD* gene product. EMBO J. 6:841-848.
62. Spaink, H. P., C. A. Wijffelman, E. Pees, R. J. H. Okker, and B. J. J. Lugtenberg. 1987. *Rhizobium* nodulation gene *nodD* as a determinant of host specificity. Nature 328:337-340.
63. Spaink, H. P., R. J. H. Okker, C. A. Wijffelman, T. Tak, L. Goosen-de Roo, E. Pees, A. A. N. van Brussel, and B. J. J. Lugtenberg. 1987. Symbiotic properties of rhizobia containing a flavonoid-independent hybrid *nodD* product. J. Bacteriol. 171:4045-4053.
64. Bender, G. L., M. Nayudu, K. K. L. Strange, and B. G. Rolfe. 1988. The *nodD*1 gene from *Rhizobium* strain NGR234 is a key determinant in the extension of host range to the nonlegume *Parasponia*. Mol. Plant-Microbe Interact. 1:259-266.
65. van Rhijn, P., and J. Vanderleyden. 1995. The *Rhizobium*-plant symbiosis. Microbiol. Rev. 59:124-142.
66. McIver, J., M. A. Djordjevic, J. J. Weinman, G. L. Bender, and B. G. Rolfe. 1989. Extension of host range of *Rhizobium leguminosarum* bv. *trifolii* caused by point mutations in *nodD* that result in alterations in regulatory function and recognition of inducer molecules. Mol. Plant-Microbe Interact. 2:97-106.
67. Györgypal, Z., E. Kondorosi, and A. Kondorosi. 1991. Diverse signal sensitivity of NodD protein homologs from narrow and broad host range rhizobia. Mol. Plant-Microbe Interact. 4:356-364.
68. Le Strange, K. K. L., G. L. Bender, M. A. Djordjevic, B. G. Rolfe, and J. W. Redmond. 1990. The *Rhizobium* strain NGR234 *nodD*1 gene product responds to activation by the simple phenolic compounds vanillin and isovanillin present in wheat seedling extracts. Mol. Plant-Microbe Interact. 3:214-220.
69. Downie, J. A., C. D. Knight, A. W. B. Johnston, and L. Rossen. 1985. Identification of genes and gene products involved in the nodulation of peas by *Rhizobium leguminosarum*. Mol. Gen. Genet. 198:255.
70. van Rhijn, P., J. Desair, K. Vlassak, and J. Vanderleyden. 1994. Functional analysis of *nodD* genes of *Rhizobium tropici* CIAT899. Mol. Plant-Microbe Interact. 7:666-676.

71. Honma, M. A., and F. M. Ausubel. 1987. *Rhizobium meliloti* has three functional copies of the *nodD* symbiotic regulatory gene. Proc. Natl. Acad. Sci. USA 84: 8558-8562.
72. Perret, X., W. J. Broughton, and S. Brenner. 1991. Canonical ordered cosmid library of the symbiotic plasmid of *Rhizobium species* NGR234. Proc. Natl. Acad. Sci. USA 88:1923-1927.
73. Mulligan, J. T., and S. R. Long. 1989. A family of activator genes regulates expression of *Rhizobium meliloti* nodulation genes. Genetics 122:7-18.
74. Mulligan, J. T., and S. R. Long. 1985. Induction of *Rhizobium meliloti nodC* expression by plant exudate requires *nodD*. Proc. Natl. Acad. Sci. USA 82:6609-6613.
75. Swanson, J. A., J. T. Mulligan, and S. R. Long. 1993. Regulation of *syrM* and *nodD3* in *Rhizobium meliloti*. Genetics 134:435-444.
76. Relic', B., R. Fellay, A. Lewin, X. Perret, N. P. J. Price, P. Rochepeau, and W. J. Broughton. 1993. *nod* genes and Nod-factors of *Rhizobium* species NGR234, p. 183-189. *In* R. Palacios, J. Mora, and W. E. Newton (ed.), New horizons in nitrogen fixation. Kluwer Academic Publishers. Dordrecht, The Netherlands.
77. Hanin, M., S. Jabbouri, R. Fellay, and W. B. Broughton. SyrM1 of *Rhizobium* sp. NGR234 controls the level of sulphated Nod factors and activates transcription of several symbiotic loci. in preparation.
78. Göttfert, M., P. Groβ, and H. Hennecke. 1990. Proposed regulatory pathway encoded by the *nodV* and *nodW* genes, determinants of host specificity in *Bradyrhizobium japonicum*. Proc. Natl. Acad. Sci. U. S. A. 87:2680-2684.
79. Sanjuan, J., P. Groβ, M. Göttfert, H. Hennecke, and G. Stacey. 1994. *nodW* is essential for full expression of the common nodulation genes in *Bradyrhizobium japonicum*. Mol. Plant-Microbe Interact. 7:364-369.
80. Loh, J., M. Garcia, and G. Stacey. 1997. NodV and NodW, a second flavonoid recognition system regulating *nod* gene expression in *Bradyrhizobium japonicum*. J. Bacteriol. 179:3013-3020.
81. P. Groβ, P. Michel, H. Henneke, and M. Göttfert. 1993. A novel response-regulator is able to suppress the nodulation defect of a *Bradyrhizobium japonicum nodW* mutant. Mol. Gen. Genet. 241:531-541.
82. Ninfa, A. J., E. G. Ninfa, A. N. Lupas, A. Stock, B. Magasanik, and J. Stock. 1988. Crosstalk between bacterial chemotaxis signal transduction proteins and regulators of transcription of the Ntr regulon: Evidence that nitrogen assimilation and chemotaxis are controlled by a common phosphotransfer mechanism. Proc. Natl. Acad. Sci. USA 85:5492-5496.
83. Schlaman, H. R. M., B. Horvath, E. Vijgenboom, R. J. H. Okker, and B. J. J. Lugtenberg. 1991. Suppression of nodulation gene expression in bacteroids of *Rhizobium leguminosarum* biovar *viciae*. J. Bacteriol. 173:4277-4287.
84. Knight, C. D., L. Rossen, J. G. Robertson, B. Wells, and J. A. Downie. 1986. Nodulation inhibition by *Rhizobium leguminosarum* multiciopy *nodABC* genes and analysis of early stages of plant infection. J. Bacteriol. 166:552-558.
85. Sadowsky, M. J., P. B. Cregan, M. Gottfert, A. Sharma, D. Gerhold, F. Rodriguez-Quinones, H. H. Keyser, H. Hennecke, and G. Stacey. 1991. The *Bradyrhizobium japonicum nolA* gene and its involvement in the genotype-specific nodulation of soybeans. Proc. Natl. Acad. Sci. USA 88:637-641.
86. Dockendorff, T. C., J. Sanjuan, P. Groβ, and G. Stacey. 1994. NolA represses *nod* gene expression in *Bradyrhizobium japonicum*. Mol. Plant-Microbe Interact. 7: 596-602.

87. Garcia, M., J. Dunlap, J. Loh, and G. Stacey. 1996. Phenotype characterization and regulation of the *nolA* gene of *Bradyrhizobium japonicum*. Mol. Plant-Microbe Interact. 9:625-635.
88. Kondorosi, E., M. Pierre, M. Cren, U. Haumann, J. Schell, and A. Kondorosi. 1991. Identification of NolR, a negative transacting factor controlling the *nod* regulon in *Rhizobium meliloti*. J. Mol. Biol. 222:885-896.
89. Fellay, R., M. Hanin, G. Montorzi, C. Friberg, M. Ferro, W. J. Broughton, and S. Jabbouri. *nodD2* of *Rhizobium* sp. NGR234 is involved in repression of the *nodABC* operon. In preparation.
90. Downie, J. A. 1994. Signalling strategies for nodulation of legumes by rhizobia. Trends Microbiol. 2:318-324.
91. Dénarié, J., F. Debellé, and J.-C. Promé. 1996. *Rhizobium* lipo-chitooligosaccharide nodulation factors: Signalling molecules mediating recognition and morphogenesis. Annu. Rev. Biochem. 65:503-535.
92. Spaink, H. P. 1996. Regulation of plant morphogenesis by lipo-chitin oligosaccharides. Crit. Rev. Plant Sci. 15:559-582.
93. Demont, N., M. Ardourel, F. Maillet, D. Promé, M. Ferro, J-C. Promé, and J. Dénarié. 1994. The *Rhizobium meliloti* regulatory *nodD3* and *syrM* genes control the synthesis of a particular class of nodulation factors *N*-acylated by (w-1)-hydroxylated fatty acids. EMBO J. 13:2139-2149.
94. Downie, J. A., C. Marie, A. K. Sheu, J. L. Firmin, K. E. Wilson, A. E. Davies, T. M. Cubo, and A. Mavridou. 1991. Genetic and biochemical analysis studies of the nodulation genes of *Rhizobium leguminosarum* bv. *viciae*, p. 134-141. *In* H. Hennecke and D. P. S. Verma (ed.), Advances in Molecular Genetics of Plant-Microbe Interactions. Kluwer Academic Publishers, Dordrecht, The Netherlands.
95. Baev, N., G. Endre, G. Petrovics, Z. Banfalvi, and A. Kondorosi. 1991. Six nodulation genes of *nod* box locus 4 in *Rhizobium meliloti* are involved in nodulation signal production: *nodM* codes for D-glucosamine synthetase. Mol. Gen. Genet. 228:113-124.
96. Banfalvi, Z., and A. Kondorosi. 1989. Production of root hair deformation factors by *Rhizobium meliloti* nodulation genes in *Escherichia coli*: hsnD(*nodH*) is involved in the plant host-specific modification of the NodABC factor. Plant Mol. Biol. 13:1-12.
97. Spaink, H. P, A. Aarts, G. Stacey, G. V. Bloemberg, B. J. J. Lugtenberg, and E. P. Kennedy. 1992. Detection and separation of *Rhizobium* and *Bradyrhizobium* Nod metabolites using thin layer chromatography. Mol. Plant-Microbe Interact. 5:72-80.
98. Atkinson, E. M, and S. R. Long. 1992. Homology of *Rhizobium meliloti* NodC to polysaccharide polymerizing enzymes. Mol. Plant-Microbe Interact. 5:439-442.
99. Bulawa, C. E. 1992. *CSD1, CSD2* and *CSD3,* genes required for chitin synthesis in *Saccharomyces cerevisiae*: the *CSD2* gene product is related to chitin synthases and to developmentally regulated proteins in *Rhizobium* species and *Xenopus laevis*. Mol. Cell Biol. 12:1764-1776.
100. Kafetzopoulos, D., G. Thireos, J. N. Vournakis, and V. Bouriotis. 1993. The primary structure of a fungal chitin deacetylase reveals the function for two bacterial gene products. Proc. Natl. Acad. Sci. USA 90:8005-8008.
101. Geremia, R. A., P. Mergaert, D. Geelen, M. van Montagu, and M. Holsters. 1994. The NodC protein of *Azorhizobium caulinodans* is an *N*-acetylglucosaminyl-transferase. Proc. Natl. Acad. Sci. USA 91:2669-2673.

102. Debellé, F., P. Roche, C. Plazanet, F. Maillet, C. Pujol, M. Ardourel, N. Demont, C. Rosenberg, G. Truchet, J-C. Promé, and J. Dénarié. 1995. The genetics of *Rhizobium* host range control: allelic and non-allelic variation, p. 275-280. *In* I. A. Tikhonovich, N. A. Provorov, V. I. Romanov, W. E. Newton (ed.), Nitrogen Fixation: Fundamentals and Applications. Netherlands Kluver Academic Publishers.
103. de Iannino, N. I. D., S. G. Pueppke, and R. A. Ugalde. 1995. Biosynthesis of the Nod factor chitooligosaccharide backbone in *Rhizobium fredii* is controlled by the concentration of UDP-N-Acetyl-D-Glucosamine. Mol. Plant-Microbe Interact. 8:292-301.
104. Kamst, E., J. Pilling, L. M. Raamsdonk, B. J. J. Lugtenberg, and H. P. Spaink. 1997. *Rhizobium* nodulation protein NodC is an important determinant of chitin oligosaccharide chain length in Nod factor biosynthesis. J. Bacteriol. 179:2103-2108.
105. Spaink, H. P., A. H. M. Wijfjes, K. M. G. M. van der Drift, J. Haverkamp, J. E. Thomas-Oates, and B. J. J. Lugtenberg. 1994. Structural identification of metabolites produced by the NodB and nodC proteins of *Rhizobium leguminosarum*. Mol. Microbiol. 13:821-831.
106. John, M., H. Röhrig, J. Schmidt, U. Wieneke, and J. Schell. 1993. *Rhizobium* NodB protein involved in nodulation signal synthesis is a chitooligosaccharide deacetylase. Proc. Natl. Acad. Sci. USA 90:625-629.
107. Atkinson, E. M., M. M. Palcic, O. Hindsgaul, and S. Long. 1994. Biosynthesis of *Rhizobium meliloti* lipooligosaccharide Nod factors: NodA is required for an N-acyltransferase activity. Proc. Natl. Acad. Sci. USA 91:8418-8422.
108. Röhrig, H., J. Schmidt, U. Wieneke, E. Kondorosi, I. Barlier, J. Schell. and M. John. 1994. Biosynthesis of lipooligosaccharide nodulation factors-*Rhizobium* NodA protein is involved in *N*-acylation of the chitooligosaccharide backbone. Proc. Natl. Acad. Sci. USA 91:3122-3126.
109. Debellé, F., C. Plazanet, P. Roche, C. Pujol, A. Savagnac, C. Rosenberg, J.-C. Promé, and J. Dénarié. 1996. The NodA proteins of *Rhizobium meliloti* and *Rhizobium tropici* specify the *N*-acylation of Nod factors by different fatty acids. Mol. Microbiol. 22:303-314.
110. Roche, P., F. Maillet, C. Plazanet, F. Debellé, M. Ferro, G. Truchet, J.-C. Promé, and J. Dénarié. 1996. The common *nodABC* genes of *Rhizobium meliloti* are host-range determinants. Proc. Natl. Acad. Sci. USA 93:15305-15310.
111. Demont, N., F. Debellé, H. Aurelle, J. Dénarié, and J-C. Promé. 1993. Role of the *Rhizobium meliloti nodF* and *nodE* genes in the biosynthesis of lipooligosaccharidic nodulation factors. J. Biol. Chem. 268:20134-20142.
112. Spaink, H. P., D. M. Sheeley, A. A. N. van Brussel, J. Glushka, W. S. York, T. Tak, O. Geiger, E. P. Kennedy, V. N. Reinhold, and B. J. J. Lugtenberg. 1991. A novel highly unsaturated fatty acid moiety of lipo-oligosaccharide signals determines host specificity of *Rhizobium*. Nature 354:125-130.
113. Shearman, C. A., L. Rossen, A. W. B. Johnston, and J. A. Downie. 1986. The *Rhizobium leguminosarum* nodulation gene *nodF* encodes a polypeptide similar to acyl-carrier protein and is regulated by *nodD* plus a factor in pea root exudate. EMBO J. 5:647-652.
114. Geiger, O., H. P. Spaink, and E. P. Kennedy. 1991. Isolation of the *Rhizobium leguminosarum* NodF nodulation protein: NodF carries a 4'-phosphopantetheine prosthetic group. J. Bacteriol. 173:2872-2878.
115. Bibb, M. J., S. Biro, H. Motamedi, J. F. Collins, and C. R. Hutchinson. 1989. Analysis of the nucleotide sequence of the *Streptomyces glaucescens tcmI* genes

provides key information about the enzymology of polyketide antibiotic synthesis. EMBO J. 8:2727-2736.

116. Göttfert, M., S. Hitz, and H. Hennecke. 1990. Identification of *nodS* and *nodU*, two inducible genes inserted between the *Bradyrhizobium japonicum nodYABC* and *nodIJ* genes. Mol. Plant-Microbe Interact. 3:308-316.

117. Lewin, A., E. Cervantes, C.-H. Wong, and W. J. Broughton. 1990. *nodSU*, two new nod-genes of the broad host range *Rhizobium* strain NGR234 encode host-specific nodulation of the tropical tree *leucaena leucocephala*. Mol. Plant-Microbe Interact. 3:317-326.

118. Geelen, D., P. Mergaert, R. A. Geremia, S. Goormachtig, M. van Montagu, and M. Holsters. 1993. Identification of *nodSUIJ* genes on Nod *locus1* of *Azorhizobium caulinodans*: evidence that *nodS* encodes a methyltransferase involved in Nod factor modification. Mol. Microbiol. 9:145-154.

119. Krishnan, H. B., A. Lewin, R. Fellay, W. J. Broughton, and S. G. Pueppke. 1992. Differential expression of *nodS* accounts for the varied abilities of *Rhizobium fredii* USDA257 and *Rhizobium* sp. strain NGR234 to nodulate *leucaena* spp. Mol. Microbiol. 6:3321-3330.

120. Geelen, D., B. Leyman, P. Mergaert, K. Klarskov, M. van Montagu, R. Geremia, and M. Holsters. 1995. NodS is an *S*-adenosyl-L-methionine-dependent methyltransferase that methylates chitooligosaccharides deacetylated at the non-reducing end. Mol. Microbiol. 17:387-397.

121. Mergaert, P., W. D'Haeze, D. Geelen, D. Promé, M. van Montagu, R. Geremia, J.-C. Promé, and M. Holsters. 1995. Biosynthesis of *Azorhizobium caulinodans* Nod factor: study of the activity of the NodABC proteins by expression of the genes in *Escherichia coli*. J. Biol. Chem. 270:29217-29223.

122. Stokkermans, T. J. W., R. Orlando, V. S. K. Kolli, R. W. Carlson, and N. K. Peters. 1996. Biological activities and structures of *Bradyrhizobium elkanii* low abundance lipo chitin-oligosaccharides. Mol. Plant-Microbe Interact. 9:298-304.

123. Roche, P., P. Lerouge, C. Ponthus, and J-C. Promé. 1991. Structural determination of bacterial nodulation factors involved in the *Rhizobium meliloti*-alfalfa symbiosis. J. Biol. Chem. 266:10933-10940 .

124. Bloemberg, G. V., J. E. Thomas-Oates, B. J. J. Lugtenberg, and H. P. Spaink. 1994. Nodulation protein NodL of *Rhizobium leguminosarum* O-acetylates lipo-oligosaccharides, chitin fragments and *N*-acetylglucosamine *in vitro*. Mol. Microbiol. 11:793-804.

125. Price, N. P. J., B. Relič¢, F. Talmont, A. Lewin, D. Promé, S. G. Pueppke, F. Maillet, J. Dénarié, J-C. Promé, and W. J. Broughton. 1992. Broad-host-range *Rhizobium* species strain NGR234 secretes a family of carbamoylated and fucosylated nodulation signals that are O-acetylated or sulphated. Mol. Microbiol. 6:3575-3584.

126. D'Haeze, W., P. Mergaert, M. Fernandez-Lopez, M. Gao, M. van Montagu, J-C. Promé, and M. Holsters. 1996. Nod factors of *Azorhizobium caulinodans*: structure and secretion. 2^(nd) European Nitrogen Fixation Conference and NATO Advanced Research Workshop. Poznan: Poland. Unpublished.

127. Folch-Mallol, J. L., S. Marroqui, C. Sousa, H. Manyani, I. M. Lopez-Lara, K. M. G. M van der Drift, H. P. Spaink, and M. Megias. 1996. Characterization of *Rhizobium tropici* CIAT899 nodulation factors: the role of *nodH* and *nodPQ* genes in their sulfation. Mol. Plant-Microbe Interact. 9:151-163.

128. Mergaert, P., W. D'Haeze, M. Fernandez-lopez, D. Geelen, K. Goethals, J-C. Promé, M. van Montagu, and M. Holsters. 1996. Fucosylation and arabinosylation

of Nod factors in *Azorhizobium caulinodans*: involvement of *nolK*, *nodZ* as well as *noeC* and/or downstream genes. Mol. Microbiol. 21:409-419.

129. Schultze, M., C. Staehelin, H. Röhrig, M. John, J. Schmidt, E. Kondorosi, J. Schell, and A. Kondorosi. 1995. *In vitro* sulfotransferase activity of *Rhizobium meliloti* NodH protein: lipochitooligosaccharide nodulation signals are sulfated after synthesis of the core structure. Proc. Natl. Acad. Sci. USA 92:2706-2709.

130. Ehrhardt, D. W., E. M. Atkinson, K. F. Faull, D. I. Freedberg, D. P. Sutherlin, R. Armstrong, and S. R. Long. 1995. *In vitro* sulfotransferase activity of NodH, a nodulation protein of *Rhizobium meliloti* required for host-specific nodulation. J. Bacteriol. 177:6237-6245.

131. Schwedock, J., and S. R. Long. 1990. ATP sulphurylase activity of the *nodP* and *nodQ* gene products of *Rhizobium meliloti*. Nature 348:644-647.

132. Schwedock, J., C. Liu, T. Leyh, and S. R. Long. 1994. *Rhizobium meliloti* NodP and NodQ form a multifunctional sulfate-activating complex requiring GTP for activity. J. Bacteriol. 176:7055-7064.

133. Laeremans, T., I. Caluwaerts, C. Verreth, J. Vanderleyden, and E. Martinez-Romero. 1996. Isolation and characterization of the *Rhizobium tropici* Nod factor sulfation genes. Mol. Plant-Microbe Interact. 9:492-500.

134. Cloutier, J., S. Laberge, Y. Castonguay, and H. Antoun. 1996. Charaterization and mutational analysis of *nodHPQ* genes of *Rhizobium* sp. strain N33. Mol. Plant-Microbe Interact. 9:720-728.

135. Poupot, R., E. Martinez-Romero, F. Maillet, and J.-C. Promé. 1995. *Rhizobium tropici* nodulation factor sulfation is limited by the quantity of activated form of sulfate. FEBS Letters 368:536-540.

136. Davis, E. O., I. J. Evans, and A. W. B. Johnston. 1988. Identification of *nodX*, a gene that allows *Rhizobium leguminosarum* biovar *viciae* strain TOM to nodulate Afghanistan peas. Mol. Gen. Genet. 212:531-535.

137. Firmin, J. L., K. E. Wilson, R. W. Carlson, A. E. Davies, and J. A. Downie. 1993. Resistance to nodulation of cv. Afghanistan peas is overcome by *nodX*, which mediates an *O*-acetylation of the *Rhizobium leguminosarum* lipo-oligosaccharide nodulation factor. Mol. Microbiol. 10:351-360.

138. Lopez-Lara, I. M., L. Blok-Tip, C. Quinto, M. L. Garcia, G. V. Bloemberg, G. E. M. Lamers, D. Kafetzopoulos, G. Stacey, B. J. J. Lugtenberg, J. E. Thomas-Oates, and H. P. Spaink. 1996. NodZ of *Bradyrhizobium* extends the nodulation host range of *Rhizobium* by adding a fucosyl residue to nodulation factors. Mol. Microbiol. 21:397-408.

139. Quesada-Vincens, D., R. Fellay, T. Nassim, V. Viprey, U. Burger, J.-C. Promé, W. J. Broughton, and S. Jabbouri. 1997. *Rhizobium* sp. NGR234 NodZ protein is a fucosyltransferase. J. Bacteriol. 179:5087-5093.

140. Chang, S., B. Duerr, and G. Serif. 1988. An epimerase-reductase in L-fucose synthesis. J. Biol. Chem. 263:1693-1697.

141. Scott, D. B., C. A. Young, J. E. Collins-Emerson, E. A. Terzaghi, E. S. Rockman, P. E. Lewis, and C. E. Pankhurst. 1996. Novel and complex chromosomal arrangement of nodulation genes. Mol. Plant-Microbe Interact. 9:187-197.

142. Hanin, M., S. Jabbouri, D. Quesada-Vincens, C. Freiberg, X. Perret, J.-C. Promé, W. J. Broughton, and R. Fellay. 1997. Sulphation of *Rhizobium* sp. NGR234 Nod factors is dependent on *noeE*, a new host-specificity gene. Mol. Microbiol. 24:1119-1129.

143. Carlson, R. W., N. P. J. Price, and G. Stacey. 1994. The biosynthesis of rhizobial lipo-oligosaccharide nodulation signal molecules. Mol. Plant-Microbe Interact. 7:684-695.

144. Orgambide, G. G., J. I. Lee, R. I. Hollingsworth, and F. B. Dazzo. 1995. Structurally diverse chitolipooligosaccharide Nod factors accumulate primarily in membranes of wild-type *Rhizobium leguminosarum* biovar *trifolii*. Biochemistry 34: 3832-3840.

145. Spaink, H. P., A. H. M. Wijfjes, and B. J. J. Lugtenberg. 1995. *Rhizobium* NodI and NodJ proteins play a role in the efficiency of secretion of lipochitin oligosaccharides. J. Bacteriol. 177:6276-6281.

146. Fernandez-Lopez, M., W. D'Haeze, P. Mergaert, C. Verplancke, J-C. Promé, M. van Montagu, and M. Holsters. 1996. Role of *nodI* and *nodJ* in lipochito-oligosaccharide secretion in *Azorhizobium caulinodans* and *Escherichia coli*. Mol. Microbiol. 20:993-1000.

147. McKay, I. A., and M. A. Djordjevic. 1993. Production and excretion of Nod metabolites by *Rhizobium leguminosarum* bv. *trifolii* are disrupted by the same environmental factors that reduce nodulation in the field. App. Environ. Microbiol. 59:3385-3392.

148. Cardenas, L., J. Dominguez, O. Santana, and C. Quinto. 1996. The role of *nodI* and *nodJ* genes in the transport of Nod metabolites in *Rhizobium etli*. Gene 173:183-187.

149. Vàzquez, M., O. Santana, and C. Quinto. 1993. The NodI and NodJ proteins from *Rhizobium* and *Bradyrhizobium* strains are similar to capsular polysaccharide secretion proteins from Gram-negative bacteria. Mol. Microbiol. 8:369-377.

150. Reizer, J., A. Reizer, and J. M. H. Saier. 1992. A new family of bacterial ABC-type transport systems catalysing export of drugs and carbohydrates. Protein Science 1:1326-1332.

151. Schlaman, H. R. M., R. J. H. Okker, and B. J. J. Lugtenberg. 1990. Subcellular location of the *Rhizobium leguminosarum nodI* gene product. J. Bacteriol.. 172: 5486-5489.

152. Saier, M. H., R. Tam, A. Reizer, and J. Reizer. 1994. Two novel families, of bacterial membrane proteins concerned with nodulation, cell division and transport. Mol. Microbiol. 11:841-847.

153. Horvath, B., E. Kondorosi, M. John, J. Schmidt, I. Török, Z. Györgypal, I. Barbaras, U. Wieneke, J. Schell, and A. Kondorosi. 1986. Organisation, structure and symbiotic function of *Rhizobium meliloti* nodulation genes determining host specificity for alfalfa. Cell 46:335-343.

154. Debellé, F., F. Maillet, J. Vasse, C. Rosenberg, F. De Billy, G. Truchet, J. Dénarié, and F. M. Ausubel. 1988. Interference between *Rhizobium meliloti* and *Rhizobium trifolii* nodulation genes: Genetic basis of *R. meliloti* dominance. J. Bacteriol. 170:5718-5727.

155. Ardourel, M., N. Demont, F. Debellé, F. Maillet, F. DeBilly, J-C. Promé, J. Dénarié, and G. Truchet. 1994. *Rhizobium meliloti* lipooligosaccharide nodulation factors: different structural requirements for bacterial entry into target root hair cells and induction of plant symbiotic developmental responses. Plant Cell 6:1357-1374.

156. Waelkens, F., T. Voets, K. Vlassak, J. Vanderleyden, and P. van Rhijn. 1995. The *nodS* gene of *Rhizobium tropici* CIAT899 is necessary for nodulation of *Phaseolus vulgaris* and *Leucaena leucocephala*. Mol. Plant-Microbe Interact. 8:147-154.

157. van Rhijn, P., E. Luyten, K. Vlassak, and J. Vanderleyden. 1996. Isolation and characterization of a pSym locus of *Rhizobium* sp. BR816 that extends nodulation ability of narrow host range *Phaseolus vulgaris* symbionts to *Leucaena leucocephala*. Mol. Plant-microbe Interact. 9:74-77.

158. Downie, J. A., and B. Surin. 1990. Either of two *nod* gene loci can complement the nodulation defect of a *nod* deletion mutant of *Rhizobium leguminosarum* bv. *viciae*. Mol. Gen. Genet. 222:81-86.

159. Economou, A., A. E. Davies, A. W. B. Johnston, and J. A. Downie. 1994. The *Rhizobium leguminosarum* biovar *viciae nodO* gene can enable a *nodE* mutant of *Rhizobium leguminosarum* biovar *trifolii* to nodulate vetch. Microbiology 1994: 2341-2347.

160. Cardenas, L., J. Dominguez, C. Quinto, I. M. Lopez-lara, B. J. J. Lugtenberg, H. P. Spaink, G. J. Rademaker, J. Haverkamp, and J. E. Thomas-Oates. 1995. Isolation, chemical structure and biological activity of the lipo-chitin oligosaccharide nodulation signals from *Rhizobium etli*. Plant Mol. Biol. 29:453-464.

161. Poupot, R., E. Martinez-Romero, and J.-C. Promé. 1993. Nodulation factors from *Rhizobium tropici* are sulfated or non-sulfated chitopentasaccharides containing an *N*-methyl-*N*-acylglucosamine terminus. Biochemistry 32:10430-10435.

162. Poupot, R., E. Martinez-Romero, N. Gautier, and J.-C. Promé. 1995. Wild-type *Rhizobium etli*, a bean symbiont, produces acetyl-fucosylated, *N*-methylated, and carbamoylated nodulation factors. J. Biol. Chem. 270:6050-6055.

163. Relic', B., C. Staehelin, R. fellay, S. Jabbouri, T. Boller, and W. J. Broughton. 1994. Do Nod-factor levels play a role in host-specificity?, p. 69-75. *In* G. B. Kiss and G. Endre (ed.), Proc. 1st European Nitrogen Fixation Conf. Officina Press, Szeged, Hungary.

164. Mergaert, P. , M. van Montagu, J.-C. Promé, and M. Holsters. 1993. Three unusual modifications, a D-arabinosyl, a *N*-methyl, and a carbamoyl group, are present on the Nod factors of *Azorhizobium caulinodans* strain ORS571. Proc. Natl. Acad. Sci. USA 90:1551-1555.

165. Carlson, R. W., S. J. Juan, U. R. Bhat, J. Glushka, H. P. Spaink, A. H. M. Wijfjes, A. N. N. van Brussel, T. J. W. Stokkermans, N. K. Peters, and G. Stacey. 1993. The structures and biological activities of the lipo-oligosaccharide nodulation signals produced by type-1 and type-2 strains of *Bradyrhizobium japonicum*. J. Biol. Chem. 268:18372-18381.

166. Sanjuan, J., R. W. Carlson, H. P. Spaink, U. R. Bhat, W. M. Barbour, J. Glushka, and G. Stacey. 1992. A 2-*O*-methylfucose moiety is present in the lipo-oligo-saccharide nodulation signal of *Bradyrhizobium japonicum*. Proc. Natl. Acad. Sci. USA 89:8789-8793.

167. Bec-Ferté, M. P., H. B. Krishnan, D. Promé, A. Savagnac, S. G. pueppke, and J.-C. Promé. 1994. Structures of nodulation factors from the nitrogen-fixing soybean symbiont *Rhizobium fredii* USDA257. Biochemistry 33:11782-11788.

168. van Brussel, A. A. N., R. Bakhuizen, P. C. van Spronsen, H. P. Spaink, T. Tak, B. J. J. Lugtenberg, and J. W. Kijne. 1992. Induction of pre-infection thread structures in the host plant by lipo-oligosaccharides of *Rhizobium*. Science 257:70-72.

169. Truchet, G., D. G. Barker, S. Camut, F. de Billy, J. Vasse, and T. Huguet. 1989. Alfalfa nodulation in the absence of *Rhizobium*. Mol. Gen. Genet. 219:65-68.

170. Relic', B., X. Perret, M. T. Estrada-Garcia, J. Kopcinska, W. Golinowski, H. B. Krishnan, S. G. Pueppke, and W. J. Broughton. 1994. Nod factors of *Rhizobium* are a key to the legume door. Mol. Microbiol. 13:171-178.

171. Ehrhardt, D. W., E. M. Atkinson, and S. R. Long. 1992. Depolarization of alfalfa root hair membrane potential by *Rhizobium meliloti* Nod factors. Science 256:998-1000.

172. Felle, H. H., E. Kondorosi, A. Kondorosi, and M. Schultze. 1995. Nod signal-induced plasma membrane potential changes in alfalfa root hairs are differentially

sensitive to structural modification of the lipochitooligosaccharide. Plant J. 7:939-947.

173. Kurkdjian, A. C. 1995. Role of the differentiation of root epidermal cells in Nod factor (from *Rhizobium meliloti*)-induced root-hair depolarization of *Medicago sativa*. Plant Physiol. 107:783-790.

174. Allen, N. S., M. N. Bennet, D. N. Cox, A. Shipley, D. W. Ehrhardt, and S. R. Long. 1994. Effects of Nod factors on alfalfa root hair Ca++, and H+ currents and on cytoskeletal behavior, p. 107-114. *In* M. J. Daniels, J. A. Downie, and A. E. Obsourne (ed.), Advances in Molecular Genetics of Plant-Microbe Interactions. Kluwer: Dordrecht , The Netherlands.

175. Gehring, C. A., H. R. Irving, A. A. Kabbara, R. W. Parish, N. M. Boukli, W. J. Broughton. 1997. Rapid, plateau-like increases in intra-cellular free calcium are associated with Nod-factor induced root-hair deformation. Mol. Plant-Microbe Interact., submitted.

176. Ehrhardt, D. W., R. Wais, and S. R. Long. 1996. Calcium spiking in plant root hairs responding to *Rhizobium* nodulation signals. Cell 85:673-681.

177. Economou, A., W. D. O. Hamilton, A. W. B. Johnston, and J. A. Downie. 1990. The *Rhizobium* nodulation gene *nodO* encodes a Ca²⁺-binding protein that is exported without *N*-terminal cleavage and is homologous to haemolysin and related proteins. EMBO J. 9:349-354.

178. Sutton, J. M., E. J. A. Lea, and J. A. Downie. 1994. The nodulation-signaling protein NodO from *Rhizobium leguminosarum* biovar *viciae* forms ion channels in membranes. Proc. Natl. Acad. Sci. USA 91:9990-9994.

179. Kijne, J. W., R. Bakhuizen, A. A. N. van Brussel, H. C. G. Canter-Cremers, C. L. Diaz, B. S. de Pater, G. Smit, H. P. Spaink, S. Swart, C. A. Wijffelman, and B. J. J. Lugtenberg. 1992. The *Rhizobium* trap: root hair curling in root-nodule symbiosis, p267-284. *In* Perspectives in Plant Cell Recognition, Society for Experimental Biology Seminar Series. Cambridge University Press.

180. Dénarié, J., and J. Cullimore. 1993. Lipo-oligosaccharide nodulation factors: A new class of signaling molecules mediating recognition and morphogenesis. Cell 74:951-954.

181. Hirsh, A. M. 1992. Developmental biology of legume nodulation. New Phytol. 122:211-237.

182. Schultze, M., E. Kondorosi, P. Ratet, M. Buire, and A. Kondorosi. 1994. Cell and molecular biology of *Rhizobium*-plant interactions. Int. Rev. Cytol. 156:1-75.

183. Spaink, H. P., G. V. Bloemberg, A. A. N. van Brussel, B. J. J. Lugtenberg, K. M. G. M. van der Drift, J. Haverkamp, and J. E. Thomas-Oates. 1995. Host specificity of *Rhizobium leguminosarum* is determined by the hydrophobicity of highly unsaturated fatty acyl moieties of the nodulation factors. Mol. Plant-Microbe Interact. 8:155-164.

184. Savouré, A., Z. Magyar, M. Pierre, S. Brown, M. Schultze, D. Dudits, A. Kondorosi, and E. Kondorosi. 1994. Activation of the cell cycle machinery and the isoflavonoid biosynthesis pathway by active *Rhizobium meliloti* Nod signal molecules in *Medicago* microcallus suspension. EMBO J. 13:1093-1102.

185. Bono, J. J., J. Riond, K. C. Nicolaou, N. J. Bockovich, V. A. Estevez, J. V. Cullimore, and R. Ranjeva. 1995. Characterization of binding site for chemically synthesized lipo-oligosaccharide NodRm factors in particulate fractions prepared from roots. Plant J. 7:253-260.

186. Staehelin, C., J. Granado, J. Müller, A. Wiemken, R. B. Mellor, G. Felix, M. Regenass, W. J. Broughton, and T. Boller. 1994. Perception of *Rhizobium*

nodulation factors by tomato cells and inactivation by root chitinases. Proc. Natl. Acad. Sci. USA 91:2196-2200.

187. Staehelin, C., M. Schultze, E. Kondorosi, R. B. Mellor, T. Boller, and A. Kondorosi. 1994. Structural modifications in *Rhizobium meliloti* Nod factors influence their stability against hydrolysis by root chitinases. Plant J. 5:319-330.

188. Staehelin, C., M. Vanney, F. Foucher, E. Kondorosi, M. Schultze, and A. Kondorosi. 1997. Degradation of nodulation signals from *Rhizobium meliloti* by its host plants, p. 43. *In* A. Legocki, H. Bothe, and A. Pühler (ed.), Biological fixation of nitrogen for ecology and sustainable agriculture. Springer-Verlag, Berlin.

189. Meinhardt, L. W., H. B. Krishnan, P. A. Balatti, and S. G. Pueppke. 1993. Molecular cloning and characterization of a sym plasmid locus that regulates cultivar-specific nodulation of soybean by *Rhizobium fredii* USDA257. Mol. Microbiol. 9:17-29.

190. Balatti, P. A., L. G. Kovacs, H. B. Krishnan, and S. G. Pueppke. 1995. *Rhizobium* sp. NGR234 contains a functional copy of the soybean cultivar specificity locus, *nolXWBTUV*. Mol. Plant-Microbe Interact. 8:693-699.

191. Dazzo, F. B., R. Hollingsworth, S. Philip-Hollingsworth, M. Robeles, T. Olen, J. Salzwedel, M. Djordjevic, and B. Rolfe. 1988. Recognition process in the *Rhizobium trifolii*-white clover symbiosis, p. 431-435. *In* H. Bothe, F. J. D. Bruijn, W. E. Newton (ed.), Nitrogen Fixation: Hundred Years After. Gustav Fischer. Stuttgart.

192. Davis, E. O., and A. W. B. Johnston. 1990. Analysis of three *nodD* genes in *Rhizobium leguminosarum* biovar *phaseoli*, *nodD1* is preceded by *nolE*, a gene whose product is secreted from the cytoplasm. Mol. Microbiol. 4:921-932.

193. Djordjevic, M. A., and J. J. Weinman. 1991. Factors determining host recognition in the clover-*Rhizobium* symbiosis. Aust. J. Plant Physiol. 18:543-557.

194. Marie, C., M. A. Barny, and J. A. Downie. 1992. *Rhizobium leguminosarum* has two glucosamine synthases, GlmS and NodM, required for nodulation and development of nitrogen-fixing nodules. Mol. Microbiol. 6:843-851.

195. Bloemberg, G. V., R. M. Lagas, S. van Leeuwen, G. A. van der Marel, J. H. van Boom, B. J. J. Lugtenberg, and H. P. Spaink. 1995. Substrate specificity and kinetic studies of nodulation protein NodL of *Rhizobium leguminosarum*. Biochemistry 34:12712-12720.

196. Ardourel, M., G. Lortet, F. Maillet, P. Roche, G. Truchet, J-C. Promé, and C. Rosenberg. 1995. In *Rhizobium meliloti*, the operon associated with the *nod* box n5 comprises *nodL*, *noeA* and *noeB*, three host-range genes specifically required for the nodulation of particular *Medicago* species. Mol. Microbiol. 17:687-699.

197. Jabbouri, S., M. Hanin, R. Fellay, D. Quesada-Vincens, B. Reuhs, R. W. Carlson, X. Perret, C. Feiberg, A. Rosenthal, D. Leclerc, W. J. Broughton, and B. Relic'. 1996. *Rhizobium* species NGR234 host-specificity of nodulation locus III contains *nod*- and *fix*-genes, p. 319-324. *In* G. Stacey, B. Mullin, and P. M. Gresshoff (ed.), Biology of Plant-Microbe Interactions. International Society for Molecular Plant-Microbe Interactions. St. Paul, MN, USA.

198. Fellay, R., X. Perret, V. Viprey, W. J. Broughton, and S. Brenner. 1995. Organisation of host-inducible transcripts on the symbiotic plasmid of *Rhizobium* sp. NGR234. Mol. Microbiol. 16:657-667.

199. Surin, B. P., J. M. Watson, W. D. O. Hamilton, A. Economou, and J. A. Downie. 1990. Molecular characterization of the nodulation gene, *nodT*, from two biovars of *Rhizobium leguminosarum*. Mol. Microbiol. 4:245-252.

200. Rivilla, R., J. M. Sutton, and J. A. Downie. 1995. *Rhizobium leguminosarum* NodT is related to a family of outer-membrane transport proteins that includes TolC, PrtF, CyaE and AprF. Gene 161:27-31.
201. Cren, M., A. Kondorosi, and E. Kondorosi. 1995. NolR controls expression of the *Rhizobium meliloti* nodulation genes involved in the core Nod factor synthesis. Mol. Microbiol. 15:733-747.
202. Lopez-Lara, I. M., J. D. J. Van den Berg, J. E. Thomas-Oates, J. Glushka, B. J. J. Lugtenberg, and H. P. Spaink. 1995. Structural identification of the lipo-chitin oligosaccharide nodulation signals of *Rhizobium loti*. Mol. Microbiol. 15:927-638.
203. Stokkermans, J. J. W., and N. K. Peters. 1994. *Bradyrhizobium elkanii* lipo-oligosaccharide signals induce complete nodule structures on *Glycine soja* Siebold & Zucc. Planta 193:413-420.
204. Horvath, B, R. Heidstra, M. Lados, M. Moerman, H. P. Spaink, J.-C. Promé, A. van Kammen, and T. Bisseling. 1993. Lipo-oligosaccharides of *Rhizobium* induce infection-related early nodulin gene expression in pea root hairs. Plant J. 4:727-733.
205. Scheres, B. C. van de Wiel, A. Zalenski, B. Horvath, H. P. Spaink, H. van Eck, F. Zwartkruis, A.-M. Wolters, T. Gloudemans, A. van Kammen, and T. Bisseling. 1990. The *Enod12* gene product is involved in the infection process during the pea-*Rhizobium* interaction. Cell 60:281-294.
206. Scheres, B., H. I. McKhan, A. Zalensky, M. Löbler, T. Bisseling, and A. M. Hirsch. 1992. The *PsENOD12* gene is expressed at two different sites in Afghanistan pea pseudonodules induced by auxin transport inhibitors. Plant Physiol. 100:1649-1655.
207. Pichon, M., E.-P. Journtem, A. Dedieu, F. de Billy, G. Truchet, and D. G. Barker. 1992. *Rhizobium meliloti* elicits transient expression of the early nodulin gene ENOD12 in the differentiating root epidermis of transgenic alfalfa. Plant Cell 4:1199-1211.
208. Csanadi, G., J. Szecsi, P. Kalo, G. Endre, A. Kondorosi, E. Kondorosi, and G. B. Kiss. 1994. ENOD12, an early nodulin gene, is not required for nodule formation and efficient nitrogen fixation in alfalfa. Plant Cell 6:201-213.
209. Allison, L. A., G. B. Kiss, P. Bauer, M. Poiret, M. Pierre, A. Savouré, E. Kondorosi, and A. Kondorosi. 1993. Identification of two alfalfa early nodulin genes with homology to members of the pea *Enod12* gene family. Plant Mol. Biol. 21:375-380.
210. McKhann, H. I., N. L. Paiva, R. A. Dixon, and A. M. Hirsch. 1997. Chalcone synthase transcripts are detected in alfalfa root hairs following inoculation with wild-type *Rhizobium meliloti*. Mol. Plant-Microbe Interact. 10:50-58.
211. Arsenijevic'-Maksimovic', I., W. J. Broughton, and A. Krause. 1997. Rhizobia modulate root-hair-specific expression of extensin genes. Mol. Plant-Microbe Interact. 10:95-101.
212. Krause, A., C. J. A. Sigrist, I. Dehning, H. Sommer, and W. J. Broughton. 1994. Accumulation of transcripts encoding a lipid transfer-like protein during deformation of nodulation-competent *Vigna unguiculata* root hairs. Mol. Plant-Microbe Interact. 7:411-418.
213. Krause, A., Vo T. T. Lan, and W. J. Broughton. 1997. Induction of chalcone synthase expression by rhizobia and Nod factors in root hairs and roots. Mol. Plant-Microbe Interact. 10:388-393.

CHAPTER 2

The *Xanthomonas avr/pth* Gene Family

Dean W. Gabriel

There have been several recent comprehensive reviews of bacterial avirulence (*avr*) genes, reviews including the *Xanthomonas avr/pth* gene family (27,41,42). However, much new information about microbial pathogenesis, including studies of animal pathogens, has converged to provide a more comprehensive picture of pathogenesis and a better perspective on the role played by *avr* genes in plant pathogenesis. Avirulence genes in plant pathogenic microbes trigger programmed plant defense responses (7), and limit virulence. Most *avr* genes are individually dispensable and many appear to be gratuitous. Why are they there? The purpose of this review is to present a perspective on the largest group of *avr* genes cloned to date, members of the *Xanthomonas avr/pth* gene family, in the light of recent discoveries in the field of bacterial pathogenesis.

Avirulence genes were discovered by classical genetic analyses of naturally occurring resistance of plant host species to races of a pathogen. Resistance in hosts was nearly always found to be due to single resistance (*R*) genes, and avirulence in the pathogen was nearly always found to be due to single *avr* genes (15,24). All classical genetic analyses were limited to studies of intraspecific variation in host/pathogen interactions, not plant/ microbe interactions generally. Until the advent of molecular genetics, the question of how a plant species became a host to a given pathogen could not be addressed. Are *avr* genes involved in limiting host range in a negative sense, or are they positively required in conditioning pathogenicity and/or virulence, with avirulence being an unusual phenotype (albeit a useful one exploited by breeders)? What adaptations are required on the part of microbes to become pathogens and what is required to attack a specific plant species—that is, to make the species a host? These questions are now being addressed, and the answers provide some perspective on the potentially different roles played by *avr* genes in pathogenesis. The genetic data on pathogenicity and avirulence are considered below and may be conceptually organized into four general requirements for a successful pathogenic lifestyle: 1) overcome preformed defenses; 2) derive nutrition from the plant; 3) move to new infection sites, and 4) avoid recognition by plant *R* genes.

Ability to Overcome Preformed Defenses

It is clear that preformed structural barriers of plants must be breached and plant-specific antimicrobial compounds must be either tolerated or degraded by a microbial pathogen (31). It is practical in some cases to directly screen plants for mutations with enhanced disease susceptibility, at least to opportunistic plant pathogens (18). This sort of analysis should be useful in identifying both preformed as well as active defense compounds and mechanisms, as it makes no assumptions regarding pathogen specificity. It may even be useful for increasing resistance against all races of a given pathogen generally (e.g., "horizontal" resistance *sensu* [38]).

Many pathogens require wounding of the plant prior to gaining entry, while others utilize a combination of degradative enzymes and/or specialized penetration structures to breach the plant cuticular layer and outer cell wall. Still others, such as the rusts and many bacterial biotrophs, enter the plant through stomata. The degradative enzymes are part of the normal assemblage of tools required by saprophytes to derive nutrition, and many may not be needed for pathogenic purposes. However, some of these may be adapted to pathogenic purposes (26). The process of wounding a plant mechanically, enzymatically or both may release a variety of preformed defense compounds effective against a broad spectrum of microbes. Gram-negative bacterial plant pathogens appear to utilize the lipopolysaccharide (LPS) cell wall component as a selective barrier to exclude specific antimicrobial compounds, and such avoidance may help determine the host range of bacteria (9,25). To date, there is no evidence that pathogen genes involved in overcoming preformed defenses are also *avr* genes.

Ability to Derive Nutrition

Necrotrophs and biotrophs appear to utilize an entirely different attack strategy to derive nutrition from plants (23). Necrotrophs produce abundant degradative enzymes and tend to have limited host cell contact. They derive nutrition by killing living cells and then living off dead and dying host cells. By contrast, biotrophs produce few degrading enzymes and have extensive contact with living host cells. Although biotrophs cause cell death, they appear to avoid this as long as possible. Avirulence genes have not been found in necrotrophs. Instead, all *avr* genes identified to date are found in biotrophs. Could *avr* genes contribute to the pathogenicity of biotrophs by assisting in releasing nutrients from plants?

The finding that mutations in nearly all genes affecting pathogenicity also affect the hypersensitive response (HR; 29) was surprising. However, since *avr* genes were known to be required for the HR, this provided a clue that *avr* gene function was linked to pathogenicity gene function. In fact, these pathogenicity genes were named *hrp* because they affected both the h̲ypersensitive r̲esponse and p̲athogenicity of strains that carried them. Not all microbial pathogenicity genes are *hrp* genes, but most are. Among the biotrophs, most pathogenicity genes that are not *hrp* genes are also *avr* genes (see below). Necrotrophic and biotrophic bacterial pathogens in the Rhizobiaceae family and *Erwinia* genus have *hrp* genes, which encode a novel type III protein secretion system found only in bacterial pathogens (1,32,39). A type III protein secretion system is also essential for diseases caused by some animal pathogens. Most mutations that abolish pathogenicity in biotrophic plant pathogenic *Erwinia, Pseudomonas* and *Xanthomonas* involve these type III export genes. There is a difference between necrotrophs and biotrophs in this regard; *hrp*[!] mutations in necrotrophs merely reduce infectivity, whereas *hrp*[!] mutations in biotrophs usually abolish all pathogenic phenotypes (1). Evidently, type III secretion is utilized by both necrotrophic and biotrophic bacterial species, but is much more important to the biotrophs.

Erwinia, Pseudomonas and *Xanthomonas* plant pathogens all have highly similar *hrp* genes (at least 16 in number), and they are similarly organized (40). Interestingly, the genes encoding these protein secretion systems in various animal and plant pathogens are all clustered and the codon usage and %GC content of the genes encoding these type III systems are typically different from that of the rest of the genome (30). This indicates that the hrp type III secretion system may have been (and still can be) acquired as a pathogenicity island in a single horizontal gene transfer event. The type III system is activated when a pathogen comes in close contact with host cells and is therefore sometimes called a contact-dependent secretion system (17). A common export system might imply a common effector, except that the host organisms can be different. Certainly with biotrophs, different effectors might be expected to reflect the level of host specificity; with necrotrophs, a few, more generalized effectors might be expected.

Only a few effector molecules have been identified as being secreted by the *hrp* genes. In both necrotrophic *Erwinia* species and biotrophic *Pseudomonas* and *Erwinia* species, harpins have been identified as effector proteins that are required for full pathogenicity. Harpins directly elicit plant symptoms on both host and nonhost plants, when applied externally to plant cells. Harpins are glycine rich proteins that do not appear to function as enzymes but that may cause direct nutrient leakage from plant cells by alkalinization of the apoplast (1,21,43).

With biotrophs, it is clear that there must be a delicate mechanism by which the plant cell gives up nutrients without immediately dying. The molecular details of slow nutrient release by harpins are unclear, and more specialized factors might be expected to also be involved in nutrient release by biotrophs—even those with harpins—according to their level of host-specificity. Interestingly, no harpin-like elicitors (i.e., an elicitor that functions when applied externally to plant cells) have been found in *Xanthomonas*, which is the most biotrophic of all *hrp*-carrying pathogens. The critical role played by the *hrp* protein secretion system in *Xanthomonas* (11) makes it likely that a protein signal is secreted that causes or contributes to nutrient leakage. Is it possible that some Avr proteins, most likely those encoded by *avr* genes known to be needed for growth *in planta*, are among the signals secreted for the purpose of programmed nutrient leakage?

The surprising discovery of functional nuclear localizing signal (NLS) sequences in *avrb6* and *pthA*, both members of the *Xanthomonas avr/pth* gene family (45) opened an unexpected avenue of research. (This gene family is described in some detail below). This discovery provided the first indirect evidence not only that some Avr proteins themselves might be secreted, but also that they might function inside the plant cell (reviewed by Gabriel [12]). Site-directed mutagenesis was used to confirm that these NLS sequences were essential for the pathogenicity phenotypes expressed by several of these genes (16,37). Strong indirect evidence that some Avr proteins were secreted from the bacterial cell and functioned inside the plant cell was provided by Gopalan et al. (19), working with AvrB of *P. syringae*. When *avrB* was biolistically delivered into *Arabidopsis* plant cells carrying resistance gene RPM1, a gene-for-gene HR resulted. These results were confirmed when *avrBs3*, a member of the *Xanthomonas avr/pth* gene family, was subsequently expressed and was active inside the plant cell (37).

Some *avr* genes, for example *avrBs2* of *X. campestris* pv. *vesicatoria* (22), are needed for growth *in planta*. Interestingly, *avrBs2* is widely conserved in *X. campestris*. The finding that the *hrp* type III secretion system can secrete Avr proteins to the plant cell wall and/or through the plant cell wall indicates that some of these proteins may be among the required signals or effector molecules that cause the plant cell to release nutrients. Such a role would be logical for genes such as *avrBs2*.

Ability to Move to New Infection Sites

Pathogenic symptoms are a host response phenotype. The type III protein secretion system provides an ideal mechanism to deliver signal proteins to host cells, and Avr proteins are among the few signal proteins known to be so delivered. By far the largest group of genes known to affect pathogenicity are *avr* genes; more than 40 have been cloned to date. Since the HR is a manifestation of a programmed plant cell death, is it possible that at least some *avr* genes serve as a signal to condition pathogenicity in ways other than by a contribution to growth *in planta*?

The *Xanthomonas avr/pth* Gene Family

The Name

By far the largest single group of *avr* genes cloned to date is the *Xanthomonas avr/pth* or *avrBs3* gene family. The first member cloned was *avrBn* (13), but the physical structure was not described until later (47). The first member physically described and sequenced was *avrBs3* (3) but its pathogenicity function is not required by *X. campestris* pv. *vesicatoria* (see below). The first member physically described, sequenced and shown to be required for pathogenicity was *pthA*; it was cloned by screening for pathogenicity, not avirulence (36). Most of the members of this gene family cloned to date are either required for, or can contribute to, pathogenicity on all hosts of the strain from which they were originally cloned. That is, most members have detectable selective value to the xanthomonads that carry them (Table 2-1 and below). Therefore, this gene group is a family of pathogenicity genes and is here termed the *Xanthomonas avr/pth* gene family to emphasize their primary role in conditioning pathogenicity.

Required for Some Diseases

Currently, 17 members of this gene family have been published and at least 22 members have been cloned (Table 2-1). Members of the gene family are known to be required for pathogenicity of *X. campestris* pv. *malvacearum* (cotton blight), *X. citri* (Asiatic citrus canker) and *X. campestris* pv. *aurantifolii* (false citrus canker and Mexican lime cancrosis). In *X. campestris* pv. *malvacearum*, at least 7 of the 10 members conferred the ability to water soak cotton leaves (47). The contribution of each gene is additive, and 5 of the 10 members appear to be redundant. All cotton blight strains examined carry multiple members of the gene family. In *X. citri, pthA* is required for Asiatic citrus canker disease, and *pthA* confers ability to elicit cankers on citrus to other xanthomonads (35). Similar results were obtained for *pthB* and *pthC* from *X. campestris* pv. *aurantifolii* (Yuan and Gabriel, *unpublished*). All citrus canker strains examined carry multiple members of the gene family. In *X. oryzae* (rice blight), *avrXa7* increases both bacterial growth *in planta* and lesion length in rice (28). As with cotton blight and citrus canker xanthomonads, all rice blight stains examined carry multiple members of the gene family, but it is not yet known if members are required for rice blight disease. A similar story may be

Table 2-1. Cloned members of the *Xanthomonas avr/pth* gene family

#	Bacterial Source and Strain	Name	Number of Repeats	Pathogenic Reaction Phenotype[a]	Reference
1.	*X. campestris* pv. *malvacearum* XcmH	*avrBn*	N.D.[b]	N.D.	13
2.		*avrb6*	13.5	Water soaking on cotton	8
3.		*avrB4*	18.5	N.D.	8
4.		*avrb7*	18.5	Water soaking on cotton	8
5.		*avrBIn*	20.5	Water soaking on cotton	8
6.		*avrB101*	22.5	Water soaking on cotton	8
7.		*avrB102*	17.5	Water soaking on cotton	8
8.		*avrB103*	N.D.	N.D.	47
9.		*avrB104*	N.D.	Water soaking on cotton	47
10.		*avrB5*	N.D.	Water soaking on cotton	47
11.	*X. campestris* pv. *malvacearum* XcmN	*pthN*	13.5	Water soaking on cotton	Chakrabarty and Gabriel, *unpublished*
12.		*pthN2*	N.D.	Water soaking on cotton	Chakrabarty and Gabriel, *unpublished*
13.	*X. campestris* pv. *vesicatoria* 71-21	*avrBs3*	17.5	Hypertrophy of pepper and tomato	4
14.	*X. campestris* pv. *vesicatoria* 82-8	*avrBs3-2* (*avrBsP*)	17.5	N.D.	2a, 20a
15.	*X. citri* 3213	*pthA*	17.5[c]	Cankers on citrus	36
16.	*X. campestris* pv. *aurantifolii* XC69	*pthB*	17.5	Cankers on citrus	Yuan and Gabriel, *unpublished*
17.	*X. campestris* pv. *aurantifolii* XC340	*pthC*	17.5	Cankers on citrus	Yuan and Gabriel, *unpublished*
18.	*X. oryzae* pv. *oryzae*	*avrxa5*	ND	N.D.	20a
19.		*avrXa7*	24.5	Elongated lesions on rice	20a
20.		*avrXa10*	17.5	N.D.	20a
21.	*X. phaseoli*	*avrXp1*	ND	N.D.	34
22.	*X. phaseoli* var. *fuscans*	*pthCBB1*	ND	Blight symptoms on bean common bean	Castaneda and Gabriel, *unpublished*

[a] Pathogenicity phenotypes on known host species of the source strain. Most *X. campestris* pv. *malvacearum* water soaking phenotypes are reported by Yang et al (47); the pathogenicity phenotype of *avrXa7* is reported by Leach et al. (28) and that of *avrBs3* is reported by Bonas et al (3).
[b] Not determined.
[c] The repeat number of *pthA* is taken from the sequence published by Yang and Gabriel (46).

emerging for *X. phaseoli* (common bean blight). *pthCBB1*, cloned from *X. phaseoli* var. *fuscans*, conferred increased pathogenicity to *X. campestris* pv. *alfalfae* on common bean (Castaneda and Gabriel, *unpublished*). As with cotton blight, citrus canker and rice blight, all common bean blight strains examined carry members of the gene family, but it is not yet known if members are required for common bean blight.

At least 14 of the 17 described members of the *avr/pth* gene family are known to either be required for, or contribute to, pathogenicity of the strain from which they were cloned (Table 2-1). Since members of the *Xanthomonas avr/pth* gene family provide critical functions for cotton blight and citrus canker diseases, these *avr/pth* genes have been examined in some detail.

avrb6 Alleles

Cotton blight disease is caused by at least two phylogenetically distinct subgroups of *X. campestris* pv. *malvacearum*: the N. American and African groups (8). Gene *avrb6* was cloned from a N. American strain, XcmH, that carries no fewer than 10 members of the *avr/pth* gene family; all 10 of these members have now been cloned, along with two hybridizing DNA fragments with no evident *avr/pth* activity (47). Although many of these 10 members exhibit at least some water soaking function, *avrb6* exhibits the highest water soaking activity on susceptible cotton plants. Strain HM2.2S is a derivative of strain XcmH that has had 7 of 12 members of the *avr/pth* gene family inactivated by marker-eviction mutagenesis, and is no longer able to water soak cotton plants. Strain HM2.2S multiplies as well as the wild type XcmH in susceptible cotton plants (47); this demonstrates that *avrb6* alleles, and the water soaking phenotype they confer, are not likely involved in obtaining nutrient release from cotton.

Based on hybridization studies, all African strains carry about half the number of *avr/pth* genes as the North American strains (5,8). Furthermore, all African strains can "defeat" a much higher number of cotton *R* genes (i.e., are virulent on a larger number of cotton cultivars) than can N. American strains, presumably because they carry fewer *avr* genes. African strain XcmN carries six *avr/pth* family members and yet has no known *avr* genes, because it can defeat the resistance of all commercial cotton cultivars (13). Because of its high level of virulence, it is a quarantine pathogen in the U.S. All six DNA fragments hybridizing to *avrb6* were recently cloned from XcmN, and four of the fragments conferred water soaking activity to HM2.2S (Chakrabarty and Gabriel, *unpublished*). Two genes with the highest level of water soaking activity were identified and partially sequenced: *pthN* and *pthN2*. Both genes are similar to *avrb6* in water soaking activity, and all three are carried on plasmids.

pthA Alleles

Citrus canker disease has been historically described as having different "forms", including Asiatic citrus canker (*X. citri* or "A form"), false citrus canker (*X. campestris* pv. *aurantifolii* form B or "B form") and Mexican lime cancrosis (*X. campestris* pv. *aurantifolii* form C or "C form"). However, these three "forms" are not distinctive in terms of disease phenotype, and have not been distinguished based upon host symptoms; the symptoms elicited on susceptible hosts are the same and the bacterial signs are the same. The taxonomic status of *X. campestris* pv. *aurantifolii* strains is unresolved. These strains are only 62-63% related to *X. citri* strains (10) form a distinct RFLP group (14), and differ serologically from *X. citri* strains (2). How then, do phylogenetically distinct strains cause identical diseases?

Gene *pthA* was first identified and cloned from *X. citri* because it conferred to other xanthomonads the ability to elicit cankers on citrus (34). DNA libraries were subsequently made from the B and C forms of *X. campestris* pv. *aurantifolii*, and clones were selected that hybridized with *pthA* (Yuan and Gabriel, *unpublished*). These were then screened in various xanthomonads for ability to elicit hyperplastic cankers on citrus. As with *X. citri*, a single fragment was identified in each *X. campestris* pv. *aurantifolii* strain that conferred the ability to cause cankers on citrus (Yuan and Gabriel, *unpublished*). The genes responsible for the activity were subcloned, partially sequenced, named *pthB* and *pthC* and identified as members of the *Xanthomonas avr/pth* gene family. Marker-exchange mutagenesis of each of these two new genes (as well as *pthA* [36]) confirmed that they were required for the source strain to elicit cankers on citrus. All three genes were functionally interchangeable in all three marker-exchanged backgrounds. Interestingly, *pthA, pthB* and *pthC* are carried on plasmids, and *pthB* was found to reside on a self-mobilizing plasmid, which may explain why these phylogenetically different xanthomonads can cause identical symptoms.

Not Required by All Xanthomonads

Members of this *Xanthomonas avr/pth* gene family are widely distributed in the genus but not all pathovars carry members of the family. Unlike *hrp* genes, these genes are not universally required for *Xanthomonas* pathogenicity. Even when found in some strains of a pathovar, they need not be conserved among all members causing the same disease. For example, members of the gene family are not required for bacterial spot of tomato and pepper caused by *X. campestris* pv. *vesicatoria*, even though some strains of this pathovar carry members (e.g., *avrBs3*). Although *avrBs3* confers hypertrophy (enlargement) of host mesophyll cells under some circumstances (3), it may be deleted from the source strain. The deletion derivative still causes bacterial spot, as do other strains of the pathovar that carry no members of the *avr/pth* gene family (4). Bacterial spot disease of pepper and tomato therefore does not depend on members of the *avr/pth* gene family, although presence of these genes may enhance the disease phenotype.

Interestingly, *avrBs3* was found on a self-mobilizing plasmid (4), and therefore its presence in some strains of this pathovar may be gratuitous. Coincidental linkage with some other factor on the plasmid with selective value to the bacterium might provide selection for this plasmid in various *Xanthomonas* strains, in addition to other *X. campestris* pv. *vesicatoria* strains. A variety of interesting genes with selective value are being discovered on plasmids carried by phytopathogens, including genes conferring resistance to toxic compounds. For example, copper resistance is widespread among *X. campestris* pv. *vesicatoria* strains in locations where copper is used, and copper resistance has been found on some *X. campestris* pv. *vesicatoria* self mobilizing plasmids recovered from strains in these areas (33).

Pathogenicity is Host Species-Specific

It is important to emphasize that when *Xanthomonas avr/pth* genes are required for pathogenicity, they are host species-specific. However, when they behave as *avr* genes, they are typically cultivar-specific (or gene-for-gene specific). For example, *pthA*, *pthB* and *pthC* are absolutely required for the citrus canker disease phenotype elicited by their respective strains (36; Gabriel lab, *unpublished*). These genes confer to other xanthomonads (in a variety of pathovars) the ability to elicit cankers on citrus, but only on citrus. If *pthA*, *pthB* or *pthC* are moved into *X. campestris* pv. *malvacearum*, the result is not the production of cankers on all cotton species attacked by *X. campestris* pv. *malvacearum*. Instead, the result is gene-for-gene avirulence on particular cotton cultivars (36; Yuan and Gabriel, *unpublished*). These genes are therefore host species-specific in their pathogenic function, and cultivar or gene-for-gene specific in avirulence function.

Involved in Dispersal

The host species-specific pathology conditioned by members of the gene family appears to condition host range by affecting dispersal. For example, *pthA*, *pthB* and *pthC* are essential for their source strains to induce hyperplastic cankers on citrus. The formation of cankers ruptures the citrus epidermis and releases abundant bacteria to the leaf surface (35). Similarly, *avrb6* is important for *X. campestris* pv. *malvacearum* to induce water soaking on cotton, since *avrb6* confers ability to release 240 times more bacteria to the leaf surface than would otherwise be released (44). As mentioned above, at least seven out of ten *avr/pth* genes tested from a single *X. campestris* pv. *malvacearum* strain confer ability to water soak cotton. When all seven of these *avr/pth* genes are destroyed in a single strain (HM2.2S, discussed below), the strain grows *in planta* to the same levels as the wild type, but is completely asymptomatic, and 1,600 times less bacteria are released to the leaf surface (47). Since both citrus canker disease and bacterial blight of cotton are spread primarily by rain splash, the presence of larger numbers of bacteria on the leaf surface undoubtedly contributes to pathogen dispersal and host range. Marker-exchange mutations of both *pthA* and *avrXa7* also resulted in reduced growth *in planta* of affected strains (28,36), but it is not clear from these studies whether or not the marker exchange events caused polar effects on other genes involved in conditioning pathogenicity.

Molecular Analyses

The most conspicuous feature of the *Xanthomonas avr/pth* gene family is the presence of a dozen or more, nearly identical, tandemly arranged, 102bp direct repeats in the central region of the genes (Figure 2-1). These repeats were shown to determine the gene-for-gene avirulence of *avrBs3* (20). By constructing chimeric genes among *pthA*, *avrB4*, *avrb6*, *avrb7*, *avrBIn*, *avrB101* and *avrB102*, the 102bp tandem repeats of the genes were found to determine not only gene-for-gene avirulence, but also the host-specific pathogenicity functions (44). That is, the repeat regions of *avrb6* and *pthA* determined water soaking of cotton and hyperplasia of citrus, respectively. The complete DNA sequence of *pthA* was determined (46) and sequence comparisons revealed that *pthA* is 97% homologous to *avrb6*

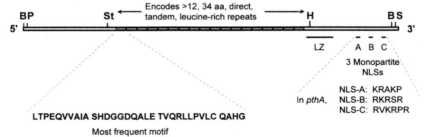

Fig. 2-1. General structure of all members of the *Xanthomonas avr/pth* gene family. The open reading frame is illustrated, along with the predicted peptide sequence of the most commonly occurring repeat, the putative leucine zipper (LZ) and the predicted nuclear localizing signal (NLS) sequences of NLS-A, NLS-B and NLS-C of *pthA*. B, *Bam*HI; P, *Pst*I; St, *Stu*I; H, *Hinc*II; and S, *Sst*I.

and *avrBs3*. All sequenced members of the gene family are flanked by nearly identical 62bp terminal inverted repeats that precisely define the limits of homology and the terminal 38 bp of these inverted repeats are highly similar to the 38 bp consensus terminal sequence of the Tn*3* family of transposons. It is therefore possible that these genes can, or once could, transpose.

By aligning the predicted amino acid sequences encoded by *avrBs3, pthA, avrBs3-2, avrb6* and *avrXa10*, the primary regions which differentiate the sequences of members of the family are found to be in specific positions within the repeats. A deletion analysis of *avrBs3* revealed the striking conclusion that not only could a few amino acid differences within the repeats change avirulence specificity, but that the *position* of a specific repeat within the repetitive region also controlled specificity (20). By a deletion analysis of *avrB102* and *pthA*, this conclusion was confirmed and also extended to include pathogenicity functions (46; De Feyter and Gabriel, *unpublished*). That is, a few amino acid differences in particular repeats changes host species-specific pathogenicity. When the predicted amino acid sequences of all five sequenced genes are compared, many unique 34 aa direct repeats are found, and some are exactly duplicated. The most frequently repeated motif is indicated in Figure 2-1, and is found 15 times in these five genes. This repeat may be a "progenitor" repeat. There were 24 repeats among the five genes which were not repeated, and there were five repeats which were repeated only within a given gene. The distribution of the unique repeats was not random in terms of position; instead, unique repeats were found clustered at positions 7 and 13:

Repeat position # 7:

{AvrBs3-2}	LIPQQVVAIA	SNIGGKQALE	TVQRLLPVLC	QDHG *
{AvrBs3}	LTPEQVVAIA	SNIGGKQALE	TVQALLPVLC	QAHG
{PthA}	LTPDQVVAIA	SHDGGKQALE	TVQRLLPVLC	QAHG **
{Avrb6}	LPPEQVVAIA	SHDGGKQALE	TVQRLLPVLC	QAHG *
{AvrXa10}	LTPDQVVAIA	SNNGGKQALE	TVQRLLPVLC	QTHG *

Repeat position #13:

{AvrBs3}	LTPEQVVAIA	SHDGGKQALE	TVQRLLPVLC	QAHG
{PthA}	LTLDQVVAIA	SNGGGKQALE	TVQRLLPVLC	QAHG *
{AvrBs3-2}	LTPQQVVAIA	SHDGGKQALE	TVQRLLPVLC	QAHG **
{Avrb6}	LTPAQVVAIA	SNNGGKQALE	TVQRLLPVLC	QAHG *
{AvrXa10}	LTPVQVVAIA	SNSGGKQALE	TVQRLLPVLC	QDHG *

* Indicates repeat is unique among repeats of all 5 *avr/pth* genes
** Indicates repeat is unique among repeats of indicated *avr/pth* gene

The repeats are leucine-rich (5/34 aa are leucine), indicating a potential role in protein-protein interactions (25a). The genes also carry leucine zipper-like heptad repeats (LESIVAQ LSRPDPA LAALTNDH LVALAC LGGRPA LDAVKKG LPHAPAL IKRTNRR IPERTSH), contiguous with the 34 amino acid tandem repeats of these proteins (Figure 2-1). Leucine zippers serve as the sites of protein-protein or protein-DNA binding (25a). It is logical to assume that these proteins may interact directly with protein receptors or directly with nuclear DNA in plants to trigger programmed cell responses.

The Role of the NLS Region

Analyses (46) of the predicted amino acid sequences encoded by *avrb6*, *pthA*, *avrXa10*, *avrBs3* and *avrBs3-2* revealed the presence of three stretches of basic residues with complete homology with the nuclear localization consensus sequences (K-R/K-X-R/K) found in many nuclear localized proteins (6). These three putative nuclear localization sequences (NLSs) are located near the C-terminus of the proteins, at positions 1020-1024 (K-R-A-K-P), 1065-1069 (R-K-R-S-R), and 1101-1106 (R-V-K-R-P-R) in PthA. (Figure 2-1). The first and third putative NLSs also contain proline residues, which are often found in NLSs. The presence of the same three putative NLSs was also found in 5 additional genes that have now been partially sequenced: *pthB* and *pthC*, required for citrus canker disease caused by *X. campestris* pv. *aurantifolii* strains, *pthN* and *pthN2*, which contribute to angular leaf spot disease caused by strain XcmN of *X. campestris* pv. *malvacearum*, and *avrB4*, an avirulence gene from strain XcmH of *X. campestris* pv. *malvacearum* (Table 2-1). In short, NLS signals are found in the predicted peptide sequences of all sequenced *avr/pth* gene family members.

The NLS signals of PthA, Avrb6 and AvrBs3 are functional and essential for the phenotypes encoded by these genes. The DNA coding sequences for the C-terminal regions of Avrb6 and PthA were independently fused to a β-glucuronidase (GUS) reporter gene. When introduced into onion cells, both of these translational fusions were transiently expressed, and GUS activity was specifically localized in the nuclei of transformed cells (45). These results demonstrate that members of the *Xanthomonas avr/pth* gene family carry functional nuclear localization signals, and these results were subsequently confirmed using AvrBs3 (37). Site-directed mutations were then used to eliminate each of the three NLSs individually and collectively in PthA, Avrb6 and AvrBs3. The three NLSs, named in order towards the C terminus, are NLS-A, NLS-B and NLS-C (Figure 2-1). The corresponding regions from *pthA, avrb6* and *avrBs3* are nearly identical in DNA sequence. Mutations affecting the NLS region in *avrBs3* resulted in confirmed loss of the HR phenotype conferred by this gene (37). Site directed mutations of

NLS-B alone, or any combination of two of the NLSs, resulted in a loss of the canker phenotype conferred by *pthA*, loss of gene-for-gene avirulence conferred by *avrb6*, but not water soaking conferred by *avrb6* or the nonhost HR conferred by *pthA* (16). When all three NLSs are destroyed in *avrb6*, the water soaking phenotype conferred by the gene is also lost (Duan and Gabriel, *unpublished*). Interestingly, another gene of *X. campestris* pv. *vesicatoria*, *avrBs3-2* (synonymous with *avrBsP*), does not require the 3' end (encoding all three NLSs) to elicit an HR on tomato. Therefore some avirulence and at least some pathogenicity phenotypes encoded by members of this gene family require nuclear localization, but others may not.

Ability to Avoid Recognition by Plant *R* Genes: One Side of the Coin

As mentioned above, the function of *avr/pth* genes in biotrophs is fully dependent upon the function of the *hrp* genes, a host contact-dependent secretion system. Importantly, when biotrophs cause host cell death due to an *avr* gene-for-*R* gene interaction, it appears to be programmed (7). Plant cell programs require specific signals and receptors, and presumably some *avr* gene products, such as those encoded by the *Xanthomonas avr/pth* gene family, function directly as the signal. Yet avirulence *per se* cannot be a function that is worth setting up a specialized, contact-dependent secretion system for. Obviously, in order to avoid recognition by plant *R* genes, the pathogen must lose or mutate the cognate *avr* gene. There are a variety of mechanisms for this, including IS element movement, point mutations, and in the case of the gene families, intergenic and intragenic recombination (refer to 46). In cases where the *avr/pth* gene is gratuitous (i.e., in situations where the *pth* function is no longer needed following horizontal gene transfer), loss of function mutations would be advantageous. On the other hand, gain of function mutations that enhanced pathogenicity would be better.

As discussed above, the *Xanthomonas avr/pth* gene family appears able to transfer horizontally to a variety of bacterial strains and possibly transpose, yet these genes have been found only in *Xanthomonas*. Perhaps due to the biotrophic nature of *Xanthomonas* and the ability of these Avr/Pth proteins to enter plant cells and affect or trigger plant programs of benefit to the pathogen, these genes have found a particular niche where they are largely conserved (*pthA, avrb6* and *avrBs3* are >97% identical in DNA sequence). The peculiar repeat structure in the central portion of these genes allows intragenic recombination to occur (46), avoiding the problem of asexual reproduction in limiting variation. The presence of these genes on plasmids allows their presence in multiple copies, allowing mismatch correction. Their presence on self-mobilizing plasmids virtually guarantees that these genes will find themselves in xanthomonads adapted to hosts in which they confer no particular function; that is, they will be gratuitous. This is likely the case with *avrBs3* and *avrBs3-2* of *X. campestris* pv. *vesicatoria*. However, intragenic recombination generates variation, and AvrBs3 is delivered into the plant cell. Any adaptive mutations would be selected; a variant disease form caused by *X. campestris* pv. *vesicatoria* may therefore be expected.

Conclusions

There are at least five new and major themes that have recently emerged from studies of bacterial pathogens. The first is that the group of genes identified as always necessary for pathogenicity of plants, the hypersensitive response and pathogenicity (*hrp*) genes, encode a novel, contact-dependent, type III protein secretion system that is unique to pathogens. The second is that the effector molecules of pathogenicity, including those determined by pathogenicity (*pth*) and avirulence (*avr*) genes, are proteins. The third is that some of these effector proteins appear to operate inside the plant cell. The fourth is that the effector molecules work by affecting pre-existing plant cell programs. The fifth is that most microbial genes conditioning pathogenicity, including *hrp, pth* and *avr* genes, are present in pathogenic strains following horizontal gene transfer. At least some of these, specifically the members of the *Xanthomonas avr/pth* gene family, likely play essential roles both in better adapting different *Xanthomonas* strains to their existing hosts, and to extending the host range of xanthomonads to plants that previously were not hosts.

References

1. Alfano, J. R., and Collmer, A. 1996. Bacterial pathogens in plants: life up against the wall. Plant Cell 8:1683-1698.
2. Bradbury, J. F. 1986. Guide to Plant Pathogenic Bacteria. CAB International, Slough. pp. 1-332.
2a. Bonas, U., Conrads-Strauch, J., and Balbo, I. 1993. Resistance in tomato to *Xanthomonas campestris* pv. *vesicatoria* is determined by alleles of the pepper-specific avirulence gene *avrBs3*. Mol. Gen. Genet. 238:261-269.
3. Bonas, U., Huguet, E., Noel, L., Pierre, M., Rossier, O., Wengelnik, K., and Van den Ackerveken, G. 1996. *Xanthomonas campestris* pv. *vesicatoria hrp* gene regulation and avirulence gene *avrBs3* recognition. Pages 203-208 in: Biology of plant-microbe interactions. G. Stacey, B. Mullin and P. M. Gresshoff, eds. ISMPMI, St. Paul, MN.

4. Bonas, U., Stall, R. E., and Staskawicz, B. 1989. Genetic and structural characterization of the avirulence gene, *avrBs3* from *Xanthomonas campestris* pv. *vesicatoria*. Mol. Gen. Genet. 218:127-136.

4a. Canteros, B., Minsavage, G., Bonas, U., Pring, D., and Stall, R. 1991. A gene from *Xanthomonas campestris* pv. *vesicatoria* that determines avirulence in tomato is related to a*vrBs3*. Mol. Plant-Microbe Interact. 4:628-632.

5. Chakrabarty, P. K., Mahadevan, A., Raj, S., Meshram, M. K., and Gabriel, D. W. 1995. Plasmid-borne determinants of pigmentation, exopolysaccharide production and virulence in *Xanthomonas campestris* pv. *malvacearum*. Can. J. Microbiol. 41:740-745.

6. Chelsky, D., Ralph, R., and Jonak, G. 1989. Sequence requirements for synthetic peptide-mediated translocation to the nucleus. Mol. Cell. Biol. 9:2487-2492.

7. Dangl, J. L., Dietrich, R. A., and Richberg, M. H. 1996. Death don't have no mercy: cell death programs in plant-microbe interactions. Plant Cell 8:1793-1807.

8. De Feyter, R., Yang, Y., and Gabriel, D. W. 1993. Gene-for-genes interactions between cotton *R* genes and *Xanthomonas campestris* pv. *malvacearum avr* genes. Mol. Plant-Microbe Interact. 6:225-237.

9. Dow, J. M., Osbourn, A. E., Wilson, T. J. G., and Daniels, M. J. 1995. A locus determining pathogenicity of *Xanthomonas campestris* is involved in lipopolysaccharide biosynthesis. Mol. Plant-Microbe Interact. 8:768-777.

10. Egel, D. S., Graham, J. H., and Stall, R. E. 1991. Genomic relatedness of *Xanthomonas campestris* strains causing diseases of citrus. Appl. Environ. Microbiol. 57:2724-2730.

11. Fenselau, S., and Bonas, U. 1995. Sequence and expression analysis of the *hrpB* pathogenicity operon of *Xanthomonas campestris* pv vesicatoria which encodes eight proteins with similarity to components of the Hrp, Ysc, Spa, and Fli secretion systems. Mol. Plant-Microbe Interact. 8:845-854.

12. Gabriel, D. W. 1997. Targeting of protein signals from *Xanthomonas* to the plant nucleus. Trends Plant Sci. 2:204-206.

13. Gabriel, D. W., Burges, A., and Lazo, G. R. 1986. Gene-for-gene recognition of five cloned avirulence genes from *Xanthomonas campestris* pv. *malvacearum* by specific resistance genes in cotton. Proc. Natl. Acad. Sci. USA 83:6415-6419.

14. Gabriel, D. W., Kingsley, M. T., Hunter, J. E., and Gottwald, T. R. 1989. Reinstatement of *Xanthomonas citri* (ex Hasse) and *X. phaseoli* (ex Smith) and reclassification of all *X. campestris* pv. *citri* strains. Int. J. Syst. Bacteriol. 39:14-22.

15. Gabriel, D. W., and Rolfe, B. G. 1990. Working models of specific recognition in plant-microbe interactions. Annu. Rev. Phytopathol. 28:365-391.

16. Gabriel, D. W., Yuan, Q., Yang, Y., and Chakrabarty, P. K. 1996. Role of nuclear localizing signal sequences in three disease phenotypes determined by the *Xanthomonas avr/pth* gene family. Pages 197-202 in: Biology of Plant-Microbe Interactions. G. Stacey, B. Mullin and P. M. Gresshoff, eds. International Society for Molecular Plant-Microbe Interactions, St. Paul, MN.

17. Galan, J. E. 1996. Molecular genetic bases of *Salmonella* entry into host cells. Mol. Microbiol. 20:263-271.

18. Glazebrook, J., Rogers, E. E., and Ausubel, F. M. 1996. Isolation of *Arabidopsis* mutants with enhanced disease susceptibility by direct screening. Genetics 143:973-982.

19. Gopalan, S., Bauer, D. W., Alfano, J. R., Loniello, A. O., He, S. Y., and Collmer, A. 1996. Expression of the *Pseudomonas syringae* avirulence protein AvrB in plant

cells alleviates its dependence on the hypersensitive response and pathogenicity (Hrp) secretion system in eliciting genotype-specific hypersensitive cell death. Plant Cell 8:1095-1105.

20. Herbers, K., Conrads-Strauch, J., and Bonas, U. 1992. Race-specificity of plant resistance to bacterial spot disease determined by repetitive motifs in a bacterial avirulence protein. Nature 356:172-174.

20a. Hopkins, C. M., White, F. F., Choi, S. H., Guo, A., and Leach, J. E. 1992. Identification of a family of avirulence genes from *Xanthomonas oryzae* pv. *oryzae*. Mol. Plant-Microbe Interact. 5:451-459.

21. Hoyos, M. E., Stanley, C. M., He, S. Y., Pike, S., Pu, X. A., and Novacky, A. 1996. The interaction of HarpinP$_{ss}$, with plant cell walls. Mol. Plant-Microbe Interact. 9:608-616.

22. Kearney, B., and Staskawicz, B. J. 1990. Widespread distribution and fitness contribution of *Xanthomonas campestris* avirulence gene *avrBs2*. Nature 346: 385-386.

23. Keen, N. 1986. Pathogenic strategies of fungi. Pages 171-188 in: Recognition in Microbe-Plant Symbiotic and Pathogenic Interactions. B. Lugtenberg, eds., Springer-Verlag, Berlin.

24. Keen, N. T. 1990. Gene-for-gene complementarity in plant-pathogen interactions. Ann. Rev. Genet. 24:447-463.

25. Kingsley, M. T., Gabriel, D. W., Marlow, G. C., and Roberts, P. D. 1993. The *opsX* locus of *Xanthomonas campestris* affects host range and biosynthesis of lipopolysaccharide and extracellular polysaccharide. J. Bacteriol. 175:5839-5850.

25a. Kobe, B., and Deisenhofer, J. 1994. The leucine-rich repeat: A versatile binding motif. Trends Biochem. Sci. 19:415-421.

26. Kolattukudy, P. E. 1985. Enzymatic penetration of the plant cuticle by fungal pathogens. Annu. Rev. Phytopathol. 23:223-250.

27. Leach, J. E., and White, F. F. 1996. Bacterial avirulence genes. Ann. Rev. Phytopathol. 34:153-179.

28. Leach, J. E., Zhu, W., Chittoor, J. M., Ponciano, G., Young, S. A., and White, F. F. 1996. Genes and proteins involved in aggressiveness and avirulence of *Xanthomonas oryzae* pv. *oryzae* to rice. Pages 191-196 in: Biology of Plant-Microbe Interactions. G. Stacey, B. Mullin and P. M. Gresshoff, eds. International Society for Molecular Plant-Microbe Interactions, St. Paul, MN.

29. Lindgren, P. B., Peet, R. C., and Panopoulos, N. J. 1986. Gene cluster of *Pseudomonas syringae* pv. *"phaseolicola"* controls pathogenicity of bean plants and hypersensitivity on nonhost plants. J. Bacteriol. 168:512-522.

30. Mecsas, J., and Strauss, E. J. 1996. Molecular mechanisms of bacterial virulence: type III secretion and pathogenicity islands. Emerging Infect. Dis. 2:271-288.

31. Osbourn, A. E. 1996. Preformed antimicrobial compounds and plant defense against fungal attack. Plant Cell 8:1821-1831.

32. Salmond, G. P. C. 1994. Secretion of extracellular virulence factors by plant pathogenic bacteria. Annu. Rev. Phytopathol. 32:181-200.

33. Stall, R. E., Loschke, D. C., and Jones, J. B. 1986. Linkage of copper resistance and avirulence loci on a self-transmissible plasmid in *Xanthomonas campestris* pv. *vesicatoria*. Phytopathology 76:240-243.

34. Swarup, S. 1991. Isolation of pathogenicity genes species and a study of their regulation. Ph. D. thesis. University of Florida, Gainesville. 108pp.

35. Swarup, S., De Feyter, R., Brlansky, R. H., and Gabriel, D. W. 1991. A pathogenicity locus from *Xanthomonas citri* enables strains from several pathovars of *X. campestris* to elicit cankerlike lesions on citrus. Phytopathology 81:802-809.

36. Swarup, S., Yang, Y., Kingsley, M. T., and Gabriel, D. W. 1992. A *Xanthomonas citri* pathogenicity gene, *pthA*, pleiotropically encodes gratuitous avirulence on nonhosts. Mol. Plant-Microbe Interact. 5:204-213.
37. Van den Ackerveken, G., Marois, E., and Bonas, U. 1996. Recognition of the bacterial avirulence protein AvrBs3 occurs inside the host plant cell. Cell 87:1307-1316.
38. van der Plank, J. E. 1963. Vertical and horizontal resistance against potato blight. Academic Press, New York. pp. 171-177.
39. Van Gijsegem, F., Genin, S., and Boucher, C. 1993. Conservation of secretion pathways for pathogenicity determinants of plant and animal bacteria. Trends Microbiol. 1:175-180.
40. Van Gijsegem, F., Gough, C., Zischek, C., Niqueux, E., Arlat, M., Genin, S., Barberis, P., German, S., Castello, P., and Boucher, C. 1995. The *hrp* gene locus of *Pseudomonas solanacearum*, which controls the production of a type III secretion system, encodes eight proteins related to components of the bacterial flagellar biogenesis complex. Mol. Microbiol. 15:1095-1114.
41. Vivian, A., and Gibbon, M. J. 1997. Avirulence genes in plant-pathogenic bacteria: signals or weapons? Microbiology 143:693-704.
42. Vivian, A., Gibbon, M. J., and Murillo, J. 1997. The molecular genetics of specificity determinants in plant pathogenic bacteria. Pages 293-328 in: Gene-for-gene relationships in plant-parasite interactions. I. R. Crute and E. B. Holub, eds. CAB International, Oxford.
43. Wei, Z. M., Laby, R. J., Zumoff, C. H., Bauer, D. W., He, S. Y., Collmer, A., and Beer, S. V. 1992. Harpin, elicitor of the hypersensitive response produced by the plant pathogen *Erwinia amylovora*. Science 257:85-88.
44. Yang, Y., De Feyter, R., and Gabriel, D. W. 1994. Host-specific symptoms and increased release of *Xanthomonas citri* and *X. campestris* pv. *malvacearum* from leaves are determined by the 102 bp tandem repeats of *pthA* and *avrb6*, respectively. Mol. Plant-Microbe Interact. 7:345-355.
45. Yang, Y., and Gabriel, D. W. 1995a. *Xanthomonas* virulence/ pathogenicity gene family encodes functional plant nuclear targeting signals. Mol. Plant-Microbe Interact. 8:627-631.
46. Yang, Y., and Gabriel, D. W. 1995b. Intragenic recombination of a single plant pathogen gene provides a mechanism for the evolution of new host specificities. J. Bacteriol. 177:4963-4968.
47. Yang, Y., Yuan, Q., and Gabriel, D. W. 1996. Watersoaking function(s) of XcmH1005 are redundantly encoded by members of the *Xanthomonas* avr/pth gene family. Mol. Plant-Microbe Interact. 9:105-113.

CHAPTER 3

Chestnut Blight and Hypovirulence

Dennis W. Fulbright

Historical Background

The introduction of chestnut blight at the beginning of the twentieth century was in many ways a preview of the turbulent times that lay ahead during the rest of the century. For quietly in the New York Zoological Gardens there began one of the worst ecological disasters to strike North America. The disaster was so complete that by the end of the century few people remember what was lost. It was also during this time that we as a civilization saw and embraced advantages of a technologically rich society; however, we are now becoming aware that technology cannot solve all of our problems. As we end the twentieth century, we now understand that we must sometimes look to nature for answers to some of our problems. Never has this fact been so amply demonstrated as it is in our frustrating attempts to understand and manage the disease known as chestnut blight.

First discovered on American chestnut trees (*Castanea dentata*) in 1904 by Merkel[1], the causal fungus, (first named *Diaporthe parasitica*, then changed to *Endothia parasitica*[2] and finally described as *Cryphonectria parasitica*[3]) moved rapidly through the New York and Pennsylvania countryside destroying most of the trees in the eastern parts of these and other nearby states by 1912.[4] Chestnut was a major component of the eastern forests and was used for rough, decay resistant timber, fine woodworking, tannin extraction, and nut harvest. After harvest, trees resprouted and never had to be replanted. Its value to the southern Appalachian region was incalculable as in some regions 25% of the trees harvested in the forest were chestnut. Large trees, up to 2.5 meters in diameter and over 30 meters in height succumbed to the advance of the pathogen as fast as younger trees.

Not long after the pathogen's introduction to North America, it was found in China. It had been suggested earlier that the pathogen was from Asia since endemic areas usually contain host populations that coexist with native pathogens[5] and various Asian chestnut species showed resistance to the pathogen. Therefore, it was thought that the fungus made its way to North America as a weak pathogen of Chinese chestnut (*C. mollissima*); part of the horticultural revolution that was sweeping America at the end of the nineteenth century.

At the end of the epidemic period, a host population estimated at between 3-4 billion mature trees was devastated, dead, or surviving as understory stump sprouts that harbored the pathogen until the sprout died and again resprouted.[4] Most of the trees were harvested as the blight swept through the Appalachian forest; small pieces of wormy chestnut can still be found at some lumber yards. The chestnut-hickory forest became the oak-hickory forest in less than 5 decades.

Our biological training tells us that there should have been some survivors of this devastation and, indeed, there were a few mature trees left that were thought to have higher levels of resistance.[6] These trees have since died or are so riddled with chestnut blight that it is difficult to survey their pathogen population. Other trees survived; either growing at the edge of or planted outside of the natural range by pioneering farmers.[7] Therefore, the American chestnut is not considered an endangered or threatened species because seeds can still be gathered and planted to provide future substrate for spores to find and infect.

Later in the twentieth century, around 1938, a time when Europe was in such disarray that few would have noticed mere plant disease, chestnut blight spread across the ocean to Italy, France and the other chestnut-rich Mediterranean countries. The European chestnut (*C. sativa*), similar to North America's chestnut tree in size and appearance, proved to be susceptible to the blight; European chestnut farmers suffered great losses. The European trees had been cultivated for nut production. Chestnuts have been part of the European diet dating back to the Roman Empire. It has even been stated that during the war years some people lived off the chestnut mast from trees that had not yet succumbed to blight.

At a meeting held in 1912[8] in Pennsylvania, the origin of the blight fungus was controversial; various hypotheses suggested that the fungus was a native pathogen which infected trees stressed by a long-term drought or that an introduced fungus had not only invaded North America but had also hybridized with native fungi making them more virulent on native vegetation. The discovery of the pathogen in China seemed to settle the matter; scientists assumed that the fungus had been accidentally imported from China. However, Milgroom et al.[9] recently demonstrated that, while the pathogenic species may have originated in China, the predominant strains that invaded North America were from Japan or another Asian country where pathogen populations were not sampled. Their evidence was based on estimates of gene flow between Asian populations and North America obtained from RFLP allele frequencies and from DNA fingerprints analyzed after hybridization with genomic clones of *C. parasitica* DNA. The origin of the European pathogen could not be determined, but they suggested that it arrived in Europe directly from Japan or via North America. Milgroom et al. suggest that these data support the work of Anagnostakis[10] who established that most chestnut importations prior to and at the time of discovery of the blight were from Japan.

Biology of the Chestnut Blight Fungus

Cryphonectria parasitica, a homothallic ascomycete (with preferential out crossing), is a filamentous, facultative saprophyte. It is a classic example of a wound infecting, stem cankering pathogen: most infections occur either at the base of stems or on stems at natural branch scars where the scar leads to the vascular cambium (Figures 3-1 and 3-2). Very little is known about the infection court and the role that ascospores or asexual conidia play in the inoculum potential. The finding that trees with more than one canker generally harbor clonal populations of the pathogen suggests that these infections occur by asexual conidia rather than by sexual outcross-produced ascospores.[11] However, it is well known that ascospores are the wind-driven dispersal agent of the pathogen. It is thought that ascospores are responsible for having moved the epidemic front by nearly 40 km per year.[5]

There are no known toxins or secondary metabolites produced by the fungus which are directly responsible for the death of the stem; previous studies suggesting toxin involvement have not been corroborated. The disease is generally thought to be a disease of the phloem, however the actual cause of the death of plant tissue and of the stem is still not known. Once infection has been initiated, it appears that mycelial fan formation, a dense mat-like growth of the fungus, is essential for enlargement of the canker.[12] These studies have focused on the colonization of the outer bark, vascular phloem and outer xylem, yet the initial symptom on the stem is chlorosis and wilting of the leaves; this fact suggests the involvement of the xylem. Ewers et al.[13] found a dramatic reduction in fluid conductance around artificially inoculated and natural infections with a strong correlation to the death of leaves distal to the infection. They also found evidence, corroborating the earlier reports of McCarroll and Thor[14], that death of the vascular cambium is involved in the death of leaves and concluded that any survival of the vascular cambium may result in the survival of the leaves.

McManus et al.[15] reported extensive colonization of the xylem rings associated with natural and artificial infections. This finding indicates that the pathogen is not restricted to the vascular cambium and outer xylem.

One of the more extensive studies to date on pathogenesis compared the histopathology of canker development on blight-resistant Chinese chestnut species and blight susceptible and resistant (survivors of the epidemic) American chestnut species[12]. It was determined that in all host-pathogen combinations a zone of lignified tissue forms 8–10 days after inoculation; the formation of this zone halts the advance of the hyphae and is followed by the formation of wound periderm next to the lignin barrier two to four days later. The formation of the mycelial fans and penetration of the wound periderm by the mycelium occurs 18 days after inoculation in the suscep-

Fig. 3-1. Typical lethal chestnut blight canker on American chestnut caused by a virulent strain of *Cryphonectria parasitica.*

tible host-pathogen combination, but occurs later, if it occurs at all, in the resistant host-pathogen combination. It was observed that mycelial fan penetration prevented the further formation of wound periderm because host tissues were killed in advance of the hyphae. No explanation was given as to why the wound periderm in the resistant hosts retarded the growth of the mycelial fans longer than in the susceptible host.

An understanding of the mechanisms of pathogenicity and resistance might lead to long-term improvement of the American chestnut and a basis for studies on susceptible and resistant interactions in the hosts. Tests for preformed inhibitors of chestnut blight in Chinese chestnut trees did not confirm the presence of such compounds, but differences in the

Fig. 3-2. Chestnut blight canker. Typical canker, as in Fig. 3-1, except the appearance of wound periderm (swelling on the left side of canker) suggests the canker may not continue to expand. Cankers in Michigan need to be viewed over time if the outcome is to be determined.

induction of β-1,3-glucanase and chitinase in American and Chinese chestnut were observed; isoforms of the enzymes also differ between the two species.[16] Protein extracts from Chinese chestnut were also shown to have more antifungal activity than protein extracts from American chestnut trees.

Recent results showed that the targeted disruption of *enpg-1*, a gene encoding the major extracellular endopolygalacturonase (endoPG), resulted in no reduction of canker size on inoculated American chestnut stems. These data indicate that this enzyme does not play a role in fungal virulence. However, two previously undetected acidic polygalacturonases, exclusively produced in planta, were identified as the major forms of polygalacturonase. This result suggests that pectin-degrading enzymes may still be involved in pathogenesis.[17]

Other compounds produced by the fungus, including oxalic acid and laccase, have also been implicated in pathogenesis and disease resistance. Oxalic acid is reactive with cations and is thought to affect host tissues by lowering the pH of the host-cell environment. Oxalic acid may also act in combination with pectic enzymes in the digestion of calcium-pectate.[18] The fungus produces large amounts of oxalic acid in culture but conflicting reports of its concentration in host tissues leave the role of the substance in doubt.

Laccase, an enzyme found in a large number of plants and fungi, may also play a role in pathogenicity.[19,20] Because saprophytes as well as pathogens produce the enzyme, its biological significance is not clear. It has been reported to be responsible for degradation of lignin and thus for penetration of the lignin barrier.[21] But these results still do not explain the lack of specificity of the host-pathogen interaction discussed above.

Our lack of knowledge regarding the genetics of the host-pathogen interaction has inhibited the understanding of this disease and its pathogenesis. New programs may remedy this situation in the future. For example, a backcross breeding program initiated in the 1980's by the American Chestnut Foundation is beginning to exploit the resistance found in the Asiatic chestnut species by introgressing resistance genes into the American chestnut. After continuous selection and back crossing to the American chestnut, it is expected that high levels of resistance to chestnut blight should be found in a population of trees with timber-like tree habit in a genetic background mostly derived from the American chestnut.[22] It is hoped that during the breeding program the segregating populations can be used to provide a better understanding of the host-pathogen interaction.

Hypovirulence

Studies on naturally-occurring strains of *C. parasitica* with reduced ability to produce disease may eventually lead to a better understanding of *C. parasitica* pathogenesis. Not long after the introduction of *C. parasitica* into Europe, the first signs of chestnut blight survival began to appear. In the early 1950's, chestnut sprouts in Italy that had died back from the disease began to show signs of healing (nonlethal) cankers associated with copious wound tissue (Figure 3-3). The fungus that was recovered from diseased bark around these nonlethal infections looked much different from the causal fungus of chestnut blight. These strains were debilitated in

Fig. 3-3. Typical nonlethal canker associated with hypovirulent strains of *Cryphonectria parasitica*. This is the same canker in Fig. 3-2 approximately four years later. Virulent strains causing lethal cankers probably become infected with hypoviruses, leading to the change from lethal canker (Fig. 3-2) to nonlethal canker. This canker is completely closed (healed over) today.

culture, grew more slowly, produced fewer spores than the wild type, and lacked the typical orange pigment. The French mycologist Grente[23] soon found that these isolates were indeed the agents of chestnut blight, and that not only had their morphology and pigmentation been altered, but they were also less aggressive in pathogenicity assays. He termed these strains hypovirulent. The characteristics of hypovirulent strains were transmissible after hyphal contact and fusion to normal virulent *C. parasitica* strains (Figure 3-4).

Not long after the discovery of hypovirulence, it was found that if a hypovirulent strain was inoculated around the outside edge of the advancing margin of a lethal canker, expansion of the canker would cease and wound periderm would appear, similar to the original nonlethal cankers on the surviving chestnut sprouts.[24] Grente proposed that a cytoplasmic factor in the fungus was responsible for the transfer of the hypovirulence-associated traits. Later, similar experiments were done in which an apparent cytoplasmic agent was transferred from European to North American strains on American chestnut trees.[25]

In 1976, reports of surviving American chestnut trees in Michigan appeared and Elliston et al.[26] isolated North America's first hypovirulent

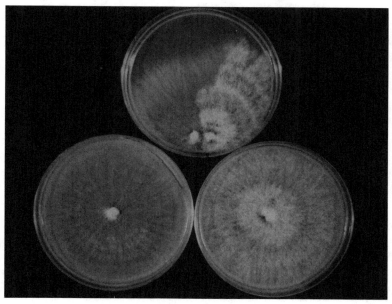

Fig. 3-4. Transmissible nature of hypovirulence. The transfer of double-stranded RNA from a hypovirulent strain to a virulent strain. As the hypovirus transfers through hyphal fusions, the morphology of the virulent strain becomes similar to the hypovirulent strain.

strains. These strains were isolated from surviving trees growing a few hundred kilometers northwest of the natural range of trees planted by pioneering orchard farmers in the late 1800's and early 1900's. The strains recovered from the non-lethal cankers on these trees were unlike the hypovirulent strains isolated in Europe; they were orange instead of white, and some grew faster and produced more spores in culture than did European hypovirulent strains.[27] Hypovirulent strains have since been found in other regions of North America, including New Jersey[28] and Ontario,[29] but none are as closely associated with chestnut survival or are as geographically widespread as the Michigan strains (Figure 3-5).

Double-Stranded RNAs

In 1977, Day et al.[29] reported that the cytoplasmic agent in the European hypovirulent strains correlated with the occurrence of large molecules of double-stranded RNA (dsRNA) and pleiomorphic vesicles.[30] Icosehedral virus-like particles, common in other fungi with dsRNA,[31] were not detected in *C. parasitica* hypovirulent strains.[32, 33] Later, it was found that the vesicles contained an RNA-dependent RNA polymerase that probably func-

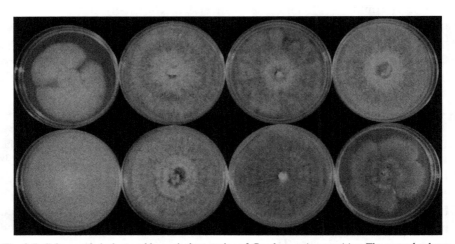

Fig. 3-5. Cultures of virulent and hypovirulent strains of *Cryphonectria parasitica*. The second column from the left contains two virulent strains from Michigan (top and bottom). The other three columns are hypovirulent strains with hypovirus-like dsRNA present in the cytoplasm. The first column represents two European hypovirulent strains (note their light color), the third column contains two hypovirulent strains from Michigan isolated from recovering trees (note how similar they appear to virulent strains), and the fourth column from the left are hypovirulent strains isolated from trees in the Appalachian Mountains.

tioned in replication of the dsRNA.[34] In the ensuing twenty years, elegant experiments finally established a direct link in *C. parasitica* between hypovirulence and the presence of dsRNA associated with the pleomorphic vesicles.[35-38] Furthermore, these experiments established the first true virus family without structural proteins, the Hypoviridae.[39]

At first, it appeared that dsRNA and its relationship to hypovirulence was a simple story: strains with dsRNA passed the hypovirulent phenotype and strains without the dsRNA were virulent. However, it soon became apparent that the presence or absence of dsRNA could not account for the dramatic morphological and pathogenic differences observed between dsRNA-containing strains. However, hypovirulent strains, regardless of their culture morphology, growth rate, or virulence, all reverted to virulent strains of normal morphology when cured of their dsRNA molecules.[40] While questions regarding phenotypic characteristics of hypovirulent strains remained unanswered, questions regarding the structure and function of the dsRNA molecules and the genetic relatedness between dsRNAs from North American hypovirulent isolates and from European hypovirulent isolates were answered.[28, 41,42]

dsRNA molecules isolated from hypovirulent strains from Europe, New Jersey and Michigan shared some features (Figure 3-6). All were large linear molecules without structural proteins and were polyadenylated at their 3′ termini.[28,43,44] The European and Michigan hypovirulent strains contain smaller dsRNA molecules which are related to the largest genomic dsRNA in each cytoplasm (Figure 3-6). These smaller molecules, or defective RNAs, appear to be the result of internal deletions of the large dsRNAs but retained their respective terminal sequences.[43-45] Alternatively, a single fungal strain could also be infected with more than one virus.[46]

Sequence analysis of the European, New Jersey and Michigan viruses has allowed a comparison of the North American and European viruses. Isolated from the hypovirulent strain EP713, the European dsRNA was the first member of the Hypoviridae and was designated the, CHV1-713.[39] CHV1-713 is 12,712 base pairs (bp) in length and contains two large open reading frames; ORF A and ORF B. ORF A and B overlap by a single basepair.[47] ORF A, 1869 bp in length, encodes two polypeptides designated p29 and p40.[48] p29 functions as a protease and is auto-catalytically released from the ORF A polypeptide during translation. Transformation of *C. parasitica* protoplasts with ORF A and the subsequent expression of p29 resulted in a strain with a phenotypic morphology similar to EP713, including loss of pigmentation, suppression of sporulation, and a reduction in the accumulation of the fungal enzyme laccase. Virulence, however, was not affected.[49]

ORF B (9498 bp) encodes a polyprotein that includes a protease (p48) with amino acid motifs similar to those found in ORF A. Two other domains have been recognized; an RNA-dependent RNA polymerase and an RNA helicase. These domains are similar to domains found in ssRNA viruses. This observation has led to speculation that the virus originated from a poty-like virus.[50]

Choi and Nuss[35] transformed *C. parasitica* with full length cDNA constructs of CHV1-713. Insertion of the cDNA occurred randomly in the nuclear DNA of the fungus; cytoplasmic forms of CHV1-713 dsRNA were produced from the nuclear DNA copies. This dsRNA was transmissible through hyphal fusion, and the integrated cDNA construct was found in sexual progeny[51] Normally, dsRNA molecules are not found in ascospores

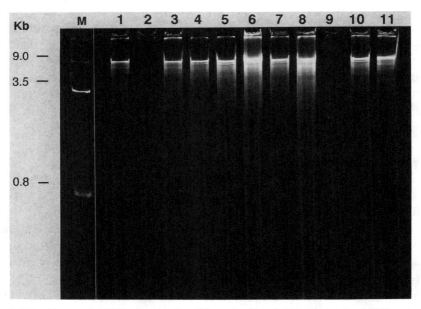

Fig. 3-6. Double stranded RNA genomes in a 5% acrylamide gel. Lane 1 contains dsRNA extracted from a Michigan hypovirulent strain of *Cryphonectria parasitica*; Lane 2 is a virulent Wisconsin strain without dsRNA; Lane 3 contains dsRNA from the culture in lane 1 that was transferred to the virulent strain in lane 2 in a manner similar to that described in Fig. 4; Lanes 4 and 5 contain dsRNA from a European strain; note the minor differences of banding patterns in lanes 1 and 3, yet these hypovirus genomes do not share significant sequence homology; also, smaller bands are present that can help identify the banding patterns of the various hypovirus-like dsRNAs including faint bands at the bottom of the gel; Lanes 6, 7 and 8 contain dsRNA from *C. parasitica* cultures doubly infected with both Michigan and European dsRNA geonomes; lanes 9, 10 and 11 contain dsRNA from *C. parasitica* cultures that grew from single asexual spores isolated from dsRNA-infected cultures; the culture in lane 9 did not have the European virus and the culture was orange, whereas the culture in lane 10 did have the European virus and its pigmentation was white. The dsRNA in lane M is from CHV3-GH2, one of the three designated hypoviruses and acts as a size standard.

because they are lost during meiosis. Integration of the virus and concomitant stability during meiosis provides an opportunity for long-distance hypovirus dispersal, as well as enhancing the ability of the virus to invade strains that might otherwise resist cytoplasmic transfer due to vegetative incompatibility.[38,52]

A hypovirulent isolate from New Jersey, NB58, harbors the second hypovirus, CHV2-NB58. The size and overall sequence of the virus are similar to those of CHV1-713.[28,37] Two open reading frames, ORF A and ORF B, are adjacent and connected by a pentanucleotide sequence. The ORF A protein product does not undergo autoproteolysis and it does not produce an active protease. The ORF B polyprotein products of CHV2-NB58 and CHV1-713 are very similar.[37]

A hypovirulent isolate from Michigan, GH2, harbors the third hypovirus, CHV3-GH2. Nearly all dsRNAs from Michigan hypovirulent strains hybridize with CHV3-GH2[42,53]. In contrast, no other dsRNAs from North American hypovirulent strains hybridize with CHV3-GH2, including CHV2-NB58. The CHV3-GH2 virus, approximately 9.0 kb in length, is found with defective dsRNA and a small dsRNA that shares no homology with the viral genome and may function as a satellite genome.[43,54] The 9.0 kb genome of CVH3-GH2 contains one ORF with replicase and helicase domains similar to CHV1-713 ORF B.[36]

Hypovirus Effects on the Fungal Host

Phenotypic variation in hypovirulent strains is not understood but is extremely important since the effect of the virus on the fitness of the fungus will ultimately decide the fate of the virus in the population. While the subtleties of the molecular interactions between the virus and its fungal host are most important, phenotypic variation may arise from large differences. For example, two strains may be infected with distinctly different viruses such as CHV1-713 and CHV3-GH2. In such a case, the fungal host would respond to the polypeptides produced by both viruses. Gene products associated with ORF A in CHV1-713 down- regulate fungal pigmentation, sporulation and the production of laccase. Because CHV2-NB58 does not have an active p29 gene product and because CHV3-GH2 does not have an analogous genel, one would not expect to see reduced pigmentation, reduced sporulation or down-regulation of laccase in strains infected with these viruses.[36] This example reinforces the need to have a good understanding of the genetic and biochemical developments in the virus-fungus interaction.

Other reasons for phenotypic variation might include hypovirulent strains infected with more than one hypovirus.[46,55] Mixed virus infections in one fungal thallus can be difficult to detect. If the viruses have genomes of different sizes, then one virus may be overlooked as a defective RNA derived from a larger genome. If two viruses are approximately the same size, it may be impossible to tell if the fungus is carrying two viruses. Elliston[55] determined that a Michigan hypovirulent strain was doubly infected by isolating individual asexual conidia that produced cultures expressing different phenotypes. He was then able to detect each of the two presumptive viruses by gel electrophoresis. If each virus in a mixed infection can be specifically detected by hybridization, both can be detected in the doubly-infected strain without the time-consuming segregation of viruses. When double infection occurs, it is possible that the fungus could be severely affected. For example, Smart and Fulbright[46] found that co-infected fungal strains were markedly reduced in virulence compared to strains infected with a single virus.

Finally, we must consider the role of viral mutation as a determinant of phenotypic variation in hypovirulent strains. The mutation rate of dsRNA viruses is not known. However, in a survey of dsRNAs associated with hypovirulent *C. parasitica* strains from New Jersey, polymerase chain reaction-generated DNA fragment banding patterns indicated all were closely related and distinct from the European virus CHV1-713. However, none of the New Jersey viruses were identical to each other.[56] Therefore, mutations appear to be common, and it will be difficult to determine if and when mutations might be lead to alterations in hypovirulent phenotypes. Understanding genetic change in the hypovirus could be important for maintaining the virus in the fungal population as a biological control agent. Information obtained from sequence analysis, studies on viral gene expression, and studies on the interactions of viral-fungal gene products, could provide insights concerning phenotypic variation in hypovirulent strains.

The ability of the hypovirulent fungus to grow in the stem tissue may be considered a separate issue from that of sporulation. The ORF A protein product in CHV1-713 can dramatically affect *C. parasitica* and its transmission by interfering with sporulation. Powell and Van Alfen[57,58] were the first to describe the effects of viruses on *C. parasitica* when they found that fungal RNA transcripts and protein products were down-regulated. While the studies by Powell and Van Alfen examined the general effects of virus infection on fungal gene expression, other research groups investigated the down-regulation of potential pathogenicity or virulence factors including cutinase,[59] laccase,[60,61] oxalic acid,[18,62] polygalacturonase[17] and others.[63]

The fact that most of these studies demonstrated that many of the targeted factors are down-regulated in the hypovirus-infected fungus led to speculation that they were virulence factors. Follow up experiments, usually involving mutation via homologous recombination with a reporter gene, in which the gene product is disrupted (e.g., as described above for the *enpg*-1 gene), showed, in each case, that these gene products were not required for virulence. In fact, natural mutants already existed in CHV2-NB58 and CHV3-GH2 in which laccase was not down-regulated by virus infection but the virus-infected strains were still hypovirulent.

Nevertheless, the continuing studies on laccase seem to have paid large dividends for other reasons. Larson et al.[63] used the *lac*-1 gene as a useful reporter gene to study the effect of hypovirus infection on gene regulation, specifically *lac*-1 transcript accumulation. These workers found that *lac*-1 was regulated by two different regulatory pathways, one a positive and the other a negative control pathway. Their studies showed that hypovirus suppression of *lac*-1 transcription was the result of alterations in the positive pathway. They suggested that their results, considered in context with the diversity observed in hypovirulent strains, imply that more studies should be focused on cellular signal transduction processes in hypovirus-infected strains.

In this regard, Nuss's laboratory initiated studies on G proteins (guanine nucleotide binding proteins) since they are a family of regulatory proteins that play an essential role in the response of eukaryotic cells to environmental stimuli.[64] Two *C. parasitica* G protein genes encoding α subunits were cloned and designated *cpg*-1 and *cpg*-2.[64] Using antibody directed against CPG-1, these workers found that CPG-1 accumulation was reduced to non-detectable levels in hypovirus-infected strains. When CPG-1 levels were reduced in virus-free strains, the strains were reduced in virulence. Therefore, these researchers concluded that reduced CPG-1 accumulation, either in hypovirus infected strains or in virus-free *cpg*-1 sense transformants, correlated with reduced virulence. (It is believed that the introduction of the sense transgene suppresses both the endogenous and introduced gene).[64] Furthermore, they suggested that at least one mechanism for hypovirus interference with virulence is by disruption of one of the key fungal signaling components.[65-69]

Mitochondrial Hypovirulence

Although we have made great strides in our understanding of dsRNA-associated hypovirulence and the role of the hypovirus in reducing virulence, it should be remembered that there are still many unanswered questions. For example, some dsRNAs that appear to be similar to hypoviruses do not appear to cause hypovirulence in *C. parasitica*[70]. Moreover, the induction of hypovirulence in *C. parasitica* is not limited to the viruses classified as Hypoviridae, since dsRNA packaged in REO virus-like protein particles have been described in hypovirulent strains.[71] Similarly, another form of dsRNA capable of cytoplasmic transmission and causing hypovirulence in *C. parasitica* has been reported to be located in the mitochondria and to be unrelated to the hypoviruses.[72] These findings make it important to emphasize that cytoplasmic factors that determine hypovirulence represent an assortment of different genetic elements, all of which are important in the biology of *C. parasitica.*

Another type of hypovirulence first described by Fulbright[53,73] in Michigan did not appear to be associated with dsRNA. Like dsRNA-associated hypovirulence, the new hypovirulent phenotype was infectious (i.e., transmissible from strain to strain via hyphal fusion). It was also maternally inherited. The only dsRNA previously known to be maternally inherited had been the dsRNA associated with the mitochondria reported by Polashock and Hillman.[72] In the dsRNA-free hypovirulent strains, most of the respiratory activity is cyanide-resistant and salicylhydroxamate-sensitive, indicating that the mitochondrial alternative oxidase was induced.[74] The respiration of virulent strains was cyanide-sensitive. The hypovirulent phenotype and altered respiratory phenotype could be transferred from the dsRNA-free hypovirulent strain to the virulent strains through hyphal anastomosis. In transmission studies, it was demonstrated that mitochondrial DNA also could move from strain to strain.[75] These early findings suggested that a cytoplasmic genetic factor other than dsRNA might be involved in the generation of the hypovirulent phenotype.

The discovery that alternative oxidase is induced in the dsRNA-free hypovirulent strains suggested that the attenuation in virulence was associated with a respiratory deficiency which could be caused by a mutation in the mtDNA. Many of the strains showing these traits begin to deteriorate upon subculturing to fresh medium (Figure 3-7). This process has been observed before in other fungi and is described as senescence.[76-78] To explore the possibility, that mitochondrial DNA (mtDNA) mutations might be involved in the hypovirulent phenotype, mutations that cause respiratory defects were generated in EP155, a stable virulent strain used in most laboratories.[79] Mutants were isolated that exhibited a cytoplasmically transmissible, hypovirulent phenotype similar to that of isolates associated with the nonlethal cankers on surviving American chestnut trees. These mitochondrial mutants, designated Mit1 and Mit2, were further implicated in hypovirulence by the demonstration of high levels of alternative oxidase, structural abnormalities in the mtDNA represented by unusual, plasmid-like DNAs, and maternal and cytoplasmic inheritance of these traits. Other mutants were obtained which showed high levels of alternative oxidase in respiration assays, but this phenotype was never cytoplasmically transmissible nor maternally inherited. Generation of a physical map for the mtDNA of strain Ep155 aided our understanding of the rearrangements in the mutant mtDNAs.[80]

Mitochondrial hypovirulent strains have been isolated from recovering American chestnut trees in Michigan and Ontario. Mitochondrial hypovirulence appears to be as effective a control of chestnut blight as hypoviruses. The discovery of these strains could provide insights into uninvestigated aspects of pathology, such as the role of the mitochondria and respiration in pathogenesis.

Biological Control

As stated earlier, hypovirulence initially appeared to be a simple system. Isolates recovered from the bark of swelling cankers usually contained dsRNA molecules. However, few thorough population studies were done in locations where chestnuts were surviving the blight or, just as importantly, were not recovering. It is now possible to find situations where aggressive, virulent strains harbor dsRNA[70] or where extremely debilitated, hypovirulent strains are without dsRNA.[74] It is important to understand that each hypovirulent strain may have its own characteristics based on the hypovirus

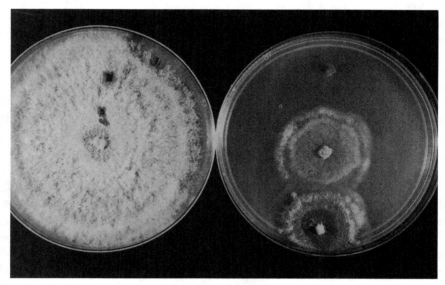

Fig. 3-7. *Cryphonectria parasitica* exhibiting senescence-like behavior. The three subcultures on the right were subcultured from the culture on the left. The small subculture on the top of the plate was taken from the growing edge—yet it would not grow. It is thought that a buildup of mutant mitochondria could be responsible for this phenotype.

or the interaction of the hypovirus with the fungus (Figure 3-6). Culture morphology and virulence may be different with each bark sample assayed or each subculture of the fungus. We are just beginning to understand the complexities of a double disease triangle in which the pathogen itself becomes a host. In Michigan, it is becoming apparent that American chestnut stands have a complex pathogen population and the interactions are more significant than first thought. In the water conductance study by McManas et al.,[15] part of the study included sampling the virulent and hypovirulent isolates obtained from the infections. No correlation was found between canker morphology (lethal vs. nonlethal) and the presence or absence of dsRNA in the pathogen recovered from the culture. These authors concluded that the cankers on surviving trees exist over many years and have the opportunity to be colonized by many strains with or without dsRNA. Studies on intracanker populations have demonstrated the presence of more than one strain per canker.[81]

The most controversial, least understood and perhaps most important aspect of hypovirulence is the mechanism of virus dissemination. Characteristics of the virus, the fungal host and the tree probably all influence the ability of the virus to move within a population of *C. parasitica*. Most attention has been paid to vegetative incompatibility as a determinant of viral dissemination (Figure 3-8). Some fungal pathogen isolates are inhibited in forming, or fail to complete hyphal fusions and, consequently, are less active in cytoplasmic transfer of hypoviruses (Figure 3-6). Therefore, it is thought that vegetative incompatibility plays an important role in the dissemination of hypoviruses. The role and importance of vegetative incompatibility (*vic*) genes are still unknown and available data are conflicting. Liu and Milgroom[82] found a negative relationship between the ease of transmission of dsRNA viruses and the number of *vic* genes that differed between isolates of the *C. parasitica*. This result supports the long held theory that the greater the number of *vic* alleles among strains in a given area, the less chance for hypoviruses to spread. In some stands of chestnut in Michigan, the *vic* gene profile of the population is narrow but in other sites the picture is more complex. Davelos et al.[83] found that isolates from their Grand Haven site had a large amount of *vic* gene diversity but that the pathogen at the County Line site showed no diversity in *vic* genes. Since both sites had similar amounts of recovery and similar frequencies of dsRNA in the population, they concluded that relatively higher *vic* gene diversity in a population does not necessarily reduce the spread of dsRNA. Evidence for natural cytoplasmic transmission among diverse strains of *C. parasitica* in the wild is supported by findings of a specific dsRNA genome in strains isolated from chestnut throughout the Appalachian Mountains[70].

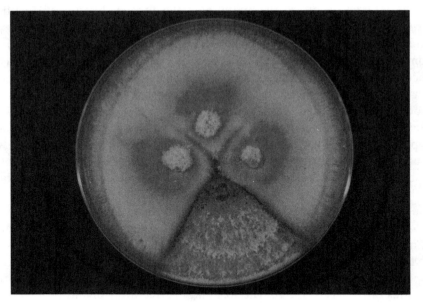

Fig. 3-8. Compatible and incompatible reactions occur as the *Cryphonectria parasitica* isolates interact along their borders. The dark line between strains is indicative of an incompatible interaction and the lack of lines between cultures suggests fusion among hypha, a result of vegetative compatibility.

It was postulated that this dsRNA was widely disseminated due to hyphal fusion among strains carrying different *vic* genes. If this particular dsRNA genome can transfer between diverse fungal populations, it should be possible for hypoviruses to disseminate in a similar manner. Our understanding of the role that *vic* genes play in reducing virus transfer is beginning to improve. Huber and Fulbright [84] have shown that the *vic* alleles each have a specific effect on dsRNA transfer, so that the *vic* genes that regulate hyphal fusions in *C. parasitica* may have either no effect or a strong negative effect depending on the specific alleles present in the strains attempting to fuse.

Acknowledgments

I would like to thank Dr. Julia Bell and Carmen Medina-Mora for reading the manuscript and offering helpful suggestions. The mitochondrial research was supported by a grant from USDA NRICGP 9200717 and the population work was supported by a grant from NSF DEB9509034.

References

1. Merkel, H. W. 1906. A deadly fungus on the American chestnut. New York Zoo. Soc. Annu. Rep. For 1905 10:97-103.

2. Shear, C. L., N. E. Stevens, and R. J. Tiller. 1917. *Endothia parasitica* and related species. USDA Bulletin 380. 82 pp.
3. Micales, J. A., and R. J. Stipes. 1987. A reexamination of the fungal genera *Endothia* and *Cryphonectria*. Phytopathology 77:650-654.
4. Hepting, G. H. 1974. Death of the American chestnut. J. For. History 18:60-67.
5. Scheffer, R. P. 1997. The Nature of Disease in Plants. Cambridge University Press, Cambridge. 325 pp.
6. Griffin, G. J., F. V. Hebard, R. A. Wendt, and J. R. Elkins. 1983. Survival of American chestnut trees; evaluation of blight resistance and virulence in *Endothia parasitica*. Phytopathology 73:1084-1092.
7. Brewer, L. G. 1982. The distribution of surviving American chestnuts in Michigan. In Proceedings USDA Forest Service American Chestnut Cooperators Meeting, eds. H. C. Smith and W. L. MacDonald, pp. 94-100. West Virginia University Books, Morgantown.
8. The Conference: Called by the governor of Pennsylvania to consider ways and means for preventing the spread of the chestnut tree bark disease. 1912. Stenographic Report of Proceedings of the Conference. The State of Pennsylvania, Harrisburg.
9. Milgroom, M. G., K. Wang, Y. Zhou, S. E. Lipari, and S. Kaneko. 1996. Intercontinental population structure of the chestnut blight fungus, *Cryphonectria parasitica*. Mycologia 88:179-190.
10. Anagnostakis, S. L. 1992. Chestnuts and the introduction of chestnut blight. Annual Rpt. North. Nut Growers' Assoc. 83:39-42.
11. Milgroom, M. G., W. L. MacDonald, and M. L. Double. 1991. Spatial pattern analysis of vegetative compatibility groups in the chestnut blight fungus, *Cryphonectria parasitica*. Can. J. Botany 69:1407-1413.
12. Hebard, F. V., G. K. Griffin, and J. R. Elkins. 1984. Developmental histopathology of cankers incited by hypovirulent and virulent isolates of *Endothia parasitica* on susceptible and resistant chestnut trees. Phytopathology 74:140-149.
13. Ewers, F. W., P. McManus, A. Goldman, R. Gucci, and D. W. Fulbright. 1989. The effect of virulent and hypovirulent strains of *Endothia parasitica* on hydraulic conductance in American chestnut. Canadian J. Bot. 67:1402-1407.
14. McCarroll, D. R., and E. Thor. 1978. Death of a chestnut; the host pathogen interaction. In Proceedings of the American Chestnut Symposium., eds. W. L. MacDonald, F. C. Cech, J. Luchok, and H. C. Smith. West Virginia University Books, Morgantown.
15. McManus, P. S., F. W. Ewers, and D. W. Fulbright. 1989. Characterization of the chestnut blight canker and the localization and isolation of the pathogen *Cryphonectria parasitica*. Can J. Botany 67:3600-3607.
16. Shain, L., J. B. Miller and R. J. Spalding. 1993. Responses of American and Chinese Chestnut to *Cryphonectria parasitica* and ethylene. In Proceedings of the International Chestnut Conference, eds. M. L. Double and W. L. MacDonald, pp. 97-101. West Virginia University Press, Morgantown, WV.
17. Gao, S., G. H. Choi, L. Shain, and D. L. Nuss. 1996. Cloning and targeted disruption of enpg-1, encoding of the major in vitro extracellular endopolygalacturonase of the chestnut blight fungus, *Cryphonectria parasitica*. Applied and Environ. Microbiology 62:1984-1990.
18. Vannini, A., C. D. Smart, and D. W. Fulbright. 1993. The comparison of oxalic acid production in vivo and in vitro by virulent and hypovirulent *Cryphonectria* (*Endothia*) *parasitica*. Physiol. and Mol. Plant Path. 43:443-451.

19. Bar Nun, N., A. Tal Lev, E. Harel and A. M. Mayer. 1988. Repression of laccase formation in *Botrytis cinerea* and its possible relation to phytopathogenicity. Phytochemistry 27:2505-2509.
20. Rigling, D., and N. K. Van Alfen. 1993. Extra- and intracellular laccase of the plant pathogenic fungus *Cryphonectria parasitica*. App. Enviro. Mirobiol. 59:3634-3639.
21. Lewis, N. G., and E. Yamamoto. 1990. Lignin: occurrence, biogenesis and bio-degradation. Annu. Rev. Plant Physiol. Plant Mol. Biol. 41:455-496.
22. Hebard, F. 1992. The American Chestnut Foundation breeding plant: Beginning and intermediate steps. In Proceedings of the International Chestnut Conference, eds. M. L. Double and W. L. MacDonald, pp.70-73. West Virginia University Press, Morgantown, WV.
23. Grente, J. 1965. Les formes hypovirulentes D'*Endothia parasitica* it les espoirs de lutte contre le chancre du chataignier, C. R. Hebd. Seances Acad. Agric. France, 51:1033-1036.
24. Grente, J., and S. Berthelay-Sauret. 1978. Biological control of chestnut blight in France. In Proceedings American Chestnut Symposium., eds. W. L. MacDonald, F. C. Cech, J. Luchok, and H. C. Smith, pp. 30-34. West Virginia University Books, Morgantown, WV.
25. Van Alfen, N. K., R. A. Jaynes, S. L. Anagnostakis, and P. R. Day. 1975. Chestnut blight: biological control by transmissible hypovirulence in *Endothia parasitica*. Science 189:890-891.
26. Elliston, J. E., R. A. Jaynes, P. R. Day, and S. L. Anagnostakis, 1977. A native American hypovirulent strain of *Endothia parasitica*. Proceedings of the American Phytopathological Society 4:83 (abstract).
27. Fulbright, D. W., W. H. Weidlich, K. Z. Haufler, C. S. Thomas, and C. P. Paul. 1983. Chestnut blight and recovering American chestnut trees in Michigan. Can. J. Botany 61:3164-3171.
28. Hillman, B. I., Y. Tian, P. J. Bedker, and M. P. Brown. 1992. A North American hypovirulent isolate of the chestnut blight fungus with a European isolate-related dsRNA. J. Gen. Virol. 73:681-686.
29. Day, P. R., J. A. Dodds, J. E. Elliston, R. A. Jaynes, and S. L. Anagnostakis.1977. Double-stranded RNA in *Endothia parasitica*. Phytopathology 67:1393-1396.
30. Dodds, J. A. 1980. Association of type 1 viral-like dsRNA with club-shaped particles in hypovirulent strains of *Endothia parasitica*. Virology 107:1-12.
31. Jensen, C. J. P., R. F. Allison, and G. C. Adams. 1995. Purification and character-ization of virus-like particles of *Leucostoma persoonii*. Mycologia 87:431-441.
32. Hansen, D. R., N. K. Van Alfen, K. Gillies, and W. A. Powell. 1985. Naked dsRNA associated with hypovirulence of *Endothia parasitica* is packaged in fungal vesicles. J. Gen. Virol. 66:2605-2614.
33. Newhouse, J. R., and W. L. MacDonald. 1990. Virus-like particles in hyphae and conidia of European hypovirulent (dsRNA-containing) strains of *Cryphonectria parasitica*. Canadian J. Bot. 68:90-101.
34. Fahima, T., Y. Wu, L. Zhang, and N. K. Van Alfen. 1993. Identification of the putative RNA polymerase of Cryphonectria hypovirus in a solubilized replication complex. J. Virology. 68:6116-6119.
35. Choi, G. H., and D. L. Nuss. 1992. Hypovirulence of chestnut blight fungus conferred by an infectious viral cDNA. Science 257:800-803.
36. Durbahn, C. M. 1992. Molecular characterization of ds-RNA associated hypoviru-lence in Michigan isolates of *Cryphonectria parasitica*. Ph.D. Dissertation, Michigan State University, East Lansing, MI.

37. Hillman, B. I., B. T. Halpern, and M. P. Brown. 1994. A viral dsRNA element of the chestnut blight fungus with a distinct genetic organization. Virology 201:241-250.
38. Chen, B., G. H. Choi, and D. L. Nuss. 1994. Attenuation of fungal virulence by synthetic infectious hypovirus transcripts. Science 264:1762-1764.
39. Hillman, B. I., D. W. Fulbright, D. L. Nuss, and N. K. Van Alfen, 1995. Family Hypoviridae. In Virus Taxonomy , eds. F. A. Murphy, et al., pp. 261-264. Springer Verlag, New York.
40. Fulbright, D. W. 1984. Effect of eliminating dsRNA in hypovirulent *Endothia parasitica*. Phytopathology 74:722-724.
41. L'Hostis, B., S. T. Hiremath, R. E. Rhoads, and S. A. Ghabrial. 1985. Lack of sequence homology between double-stranded RNA from European and American hypovirulent strains of *Endothia parasitica*. J. Gen. Virol. 66:351-355.
42. Paul, C. P., and D. W. Fulbright, 1988. Double-stranded RNA molecules from Michigan hypovirulent isolates of *Endothia parasitica* vary in size and sequence homology. Phytopathology 78:751-755.
43. Tartaglia, J., C. P. Paul, D. W. Fulbright, and D. L. Nuss. 1986. Structural properties of double-stranded RNAs associated with biological control of chestnut blight fungus. Proc. Nat. Acad. Sci. USA 83:9109-9113.
44. Hiremath, S., B. L'Hostis, S. A. Ghabrial, and R. E. Rhoads. 1986. Terminal structure of hypovirulence-associated dsRNAs in the chestnut blight fungus *Endothia parasitica*. Nuc. Acids Res. 14:9877-9896.
45. Shapira, R., G. H. Choi, B. I. Hillman, and D. L. Nuss. 1991. The contribution of defective RNAs to the complexity of viral-encoded double-stranded RNA populations present in hypovirulent strains of the chestnut blight fungus *Cryphonectria parasitica*. EMBO J. 10:731-739.
46. Smart, C. D., and D. W. Fulbright, 1995. Characterization of a strain of *Cryphonectria parasitica* doubly-infected with hypovirulence-associated dsRNA viruses. Phytopathology 85:491-494.
47. Shapira, R., G. H. Choi, and D. L. Nuss. 1991. Virus-like genetic organization and expression strategy for a double-stranded RNA genetic element associated with biological control of chestnut blight. EMBO J. 10:731-739.
48. Choi, G. H., R. Shapira, and D. L. Nuss. 1991. Cotranslational autoproteolysis involved in gene expression from a double-stranded RNA genetic element associated with hypovirulence of the chestnut blight fungus. Proc. Nat. Acad. Sci. USA 88:1167-1171.
49. Craven, M. G., D. M. Pawlyk, G. H. Choi, and D. L. Nuss. 1993. Papain-like protease p29 as a symptom determinant encoded by a hypovirulence-associated virus of the chestnut blight fungus. J. Virology 67:6513-6521.
50. Koonin, E. V., G. H. Choi, D. L. Nuss, R. Shapira, and J. C. Carrington. 1991. Evidence for common ancestry of a chestnut blight hypovirulence-associated double-stranded RNA and a group of positive-strand RNA plant viruses. Proc. Natl. Acad. Sci. USA 88:10647-10651.
51. Chen, B., G. H. Choi and D. L. Nuss. 1993. Mitotic stability and nuclear inheritance of integrated viral cDNA in engineered hypovirulent strains of the chestnut blight fungus. EMBO J. 12:2991-2998
52. Nuss, D. L. 1992. Biological control of chestnut blight: an example of virus-mediated attenuation of fungal pathogenesis. Microbiol. Rev. 56:561-576.
53. Fulbright, D. W. 1990. Molecular basis for hypovirulence and its ecological relationship. In New Directions in Biocontrol: Alternatives for Suppressing Agri

cultural Pests and Diseases, eds. R. R. Baker, and P. E. Dunn., pp. 693-702. Alan R. Liss, Inc., New York.

54. Collmer, C. W., and S. H. Howell. 1992. Role of satellite RNA in the expression of symptoms caused by plant viruses. Annu. Rev. Phytopathol. 30:419-442.

55. Elliston, J. E. 1985. Further evidence for two cytoplasmic hypovirulence agents in a strain of *Endothia parasitica* from Western Michigan. Phytopathology 75:1405-1413.

56. Chung, P., P. J. Bedker, and B. I. Hillman. 1994. Diversity of *Cryphonectria parasitica* hypovirulence-associated double-stranded RNAs within a chestnut population in New Jersey. Phytopathology 84:984-990.

57. Powell, W. A., and N. K. Van Alfen. 1987. Differential accumulation of poly (A)+RNA between virulent and double-stranded RNA-induced hypovirulent strains of *Cryphonectria (Endothia) parasitica*. Mol. Cell. Biol. 7:3688-3693.

58. Powell, W. A., and N. K. Van Alfen. 1987. Two nonhomologous viruses of *Cryphonectria (Endothia) parasitica* reduce accumulation of specific virulence-associated polypeptides. J. Bacteriol. 1169:5324-5326.

59. Varley, D. A., G. K. Podila, and S. T. Hiremath. 1992. Cutinase in *Cryphonectria parasitica*, the chestnut blight fungus: suppression of cutinase gene expression in isogenic hypovirulent strains containing double-stranded RNAs. Molec. and Cell. Biology 12:4539-4544.

60. Rigling, D., and N. K. Van Alfen. 1991. Regulation of laccase biosynthesis in the plant-pathogenic fungus *Cryphonectria parasitica* by double-stranded RNA. J. Bacteriol. 173:8000-8003.

61. Rigling, D., U. Heiniger, and H. R. Hohl. 1989. Reduction of laccase activity in dsRNA-containing hypovirulent strains of *Cryphonectria (Endothia) parasitica*. Phytopathology 79:219-223.

62. Havir, E. A., and S. L. Anagnostakis. 1983. Oxalic acid production by virulent but not by hypovirulent strains of *Endothia parasitica*. Physiol. Plant Pathol. 23:369-376.

63. Larson, T. G., G. H. Choi, and D. L. Nuss. 1992. Regulatory pathways governing modulation of fungal gene expression by a virulence-attenuating mycovirus. EMBO J. 11:4539-4548.

64. Nuss, D. L. 1996. Using hypoviruses to prove and perturb signal transduction processes underlying fungal pathogenesis. The Plant Cell 8:1845-1853.

65. Choi, G. H., B. Chen, and D. L. Nuss. 1995. Virus-mediated or transgenic suppression of a G-protein a subunit and attenuation of fungal virulence. Proc. Nat. Acad. Sci. USA 92:305-309.

66. Chen, B., C.-H. Chen, B. H. Bowman, and D. L. Nuss. 1996. Phenotypic changes associated with wild-type and mutant hypovirus RNA transfection of plant pathogenic fungi phylogenetically related to *Cryphonectria parasitica*. Phytopathology 86:301-310.

67. Chen, B., S. Gao, G. H. Choi, and D. L. Nuss. 1996. Extensive alteration of fungal gene transcript accumulation and elevation of G-protein-regulated cAMP levels by a virulence-attenuating hypovirus. Proc. Nat. Acad. Sci. USA 93:7996-8000.

68. Gao, S., and D. L. Nuss. 1996. Distinct roles for two G protein a subunits in fungal virulence, morphology, and reproduction revealed by targeted gene disruption. Proc. Nat. Acad. Sci. USA 93:14122-14127.

69. Wang, P., and D. L. Nuss. 1995. Induction of a *Cryphonectria parasitica* cellobiohydrolase I gene is suppressed by hypovirus infection and regulated by a GTP-binding-protein-linked signaling pathway involved in fugal pathogenesis. Proc. Nat. Acad. Sci. 92:11529-11533.

70. Enebak, S. A., W. L. MacDonald, and B. I. Hillman. 1994. Effect of dsRNA associated with isolates of *Cryphonectria parasitica* from the central Appalachians and their relatedness to other dsRNAs from North America and Europe. Phytopathology 84:528-533.

71. Enebak, S. A., B. I. Hillman, and W. L. MacDonald. 1994. A hypovirulent isolate of *Cryphonectria parasitica* with multiple, genetically unique dsRNA segments. Mol. Plant-Microbe Interact. 7:590-595.

72. Polashock, J. J., and B. I. Hillman. 1994. A small mitochondrial double-stranded (ds) RNA element associated with a hypovirulent strain of the chestnut blight fungus and ancestrally related to yeast cytoplasmic T and W dsRNAs. Proc. Nat. Acad. Sci. USA 91:8680-8684.

73. Fulbright, D. W. 1985. A cytoplasmic hypovirulent strain of *Endothia parasitica* without double-stranded RNA. Phytopathology 75:1328 (Abstract).

74. Mahanti, N., Bertrand, H, Monteiro-Vitorello, C. and Fulbright, D. W. 1993. Elevated mitochondrial alternative oxidase activity in dsRNA-free, hypovirulent isolates of *Cryphonectria parasitica*. Physiol. Mol. Plant Path. 42:455-463.

75. Mahanti, N., and D. W. Fulbright. 1995. Detection of mitochondrial DNA transfer between strains after vegetative contact in *Cryphonectria parasitica*. Mol. Plant-Microbe Inter. 8:465-467.

76. Bertrand, H. 1983. Aging and senescence in fungi. In Intervention in the Aging Process, Part B: Basic Research and Preclinical Screening, eds. W. Regelson and F. M. Sinex, pp. 233-251. Alan R. Liss, Inc., New York.

77. Bertrand, H. 1995. Senescence is coupled to induction of an OXPHOS stress response by mitochondrial DNA mutations in *Neurospora*. Can. J. Botany 73:S198-S204.

78. Bertrand, H., B. S-S. Chan, and A. J. F. Griffiths. 1985. Insertion of a foreign nucleotide sequence into mitochondrial DNA causes senescence in *Neurospora intermedia*. Cell 41:877-884.

79. Monteiro-Vitorello, C. B., J. A. Bell, D. W. Fulbright, H. Bertrand. 1995. A cytoplasmically-transmissible hypovirulence phenotype associated with mitochondrial DNA mutations in the chestnut blight fungus *Cryphonectria parasitica*. Proc. Nat. Acad. Sci. USA 92:5935-5939.

80. Bell, J. A., C. B. Monteiro-Vitorello, G. Hausner, D. W. Fulbright, and H. Bertrand. 1996. Physical and genetic map of the mitochondrial genome of *Cryphonectria parasitica* Ep155. Current Genet. 30:34-43.

81. Schaupp, J. K., A. M. Jarosz, and A. L. Davelos. 1997. Intra-canker dynamics of *Cryphonectria parasitica*. Phytopathology 87:S85 (abstract).

82. Liu, Y.-C., and M. G. Milgroom. 1996. Correlation between hypovirus transmission and the number of vegetative incompatibility (*vic*) genes different among isolates from a natural population of *Cryphonectria parasitica*. Phytopathology 86:79-86.

83. Davelos, A. L., J. K. Schaupp, D. H. Huber, and D. W. Fulbright. 1996. Variation in diversity of vegetative compatibility groups of *Cryphonectria parasitica* among populations in Michigan. Phytopathology 86:S11(abstract).

84. Huber, D. H., and D. W. Fulbright. 1994. Preliminary investigations on the effect of individual *vic* genes upon the transmission of dsRNA in *Cryphonectria parasitica*. In Proceedings of the International Chestnut Conference, eds., M. L. Double and W. L. MacDonald, pp. 15-19. West Virginia University Press, Morgantown, WV.

CHAPTER 4

Active Oxygen and Pathogenesis
in Plants

C. Jacyn Baker and Elizabeth W. Orlandi

As evidenced by a recent meeting devoted entirely to active oxygen (AO) in plant stress[1] it is generally accepted that AO species are involved in many different facets of stress physiology. Plants stressed not only by pathogens but by drought, salinity, heavy metal toxicity, air pollution, and UV exposure all have increased AO metabolism that, at different times, could be either beneficial or detrimental. The wide-spread involvement of AO in both plant and animal physiology should not be too surprising because we exist under a blanket of air made up of 21% O_2. While molecular oxygen is stable in its ground state, it becomes highly reactive or "active" when its electrons absorb energy from either light or energetic electrons with which it comes into contact. Active oxygen is not only produced in response to stress, it is also produced during several normal processes in healthy plants.[2] For example, during photosynthesis it has been reported that, under certain conditions, as much as 5 to 10 % of the electrons can "leak" from electron transport pathways resulting in the production of AO.[3,4] Because AO production accompanies and is even necessary for many normal plant cell · processes, the cell has evolved highly efficient antioxidant systems to control the accumulation of AO. However, during periods of stress, AO production increases and can overwhelm these antioxidant systems.

The focus of this review will be on the involvement of AO in several cellular processes that occur during plant/pathogen interactions. Currently in the literature, there is a heavy emphasis on the AO "burst" that occurs during the initial contact between plant cells and pathogens or various elicitors. However, AO species may be produced and play a significant although more subtle role at later stages of pathogenesis as well. In this review we will look at AO involvement during plant/pathogen interactions in a chronological fashion, describing its potential production and impact during three stages of pathogenesis (Figure 4-1). Stage I includes the early phase of the interaction

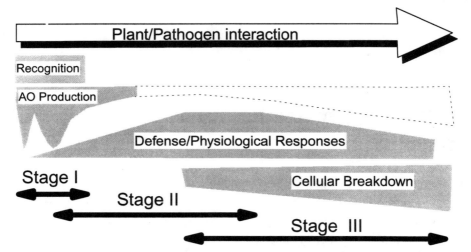

Fig. 4-1. Chronological summary of events in plant pathogenesis that involve AO metabolism. Stage I encompasses the earliest events including plant/pathogen recognition and the initial phases of AO production. Active oxygen production continues throughout pathogenesis and may even increase in later stages, however, it is scavenged and may not accumulate to the levels observed during Stage I. Stage II includes the secondary responses initiated by Stage I including those that require gene activation. Stage III is comprised of the senescent or degradative processes leading to tissue necrosis during a HR response or the development of symptoms in compatible interactions.

during which recognition of the potential pathogen occurs. During Stage II, various defense-related processes are initiated, but visible symptoms have not yet developed. Stage III is comprised of the degradative processes of pathogenesis that lead to the development of visible symptoms. The compartmentalization of pathogenesis into three distinct stages is clearly artificial, and there is considerable overlap as indicated in Figure 4-1. In addition, the occurrence and timing of the various physiological events described in each stage will vary depending on the interaction.

Background

This section will provide a brief review of the key AO species that are important in plant physiology; some of the antioxidative mechanisms plant cells employ to control the accumulation of AO; and the methods that are most commonly used to detect and/or measure AO in plant tissues.

Active Oxygen Species

What is active oxygen? Molecular oxygen (O_2) is relatively unreactive and nontoxic. This is attributed to the stable electron structure in its outer shell. The ground state of atmospheric oxygen is a triplet state with two outer electrons having the same spin. Molecules in a triplet state cannot bond with molecules in a singlet state, which is the state of most biological matter. However, once the electronic distribution of O_2 is altered the structure becomes very reactive and can influence biological systems. The AO species generated within plants result either from the excitation and "flipping" of an outer electron, forming singlet oxygen ($^1O_2$) or from the successive addition of electrons to molecular oxygen yielding superoxide (O_2^-), hydrogen peroxide (H_2O_2), and the hydroxyl radical ($OH\cdot$) respectively (Figure 4-2). These molecules are considered "active" because they will react with other molecules without the input of energy. Figure 4-2

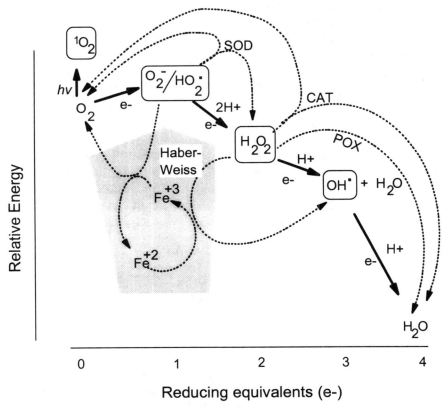

Fig. 4-2. Active oxygen species derived from O_2 and the scavenging and interconversion pathways that are likely to exist in plants. Singlet oxygen ($^1O_2$) is produced by the absorption of energy, generally from light. The other species result from the reduction of oxygen. The first one-electron reduction of O_2 requires a slight input of energy and results in the formation of the superoxide radical (O_2^-) which is in equilibrium with its perhydroxy radical form ($HO_2\cdot$). The following 3 one-electron reductions yield hydrogen peroxide (H_2O_2), the hydroxyl radical ($OH\cdot$) and finally, water. The Haber-Weiss reaction produces $OH\cdot$ and O_2 from H_2O_2 and O_2^- in the presence of Fe^{+++}. The scavenging activity of superoxide dismutase (SOD), peroxidase (POX), and catalase (CAT) can result in a reduction of AO levels in plant tissues.

demonstrates the energetic relationships of these species including some of the cellular scavenging mechanisms that will be discussed later in this section. Figure 4-3 presents much of the same information but in the cellular context, showing the organellar location of the various AO interactions.

Although we will limit our discussion in this section to those species described by Figure 4-2, it should be kept in mind that a broader definition of AO encompasses any reactive species in which the active moiety involves an oxygen atom. This includes the phenoxy radicals involved in lignin synthesis and the lipoxy radicals produced by lipoxygenase (LOX) activity or lipid peroxidation. The relevance of these free radicals to pathogenesis will be discussed later in this review.

Superoxide. Superoxide is reportedly produced in plants through several mechanisms including membrane-bound NADH-oxidases/synthases,[5] cell wall peroxidases,[6,7] LOX,[8,9] and as a result of aberrant transfer of electrons in the electron transport chains of the mitochondria and chloroplast (Figure 4-3). Under the neutral to slightly acidic conditions often found in plants, the negatively charged O_2^- would be in equilibrium with its neutral charged conjugate acid, the hydroperoxyl radical (HOO•) (equation a). Being neutral, HOO• is more lipophilic than O_2^- and, therefore, more capable of lipid peroxidation. Superoxide in either form dismutates to H_2O_2 and O_2 (equations b and c). Therefore, whenever O_2^- is produced, significant amounts of H_2O_2 will also be formed. The reaction in equation b occurs spontaneously at neutral or slightly acidic pH while the reaction in equation c is slow due to electrostatic repulsion and requires SOD as a catalyst.

$$O_2^- + H^+ \longleftrightarrow HOO• \quad (pKa = 4.8) \quad \text{(a)}$$

$$HOO• + O_2^- \xrightarrow{\ H^+\ } H_2O_2 + O_2 \quad \text{(b)}$$

$$O_2^- + O_2^- \xrightarrow{\ 2H^+\ } H_2O_2 + O_2 \quad \text{(c)}$$

Superoxide is unusual in that it can act either as an oxidant or reductant. It will oxidize a variety of organic molecules such as ascorbate or reduce transition metals such as Fe^{3+}, as in the Haber-Weiss reaction (Figure 4-2).

Hydrogen peroxide. As described above, most of the H_2O_2 in the cell probably arises from the dismutation of O_2^- catalyzed by SOD. However, another major source is the leaf peroxisome where two-electron oxidases, such as glycollate oxidase, generate H_2O_2 directly (Figure 4-3). H_2O_2 is a relatively stable oxidant and can readily cross the lipid bilayer of cell membranes due to its lack of charge. H_2O_2 can directly oxidize transition

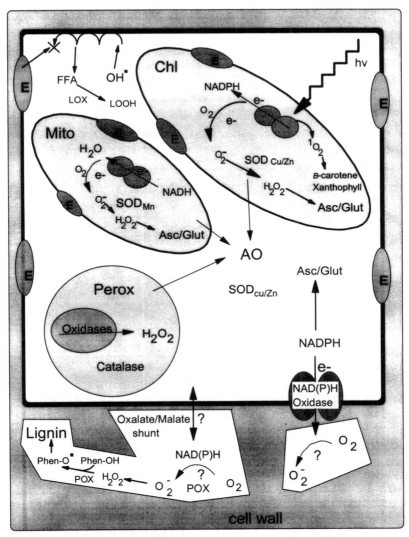

Fig. 4-3. Cellular perspective of AO metabolism. Active oxygen species are generated during normal metabolism in healthy plant cells and are quickly detoxified by scavenging mechanisms to avoid significant cytotoxic effects. The chloroplast (Chl) and mitochondria (Mito) generate O_2^- from stray electrons from the transport chains while peroxisomes (Perox) generate H_2O_2 from catabolic oxidases. The production of lignin in the cell wall may require H_2O_2, which has been hypothesized to be produced via a cell wall-bound peroxidase utilizing NAD(P)H generated by the malate/oxalate shuttle. An NAD(P)H oxidase, similar to that found in human neutrophils, has also been hypothesized to generate extracellular AO during times of stress. Under normal metabolism catalase, superoxide dismutase (SOD), and the ascorbate glutathione cycle are able to scavenge the AO species generated by these various processes. Another antioxidant, α-tocopherol (vitamin E), protects membranes by interrupting the chain reaction of lipid peroxidation that is often initiated by OH· or other species, β-carotene and the xanthophyll cycle protect against the damaging effects of $^1O_2$ generated in chloroplasts.

metals such as Fe^{2+} or organic molecules, generally with the aid of peroxidase.

Hydroxyl radical. The hydroxyl radical has a half life in the range of microseconds and, therefore, reacts with biomolecules close to its generation site. Due to its short half-life it has been difficult to do extensive, precise studies of this radical. In biological materials its production is presumed to result from the Haber-Weiss reaction (Figure 4-2) as summarized in equation (d). Because of the role of transition metals as a catalyst, the formation of the OH. appears to be determined by the location of metal ion complexes in the cell. Iron associated with the membrane would likely initiate lipid peroxidation upon production of O_2^- and subsequent H_2O_2.

$$O_2^- + H_2O_2 \xrightarrow{\quad Fe^{3+} \quad} OH. + OH^- + O_2 \quad (d)$$

Singlet oxygen. Like the OH., $^1O_2$ has a very short lifetime (in the range of µsec) and does the most damage near the site of its production. It is generated predominately in the chloroplast by the transfer of energy from photo-excited chlorophyll to the electrons of molecular oxygen (Figures 4-3 and 4-4). Singlet oxygen is converted back to stable triplet oxygen via β-carotene as mentioned below, however, in stressed plants, the β-carotene mechanism can become overwhelmed, leading to bleaching or chlorosis.[10] Singlet oxygen reacts readily with double bonds and has a high reactivity with dienes in membranes and His, Try, Met, and Cys residues in proteins [11]. In plant cells infected with the fungal pathogen *Cercospora,* cell death reportedly results from the generation $^1O_2$ and O_2^- produced upon light activation of the fungal toxin, cercosporin.[12,13]

Scavenging Mechanisms

β-carotene. β-carotene is one of the most effective scavengers of $^1O_2$ (Figure 4-4). Carotenoids are part of the chloroplast antenna system that absorbs light and transfers excitation energy on to the reaction centers. However, carotenoids can also dissipate energy during photooxidative stress. During these periods, a fraction of the chlorophyll molecules remain excited for longer periods of time and may stand a greater chance of converting to the triplet state and subsequently producing singlet oxygen. Carotenoids intervene to prevent these destructive reactions by quenching the triplet chlorophyll or any singlet oxygen that has formed, thereby reducing their ability to cause damage by bleaching the chlorophyll or reacting with membrane fatty acids or proteins.

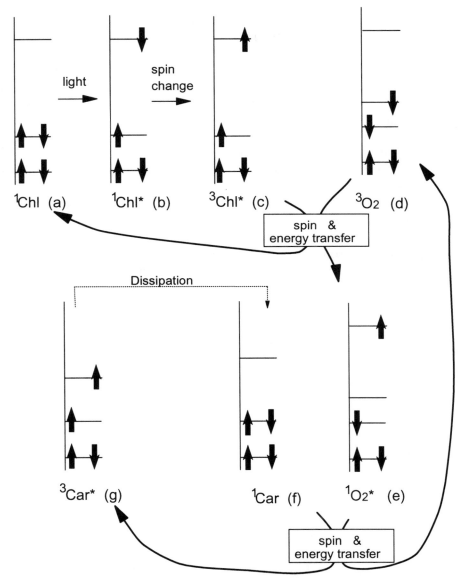

Fig. 4-4. Production and scavenging of singlet oxygen during photooxidative stress. When chlorophyll (a) absorbs light, an outer electron becomes excited (b). Some of these electrons will change their spin, creating a triplet state (c). The triplet state cannot decay to ground state until the electron spin changes again, which does not happen readily. Oxygen, which has a triplet ground state (d), can interact and accept the energy of the triplet chlorophyll molecule. This allows the chlorophyll molecule to return to its singlet ground state (a). However, the oxygen is now in its highly reactive singlet state (e). Carotene, which has a singlet ground state (f) can scavenge singlet oxygen creating triplet carotene (g). This excited molecule is harmless because its energy level is below $^1O_2$ and is able to dissipate harmlessly to ground state.

Superoxide dismutase. Superoxide is scavenged via the disproportionation reaction catalyzed by superoxide dismutase that results in the production of H_2O_2 (Figure 4-5). There are three major types of SOD that differ mainly in their prosthetic metals: Cu/Zn, Mn, and Fe. Plants usually have a Cu/Zn SOD in the cytosol, a Mn SOD in the mitochondria, and Cu/Zn and/or Fe SOD in the chloroplast (Figure 4-3). SOD plays a very important role in reducing O_2^- levels in plant cells. For example, in isolated chloroplasts SOD reportedly reduced the level of superoxide from ~22 µM to ~2.4 nM, thus significantly reducing the potential for oxidative damage.[11]

Catalase. Catalase is found predominately in the peroxisome (Figure 4-3) where it scavenges the H_2O_2 produced during the catabolic oxidation that takes place there. As evidenced by its high Km (generally >10–100 mM), catalase is very inefficient at scavenging low concentrations of H_2O_2. To compensate for this inefficiency, very high levels of catalase are present in peroxisomes, sometimes in a nearly crystalline state.[14] One unique advantage of catalase is that it does not require an energy source to scavenge H_2O_2 and, therefore, it does not affect the cellular NAD(P)H pool (Figure 4-5).

Ascorbate/glutathione cycle. H_2O_2 can be reduced/scavenged by ascorbate peroxidase (APX) using ascorbate as a reductant forming monodehydroascorbate radical (MDA·) which will dismutate to dehydroascorbate

Superoxide Dismutase:

$$O_2^- + O_2^- + 2H^+ \longrightarrow H_2O_2 + O_2$$

Catalase:

$$H_2O_2 + H_2O_2 \longrightarrow 2H_2O + O_2$$

Ascorbate Peroxidase (APX):

$$APX + H_2O_2 \longrightarrow \text{Compound I} + H_2O$$

$$\text{Compound I} + \text{Ascorbate} \longrightarrow \text{Compound II} + MDA^\bullet$$

$$\text{Compound II} + \text{Ascorbate} \longrightarrow APX + MDA^\bullet + H_2O$$

Fig. 4-5. Mechanism of key AO scavenging enzymes.

(DHA) and ascorbate (Figures 4-5 and 4-6). Relatively high levels of ascorbate exist in the cytoplasm and organelles and can reach 10 mM in the chloroplast.[11] Ascorbate peroxidase, unlike most peroxidases, demonstrates substrate specificity and appears to be unique to plants, playing a role similar to glutathione peroxidase in animals. The K_m of APX (5–100 μM) for H_2O_2 is much lower than that of catalase, allowing it to effectively scavenge low levels of H_2O_2.

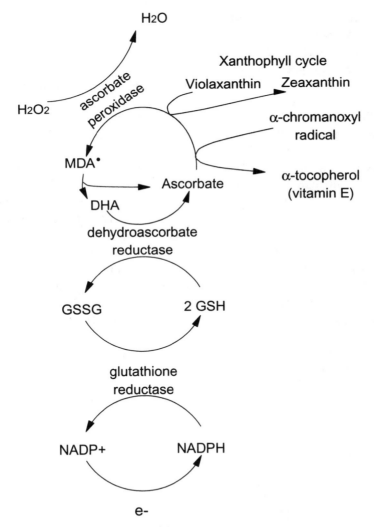

Fig. 4-6. The proposed ascorbate/glutathione cycle and associated reactions. MDA·, monodehydroascorbate radical; DHA, dehydroascorbate; GSSG, oxidized glutathione; GSH, reduced glutathione.

Ascorbate peroxidase catalyzes the first step of the ascorbate-glutathione cycle, which involves successive enzymatic oxidations and reductions of ascorbate, glutathione, and NAD(P)H, as shown in Figure 4-6. Some isozymes of APX appear to be membrane bound and help protect membranes by regenerating components of other membrane-associated antioxidant cycles such as the xanthophyll cycle components, β-carotene and α-tocopherol (vitamin E).[11]

AO Measurement During Plant/ Pathogen Interactions

The extraordinary potential impact of AO metabolism on several aspects of the plant/pathogen interaction makes it important for us to be aware of its existence and to determine its significance. However, the detection and quantification of AO in biological systems has been problematic. Active oxygen species cannot be directly measured using spectrophotometric or HPLC methods that are applicable to the study of carbon-based molecules. Therefore, the majority of AO detection techniques rely upon the oxidation or reduction of chemical probes by AO and the subsequent measurement of these changes using various methods. Quantification of AO species is difficult due to the rapid destruction and scavenging of AO species by the scavenging mechanisms described above. Inhibitors of antioxidant enzymes are often potent inhibitors of AO production, as well, making interpretation of assay results especially difficult. Table 4-1 lists several methods of AO detection that have been employed to monitor AO in plant tissues. Recent reviews have summarized the usefulness of these methods for the study of plant/pathogen interactions along with some cautions on interpreting the results.[15-17]

Table 4-1. Methods of AO detection and measurement in plant tissues

Method of Detection	Chemical Probe	References
Macroscopic/microscopic	Nitroblue tetrazolium	40,42,49,50
	Starch/potassium iodide	99,66,148,73
	Titanium chloride (Ti(IV)Cl$_4$)	66
Spectrophotometric	Nitroblue tetrazolium	40,149,150
	Cytochrome C	42,151
	Epinephrine	152
Fluorescent	Pyranine	80
	Scopoletin	73
Chemiluminescent	5-amino-2,3-dihydro-1,4-phthalazinedione (luminol)	50,153,20,117
	Luciginen	154
Electron spin resonance spectroscopy	5,5′-dimethyl-1-pyrroline-1-oxide (DMPO)	155
	4-pyridyl 1-oxide N-tert-butylnitrone (4-POBN)	156

Stage I: Initial Recognition and Early Events

When plant cells come into contact with potential plant pathogens, one of the first measurable responses is the production of AO, often called an AO "burst". In suspension cells, this first AO response occurs within minutes after the addition of bacteria. It is assumed that during this short amount of time recognition of the pathogen occurs, signal transduction events are triggered, and the mechanisms responsible for AO production are activated. In this section we will first present some generalizations about these initial processes and then put them into the context of specific plant/fungi and plant/bacteria interactions. We will also discuss the immediate effects of AO production that can occur during Stage I pathogenesis.

Elicitation, Signal Transduction, and AO Production: A Brief Overview

Elicitation. Table 4-2 provides a list of several AO elicitors from fungi and bacteria. The table summarizes studies done with whole pathogens as well as pathogen biomolecules that have been found to elicit AO production in plant cells. Many of the elicitor preparations are crude extracts from fungal hyphae and were originally isolated for use as phytoalexin elicitors. It is assumed that the carbohydrate, protein, or glycoprotein portions of these extracts are the components that are recognized by plant cells. However, it is still unclear which of these biomolecules actually triggers recognition during pathogenesis.

Signal transduction. Recognition of elicitors results in the triggering of signal transduction events that may vary depending on the elicitor. The signaling cascades triggered by different AO elicitors have recently been compared by Low and colleagues[18,19]. Several studies have indicated a common requirement for protein phosphorylation and Ca^{++} influx for a variety of AO elicitors.[20–23] However, other established signal transduction components such as G-proteins and phospholipases may not be absolutely required for all elicitors.[24–26]

AO production. The source of AO production during plant/pathogen interactions has not been isolated. For several years, researchers have proposed mechanisms centering on the involvement of either a cell wall peroxidase[6,28,29] or a membrane-bound NAD(P)H oxidase similar to that found in human neutrophils[30,31,5] (Figure 4-3). In recent years, researchers have primarily focused their attention on the latter mechanism. The similarities between the triggering of the human neutrophil NADPH oxidase and elicitation of AO production in plant suspension cells have been reviewed by Low and Dwyer.[32] The parallels they draw between the two systems include the rapid reaction of cells to elicitors (0.5 to 2 minutes), the magnitude of the response (reportedly 10^{-14} mol/cell/minute), and the cessation of AO production after approximately 15 minutes. Recently, these authors and others have also found some cross-reaction between antibodies raised against components of the neutrophil oxidase and plant proteins. The components of human neutrophil NADPH oxidase include cytochrome b_{558}, composed of the two membrane-bound components ($gp91^{phox}$ and $gp22^{phox}$), and a cytosolic complex of two proteins ($p47^{phox}$ and $p67^{phox}$) that is translocated to the plasma membrane along with a small GTP-binding protein (rac) during oxidase activation[33,34]. Proteins immunologically related to $p47^{phox}$ and $p67^{phox}$ were found in suspension cells of soybean and arabidopsis, while tobacco, cowpea, and cotton cells contained proteins immunologically related to $p47^{phox}$ but not $p67^{phox}$[35,36]. In

Table 4-2. Elicitors of AO production in plant tissues

Source	Plant Tissue	Elicitor	References
Fungi			
Phytophthora infestans	Potato (*Solanum tuberosum*) tuber protoplasts, disks, leaves	Zoospores, hyphal wall components	42, 41,44
Phytophthora mega- sperma	Rose (*Rosa damascena*) suspension cells	Mycelial cell walls	157, 158
Colletotrichum linde- muthianum	Bean (*Phaseolus vulgaris*) suspension cells	Partially purified galac- toglucomannan	159
Verticillium dahliae	Soybean suspension cells	Autoclaved mycelia	160,80
Cladosporium fulvum	Tomato (*Lycopersicon esculentum* L.) suspension cells or cotyle- dons	Intercellular fluids of infected tomato leaves	161,40
Pyricularia oryzae	Rice (*Oryza sativa*) protoplasts	Proteoglucomman	162
Verticillium albo-atrum	Cotton leaves	Endopolygalacturonase	14,163
Amanita muscaria; Hebel- oma crustuliniforme; Heterobasidion anno- sum	Spruce (*Picea abies*) suspension cells	Autoclaved mycelial cell walls	20
Phytophthora cryptogea	Tobacco (*Nicotiana tabacum*) suspension cells	Cryptogein	85,22
Phytophthora capsici	Tobacco (*Nicotiana tabacum*) suspension cells	Capsicein	85
Bacteria			
Pseudomonas syringae pv. *syringae*	Tobacco (*Nicotiana tabacum*) leaf disks	Whole, viable bacteria	49
Pseudomonas syringae pv. *syringae*	Tobacco (*Nicotiana tabacum*) suspension cells	Whole, viable bacteria	50,51
Pseudomonas corrugata	White clover (*Trifolium repens* L.)	Whole, viable bacteria or autoclaved bacteria	86
Erwinia amylovora	Tobacco (*Nicotiana tabacum*) suspension cells	Whole, viable bacteria or harpin$_{ea}$	21
Pseudomonas syringae pv. *gylcinea*	Soybean and tobacco (*Nico- tiana tabacum*) suspension cells	Whole, viable bacteria	51,153, 55,160, 73

addition, antibodies to *gp22phox* cross-reacted with a plasma membrane protein from soybean[37] and DNA sequences from rice roots and shoots were found to be homologous to the *gp91phox* of neutrophil oxidase[38]. Recently, Xing et al.[39] monitored AO production and the corresponding appearance of proteins immunologically related to *p47phox*, *p67phox*, and rac in tomato cell membranes after treatment with race-specific elicitors from *Cladosporium fulvum*.

Early Events in Plant/Pathogen Interactions

As described above, many of the insights into the mechanisms of elicitation, signal transduction, and production of AO in plant cells have been gained from studies employing crude or purified elicitors. However, these studies cannot accurately examine the complexities of the early events of plant/pathogen interactions. In elicitor studies, the production of AO is often a single, transient event triggered by a one-time addition of an elicitor. However, during pathogenesis, elicitors may be produced at several time points during the interaction by both the pathogen and the plant cell.

In addition, the production and accumulation of AO may vary over time as the physiological state of the plant cell changes. For example, changes in antioxidant activity, membrane permeability, and oxidative state of the plant may have an effect on AO production and accumulation. Current research methods still cannot give us a complete picture of AO production in plants infected with pathogens. However, by combining the results of studies employing elicitors, whole pathogens, suspension cells and intact plant tissues a more complete picture of AO production during pathogenesis will emerge. The following sections present an overview of AO production during the initial stage of a plant/fungi and plant/bacteria interaction.

Early events in Phytophthora/potato interactions. Doke and colleagues performed some of the earliest studies on the production of AO during plant/pathogen interactions. Inoculation of aged potato tuber discs with zoospores of incompatible races of *Phytophthora infestans* elicited the reduction of extracellular cytochrome c and nitroblue tetrazolium.[40,41] AO production was measured in discs treated with incompatible but not compatible races of *P. infestans*. Crude preparations of hyphal wall components of both compatible and incompatible races also elicited AO production by tuber discs, protoplasts[42], and membrane fractions.[30] However, the compatible races were also found to produce water-soluble glucans (WSC) that suppressed the stimulation of AO production[42,43] and were proposed to be the determinants of whether an interaction was incompatible or compatible.

The inoculation of potato leaf discs with zoospores of compatible and incompatible races of *Phytophthora infestans* resulted in the production of

AO approximately 1 h after inoculation.[44,45] This "step I" reaction lasted approximately 3-4 hours and was also induced by germination fluid from either compatible or incompatible races of *P. infestans* [44]. This indicated that the zoospores secreted nonspecific elicitors during germination that triggered the production of AO.

Step I AO production was followed by a larger accumulation of AO, "step II", which occurred only after inoculation with zoospores from incompatible races. The step II AO production occurred approximately 4 hours after inoculation, at the same time that the germinated zoospores began to penetrate the epidermal cells of the leaf tissues, and increased until approximately 10 hours after inoculation.

The race-specificity of the step II AO production may be attributed to the production of water-soluble glucans by compatible *P. infestans* races that suppress the production of AO[42,43]. These suppressers were isolated from *P. infestans*, *P. capsici*, and *P. nicotianae* var. *nicotiana*.[46,47] Hyphal wall components from these *Phytophthora* species elicited AO production from suspension cells of tomato, sweet pepper, and tobacco in a species-nonspecific manner. However, the water-soluble glucans isolated from the *Phytophthora* species suppressed the elicited AO production only on their respective hosts.

Studies on the signal transduction mechanisms activated during the elicitation of AO in potato cells treated with viable *Phytophthora infestans* or isolated hyphal wall components indicate the involvement of GTP-binding proteins and Ca^{++}.[48]

Early events in Pseudomonas syringae/tobacco and soybean interactions. The production of AO during plant/bacteria interactions was first monitored in tobacco leaf discs and suspension cells treated with the incompatible pathogen *Pseudomonas syringae* pv. *syringae*.[49,50] In tobacco (*Nicotiana tabacum* cv. Burley 21) leaf discs, Adam et al.[49] measured AO production using NBT. In this study, AO production began approximately 3 hours after infiltration with bacteria and continued to increase for approximately 3 more hours until the first signs of hypersensitive cell death occurred. Treatment of leaf discs with a Tn5 transposon mutant of *P. s.* pv. *phaseolicola* that did not cause hypersensitive cell death also did not induce phase II AO production.

Using luminol-dependent chemiluminescence, Keppler, Baker and co-workers monitored a two-phased production of AO in tobacco (cv. Hicks) and soybean (*Glycine max* L. cv. Mandarin) suspension cells incubated with *Pseudomonas* pathovars (Figure 4-7A and B).[50,51] Similar to AO production in the *Phytophthora*/potato system described above, AO production in suspension cells occurred in two phases. However, in plant/bacteria interactions, AO production occurred sooner after inoculation and was shorter lived than in the plant/fungi interaction described above. Phase I AO production occurred within a few minutes after treatment of suspension cells with bacteria, regardless of whether the resulting interaction was compatible, incompatible, or saprophytic. Phase II occurred 1.5–3 hours after bacterial treatment, was more sustained, and was specific to incompatible interactions.

Phase II AO production occurred in both non-host and race-specific incompatible reactions and was consistently followed by hypersensitive

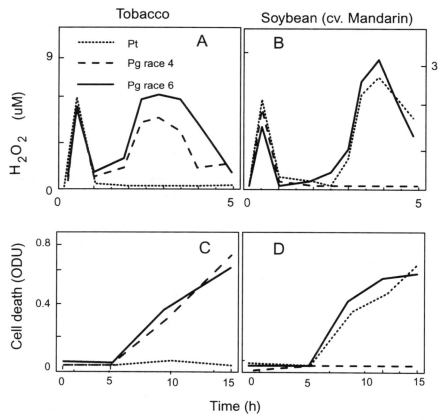

Fig. 4-7. Two-phased AO production (**A and B**) and cell death (**C and D**) in tobacco and soybean suspension cells during non-host and race-specific plant/bacteria interactions. Phase I is immediate and non-specific, occurring in both compatible and incompatible interactions within the first few minutes of each interaction. Phase II occurs only in incompatible interactions and is subsequently followed by cell death. The *Pseudomonas syringae* pathovars used are pv. *tabaci* (Pt), a pathogen of tobacco (compatible) that causes HR on soybean (non-host, incompatible); pathovar *glycinea* race 4 (Pg race 4), which causes HR on tobacco (non-host, incompatible) and is a race-specific pathogen of soybean cv. Mandarin; pathovar *glycinea* race 6 (Pg race 6), which causes HR on tobacco (non-host incompatible) and is a race-specific incompatible pathogen on soybean cultivar Mandarin.

cell death as measured by Evan's blue staining[52]. As illustrated in Figure 4-7A and C, non-host tobacco suspension cells reacted similarly to *P. s.* pv. *glycinea* races 4 and 6, whereas host soybean suspension cells reacted in a race-specific manner (Figure 4-7B and D). Similarly, *P. s.* pv. *tabaci* did not elicit phase II AO production during a compatible reaction with tobacco cells but only during the non-host incompatible interaction with soybean suspension cells. All combinations of host cells and bacteria resulted in the production of phase I AO production.

The rapidity of the phase I AO response indicated that it was triggered by a pre-formed bacterial elicitor. In addition, because the first phase was detected in suspension cells treated with a wide variety of bacteria whether compatible, incompatible, or saprophytic, it is apparent that the phase I elicitor(s) is common among bacteria. Preliminary studies in our laboratory have isolated a protein that is loosely-bound to *Pseudomonas* and is a very potent elicitor of AO[53]. It is possible that this first phase elicitor is a flagellin, as these proteins are widespread among bacteria and have recently been hypothesized to trigger ion fluxes in tomato suspension cells.[54]

The elicitation of phase II AO production during the race-specific soybean/*P. s.* pv. *glycinea* interaction was apparently due to the expression of the *avr*A gene in race 6.[55] Incorporation of this gene into the compatible race 4 resulted in an incompatible pathogen capable of inducing both hypersensitive cell death and phase II AO production in soybean suspension cells.

The genes in the *hrp* region of *P. s.* pv. *syringae* appear to be responsible for the elicitation of phase II AO production in non-host tobacco suspension cells. Transfer of the *hrp/hrm* region from *P. s.* pv. *syringae* to the saprophyte *P. fluorescens* also transferred the ability to induce phase II AO production as well as the HR in tobacco.[56,57] In addition, mutations in the *hrp* region rendered the bacteria incapable of inducing either the phase II response or the HR. Harpins, proteinaceous products of certain *hrp* genes have been isolated from *Erwinia amylovora* and *P. s.* pv. *syringae* [58,59] and have been demonstrated to elicit AO production.[21,53] Expression of these genes during interactions with suspension cells appears to correspond to phase II. However, it is still not clear if harpins are the actual elicitors of the phase II AO production or if they facilitate the secretion of other biomolecules, such as *avr* gene products, that are the true elicitors during the interaction.[60]

Both phases of AO production during plant/bacteria interactions appear to be dependent upon protein phosphorylation and Ca^{++}.[53] Pretreatment of tobacco suspension cells with the protein kinase inhibitor k-252a or the Ca^{++} channel blocker $LaCl_3$ completely inhibited the production of AO in

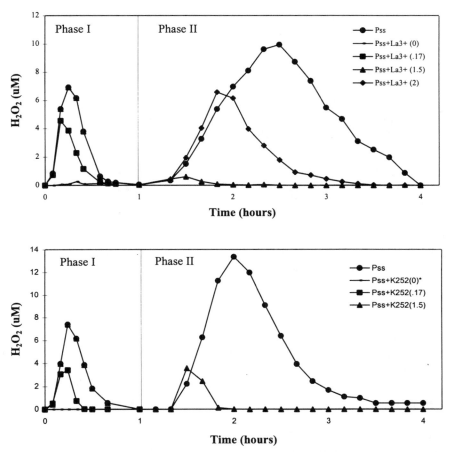

Fig. 4-8. Inhibition of Phase I and Phase II AO production by (A) 0.25 mM LaCl$_3$ or (B) 0.16 μM K-252a. Tobacco suspension cells were treated with *P. syringae* pv. *syringae* and AO was measured by luminol-dependent chemiluminescence. Inhibitors were added at indicated times after the addition of bacteria. Controls received bacteria only (●). Numbers in parentheses indicate time (in hours) when inhibitor was added to suspension cells.

tobacco suspension cells treated with *P. s.* pv. *syringae* (Figure 4-8). In addition, treatment of suspension cells with either of these inhibitors after either phase I or II AO production has been elicited resulted in an almost immediate cessation of production.

Direct Effects of AO Production

Because of the reactivity of AO, its production during the first few hours of pathogenesis could have an immediate impact on both the plant cell and the invading pathogen. In this section, we will discuss evidence for the direct participation of AO produced during Stage I in pathogenesis-related events. Because of the reactivity and short life of AO species, direct effects of their production would most likely occur within minutes and at the site of pathogen ingress.

Cell wall changes. One of the earliest events in pathogenesis that is believed to be caused by AO is the immobilization of specific cell wall proteins. After treatment with fungal elicitors, specific proteins were no longer detectable in suspensions of soybean or bean cells.[61,62] The finding that 1 mM H_2O_2 also elicited the disappearance of these same proteins led the authors to conclude that elicitation of H_2O_2 production contributed to protein cross-linking similar to the cross-linking of phenolics that leads to the formation of lignin. As in lignification, the protein cross-linking was believed to require peroxidase as a catalyst. However, unlike lignification, the immobilization of the proteins was very rapid, occurring within approximately 15 minutes after elicitor treatment of suspension cells, and did not appear to require the *de novo* synthesis of phenolics.[63] Cell wall protein immobilization was also measured in intact leaves during interactions with incompatible bacteria but not with compatible pathovars.[63] Treatment of suspension cells with elicitor or 1 mM H_2O_2 also led to a decrease in cell wall digestibility, as measured by protoplastibility of the cells. The decrease in digestibility was proposed to be due to the protein cross-linking and to play a role in protecting cells against subsequent microbial attack.

Antimicrobial effects. Another direct role for AO production that has been proposed is that of an antimicrobial against invading pathogens. AO can have deleterious effects on most biological molecules including proteins, nucleic acids, and lipid membranes.[64,65] Studies have demonstrated that artificially augmented levels of H_2O_2 or O_2^- can have deleterious effects on pathogens and can inhibit the development of symptoms in plant tissues. The in vitro generation of H_2O_2 by a mixture of peroxidase, NADH, and various cofactors inhibited the germination of sporangiospores of *Peronospora tabacina* and reduced the disease severity of blue mold on tobacco leaf disks.[28] The artificial generation of O_2^- by riboflavin in bean leaves infected by *Colletotrichum lindemuthianum* significantly decreased disease symptoms [27]. Recently, transgenic potato plants were engineered to express fungal glucose oxidase.[66] These plants had increased H_2O_2 production in leaf, tuber, and root tissues, and demonstrated increased resistance to the fungus *Phytophthora infestans* and the bacteria *Erwinia carotovora* sp. *carotovora*. However, it was not determined whether the increased resistance was due to the antimicrobial effects of the elevated H_2O_2 levels, or whether other defense responses that require H_2O_2, such as lignification, were enhanced.

Although these studies demonstrate that abnormally high AO levels can negatively impact disease development, it is unclear if unaugmented amounts of AO produced during plant/pathogen interactions accumulate to concentrations high enough to be antimicrobial. As discussed above, plant and pathogen cells are armed with effective antioxidants such as catalase, superoxide dismutase, and ascorbate peroxidase that limit the accumulation of harmful AO species. For example, during the first few hours after treatment of tobacco or soybean cell suspensions with *Pseudomonas sp.* measurable H_2O_2 concentrations did not exceed 15 μM[67]. If one were to entirely inactivate the AO scavenging activity measured in the system, the maximal accumulation of H_2O_2 in treated suspension cells was estimated to be between 40–240 nmol min^{-1} g^{-1} f.w. suspension cells. These levels of H_2O_2 appear to be easily detoxified and/or tolerated by *Pseudomonas* sp.[68] It should be kept in mind that the production of AO in intact plant tissues may be higher than in suspension cells due to increased cell density. However, the scavenging activity of intact tissues would presumably be sufficient to protect both plant and pathogen cells from the excessive accumulation of AO.

Stage II: Defense Processes

In this section we will be discussing processes or metabolic pathways that are triggered as a result of recognition during Stage I. Some of these processes, such as increases in antioxidants, may be triggered by the AO produced earlier in the interaction. Others, such as increases in LOX activity, may actually be involved in further production of AO. During this stage, the plant cells may still appear healthy but are actively responding to the pathogen. As mentioned above, the delineation between these three stages of pathogenesis is unclear and many of the metabolic processes discussed in this section continue, and may even increase, during Stage III (Figure 4-1).

Antioxidant Increases

As discussed above, plant cells have several scavenging mechanisms to limit the accumulation of harmful amounts of AO and the activity of these scavenging systems appears to increase during pathogenesis.[69,70,67,71] In *Arabidopsis thaliana* suspension cells, an increase in glutathione synthesis, an important component of the ascorbate/glutathione cycle (Figure 4-6) was measured after treatment with H_2O_2.[72] Similarly, soybean suspension cells treated with incompatible bacteria or at least 2 mM H_2O_2 showed increased levels of glutathione S-transferase, an enzyme important in the synthesis of glutathione.[73] In studies with tobacco and soybean suspension cells, AO scavenging activity increased within a few hours after contact with bacteria.[67] Recent studies with heat-killed bacteria that elicit phase I but not phase II have suggested that the increased scavenging is induced by phase I AO production,[53] thus greatly limiting the accumulation of AO during phase II and later stages of pathogenesis. However, it must be understood that although AO is scavenged and does not measurably accumulate, its production during the later stages of pathogenesis does not necessarily decrease.

Systemic Acquired Resistance

Systemic acquired resistance (SAR) describes the enhanced resistance of plants that follows an initial infection by necrotizing pathogens or treatment with elicitors. It has been hypothesized that a highly mobile signal is translocated throughout the plant after the initial infection triggering a systemic resistance response. This phenomenon has been extensively reviewed.[74,75a] A role for AO in SAR was proposed by Chai and Doke[69] who measured increased production of O_2^- in upper leaves of potato plants one to seven days after treatment of lower leaves with hyphal wall components from *P. infestans*. Subsequent challenge of the upper leaves with *P. infestans* zoospores showed enhanced production of O_2^- prior to penetration and almost complete inhibition of fungal penetration and subsequent lesion formation. The authors hypothesized that a signal for activation of O_2^- production was systematically translocated throughout the plant following the initial treatment with elicitor and that elevated levels of O_2^- inhibited the infection processes.

More recently, Chen et al.[76] proposed that systemic increases in H_2O_2 during SAR were the result of specific inhibition of catalase by salicylic acid, a known inducer of SAR. They also found that treatment of tobacco leaves with 1mM or 5mM H_2O_2 induced the expression of pathogenesis related (PR) proteins that enhance disease resistance. The authors hypothesized that H_2O_2 played the role of second messenger for salicylic acid during SAR. This hypothesis has been challenged, however. Rüffer, et al.[77] demonstrated that salicylic acid did not specifically bind to catalase, but also bound to other iron-containing enzymes in plants, animals, and fungi such as LOX and peroxidase. They further hypothesized that salicylic acid did not reach levels in the plant high enough to inhibit catalase but, rather, acted merely as a phytoalexin or antimicrobial agent. Neuenschwander, et al., [78] also challenged the role of H_2O_2 as a second messenger for salicylic acid. In their study they could not detect an increase in H_2O_2 during the initiation of SAR in tobacco leaves and found no evidence supporting a role for H_2O_2 in SAR induction. They did, however, find that exogenous addition of H_2O_2 resulted in increased production of salicylic acid and hypothesized that H_2O_2 increases may act upstream of salicylic acid. Recently, Klessig et al.[79] have also modified their original hypothesis and it is currently unclear what role H_2O_2 may play in SAR induction.

Phytoalexins

Many of the AO elicitors listed in Table 4-2 were originally characterized as phytoalexin elicitors and several laboratories have explored the connection between AO production and phytoalexin synthesis. Phytoalexin production can occur from a few to several hours after AO production. However, despite numerous studies in recent years, there is still no clear consensus on whether the two responses are related. Doke[41] found that pretreatment of potato tuber discs with SOD delayed but did not eliminate phytoalexin production. In a more recent study by Apostol, et al.[80], catalase pretreatment of soybean suspension cells decreased the accumulation of glyceollin. In addition, glyceollin production was induced by the addition of 0.5 mM or 1 mM H_2O_2. This led to the hypothesis that AO production acted as a second messenger to elicit the synthesis of phytoalexins. However, a follow-up study in this same laboratory found that the crude elicitor preparation used to elicit phytoalexins was composed of a proteinaceous elicitor of phytoalexin production and a separate carbohydrate portion that was responsible for the elicitation of AO production.[81]

In general, treatment of plant cells with exogenous AO may elicit phytoalexin production but it may not be absolutely required in the elicitation of phytoalexins during pathogenesis. The addition of H_2O_2 or the artificial generation of superoxide or hydroxyl radical triggered phytoalexin production in soybean suspension cells and hypocotyls.[82–84] However, in tobacco cell suspensions treated with the elicitor cryptogein neither catalase, SOD, nor diphenyleneiodonium, an inhibitor of oxidase AO production, inhibited phytoalexin production.[85] SOD and catalase had no effect on the production of medicarpin in white clover treated with *Pseudomonas corrugata*.[86] In addition, autoclaved *P. corrugata* did not elicit medicarpin, but was still able to elicit the production of AO. These conflicting reports indicate that the relationship between AO production and phytoalexin synthesis is not a direct causal relationship although the two responses are often correlated.

Lignification

Another metabolic process associated with pathogenesis in which AO species may play an important role is lignification of plant cell walls. Lignification occurs during normal plant development and is often enhanced during pathogenesis leading to increased disease resistance[87-89] including induced systemic resistance.[75,90] Lignin increases the structural integrity of the cell wall and may protect cell wall constituents from the degradative enzymes secreted by invading microorganisms. Some lignin precursors are antimicrobial and have been suggested to impede the growth of fungi through lignification of hyphae.[75]

During lignification, phenoxy radicals are produced which then polymerize into lignin (Figure 4-9). Two separate mechanisms have been proposed for the oxidation step resulting in the production of these radicals, which are considered to be forms of AO. The first requires O_2 and is catalyzed by a phenol oxidase, often referred to as laccase.[91] The second mechanism involves H_2O_2 and is catalyzed by cell wall-bound peroxidase.[92,93] There is compelling evidence for both mechanisms. A purified laccase from suspension culture cells of *Acer pseudoplatanus* was recently demonstrated to oxidize lignin precursors to form lignin-like polymers in vitro.[94] In addition, Bao *et al.*[95], found increased laccase activity that correlated with the time and site of lignin deposition in differentiating xylem of Loblolly pine. On the other hand, cell wall-bound peroxidases have been widely accepted as the exclusive catalyst for phenoxy radical production, in part due to their ubiquitous distribution in plant cell walls and, in particular, in lignifying tissues[96,92]. Peroxidase isozyme production was demonstrated to coincide with lignin formation in elicited suspension cells[97] and recent studies have localized H_2O_2 in tissues undergoing lignification[98,99] It is feasible that both laccase- and peroxidase-dependent

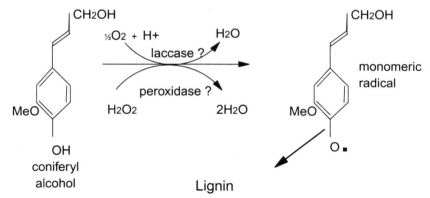

Fig. 4-9. Two reported mechanisms for the production of phenoxy radical monomers used in lignin synthesis. Although coniferyl alcohol is shown, sinapyl and coumaryl alcohols are also commonly involved in this process.

mechanisms may be responsible for lignin formation in various tissues and may be active during different periods of development.

Lipoxygenase

Lipoxygenase was first discovered in plants and was one of the first enzymes to be crystallized.[100] Numerous studies have demonstrated increased LOX activity during plant/pathogen interactions.[101–103] These increases were earlier and more pronounced in tissues undergoing an HR than in compatible interactions. Lipoxygenase activity has been hypothesized to lead to the formation of several biologically active compounds, including AO species and jasmonic acid. However, it is still not clear whether LOX plays a critical role in pathogenesis or if LOX activity increases in response to fatty acid release during cellular breakdown.

Lipoxygenase catalyzes the addition of molecular oxygen to polyunsaturated fatty acids such as linoleic (18:2) and linolenic acids (18:3) found in plant membranes and arachidonic acid (20:4) found in mammalian and fungal membranes (Figure 4-10). The enzymatic mechanism of LOX involves the transfer of high energy electrons in the presence of molecular oxygen, which means that the sporadic production of AO is likely (Figure 4-10). Lipoxy radicals (LO•, LOO•), which can be considered to be species of AO, are believed to be intermediates in the reaction of LOX with fatty acids[104,105]. In addition, in vitro experiments have found that the incubation of LOX with linoleic acid results in the formation of O_2^-.[8] Another well-cited study reports the ability of LOX and linoleic acid to produce singlet oxygen[106] via condensation of the peroxy radicals[107]. Some interesting and careful studies by Kulkarni et al.[108] and Roy et al.[109] demonstrated that, in vitro, LOX can act as both a dioxygenase and a peroxidase, generating AO from NADH by a mechanism similar to that described earlier for peroxidase (Figure 4-3). Although AO has been detected during these in vitro studies with purified LOX and linoleic acid, other studies have monitored the production of lipoxy radicals without the production of O_2^- or H_2O_2 as critical intermediates[110,111] and it is not apparent if significant amounts of AO would be generated by LOX activity during plant pathogenesis.

The production of AO by LOX may or may not be significant; however, LOX activity can lead to the formation of other biologically active compounds. In mammalian cells the hydroperoxides produced from arachidonic acid by LOX serve as precursors for regulatory molecules such as leukotrienes or lipoxins. These products mediate numerous physiological and pathological processes including tumor induction, inflammation, and cell killing in various tissues.[112–115] Similarly, LOX activity in plants is

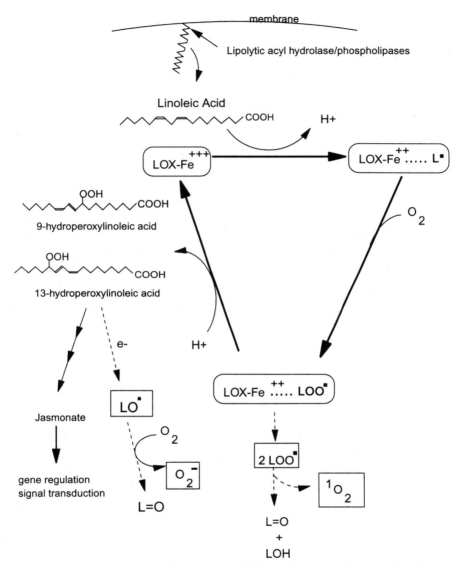

Fig. 4-10. Proposed mechanisms of lipoxygenase (LOX) activity in the lipoxygenation of poly-unsaturated fatty acids. Fatty acids such as linoleic acid, which is enzymatically released from membrane lipids, is oxidized by LOX forming 9- and 13-hydroperoxylinoleic acids. This is accomplished via redox reactions within the enzyme/lipoxyl radical complex resulting in the addition of molecular oxygen to the fatty acid. Ultimately jasmonate can be produced which is an important gene regulator and may be involved in signal transduction. In addition, lipoxy radicals (LOO·, LO·, L·) can contribute to lipid peroxidation if levels of LOX and free fatty acids increase as expected in Stage III. Superoxide (O_2^-) and singlet oxygen ($^1O_2$) are proposed to be produced via the pathway outlined in dashed lines (-------).

believed to lead to the formation of molecules that are active in signaling of various processes during normal growth and pathogenesis. Two of these products, jasmonic acid and methyl jasmonate, are potent inducers of proteinase inhibitors in various plants[116]. Methyl jasmonate has also been shown to enhance AO production by parsley (*Petroselinum crispum* L.) suspension cells treated with fungal elicitors.[117]

During plant/pathogen interactions, LOX activity increases several hours or days after inoculation depending upon the pathogen. For example, increases in LOX occurred in bean or tomato leaves within 4 to 12 hours after inoculation with *Pseudomonas* sp.,[101,103] but LOX increases were not measured until 9 days after inoculation of pepper leaves with *Xanthomonas campestris*.[118] Increases in LOX were generally induced after several days in incompatible plant/fungi interactions such as *Puccinia coronata* on oats, powdery mildew on tobacco, and leaf rust in coffee.[119–121] It must be remembered that correlations between increases in LOX activity and hypersensitive cell death do not necessarily mean that the enzyme initiates membrane breakdown. In most of the studies discussed above, it is not evident whether LOX increased before or after hypersensitive cell death was initiated. Although some isozymes appear able to directly act on linoleic acid in the membrane[114] LOX primarily uses free fatty acids as substrates. It appears likely that fatty acids are released from membrane breakdown during pathogenesis, possibly by AO-induced lipid peroxidation[122], and then serve as substrate for the constitutive LOX.

Stage III: Degradative Processes

The last stage of pathogenesis we will describe encompasses processes that are involved in cellular degradation and death. Many of these processes are active during cellular breakdown induced by environmental stresses and natural senescence. Although some of them have not been studied specifically in association with pathogenesis, they undoubtedly contribute to symptoms such as chlorosis and necrosis associated with both compatible and incompatible interactions.

Hypersensitive Cell Death

Cell death associated with compatible interactions can occur days or even weeks after pathogen ingress and is often accompanied by spreading lesions and yellowing of infected and adjacent tissues. Hypersensitive cell death, however, generally occurs within 24 hours of infection and is characterized by a rapid necrosis confined to the area surrounding the pathogen. The production of the phase II AO burst during the first few hours of incompatible plant/bacteria interactions has been correlated to hypersensitive cell death in several studies[16] and has been suggested to directly cause the subsequent HR.[73] However, a recent study demonstrated that *Pseudomonas* sp. with mutations in the *hrp* region induced phase II AO production in tobacco suspension cells but no hypersensitive cell death in suspension cells or leaves.[56] This study concluded that phase II AO may accompany or trigger subsequent responses that are necessary for HR, but may not be sufficient by itself to cause HR.

There has been much discussion and study on whether or not hypersensitive cell death is a form of programmed cell death, specifically apoptosis, and what role AO may play in the process.[123–125] In animals there have been several studies that have demonstrated a potential role for AO in apoptosis.[126,127] These studies found that specific genes, which encode components of antioxidant pathways, are able to prevent apoptosis. However, other studies questioned the role of AO in apoptosis.[128,129] Using tomato leaflets and protoplasts treated with host-selective toxins secreted by *Alternaria alternata* f. sp. *lycopersici*, Wang, et al.[130], clearly demonstrated the occurrence of apoptotic processes such as shrinkage, loss of cell-to-cell contact, and DNA laddering, similar to those documented in animal cells. Ryerson and Heath[131] also found DNA laddering in intact leaves of cowpea undergoing an incompatible reaction to cowpea rust fungus, but not in susceptible leaves. The authors were not able to detect laddering in leaves treated with 0.29 mM–2.9M H_2O_2, even when concentrations were high enough to kill the cells. Cell shrinkage, blebbing, and nuclear condensation were monitored by Levine, et al.[132] in suspension cells treated with incompatible bacterial pathogens. Large DNA fragments were also formed in cells killed by high concentrations of H_2O_2 (8 mM). As has been hypothesized by the authors of the above studies, cell death processes may vary depending upon the plant/pathogen interaction and there may not be any one "program" that is triggered during hypersensitive cell death.

Lipid Peroxidation

A key process of cellular degradation that occurs during stage III of pathogenesis is membrane deterioration leading to the loss of membrane integrity in cells and cellular organelles. There is increasing evidence that once the plant is in a degenerative stage, increased AO helps promote this deterioration through lipid peroxidation.[133] Electron microscope studies have revealed that peroxisomes, organelles with highly oxidative metabolisms, increase in number during senescence.[134] This has lead several authors to suggest that they may play a role in tissue degeneration caused by environmental stress or pathogenesis through their increased production of AO and the subsequent lipid peroxidation of membranes.[135,136]

Leaf peroxisomes contain a number of AO-producing enzymes such as glycollate oxidase involved in photorespiration (equation e)[137] and xanthine oxidase (equation f) and urate oxidase (equation g) that produce H_2O_2 and O_2^- as a result of nucleic acid breakdown.[138,139,135] These enzymes appear capable of generating substantial levels of AO during periods of senescence. In non-senescent tissue, peroxisomes contain high concentrations of catalase to scavenge excess H_2O_2. However, during senescence catalase decreases and SOD activity increases resulting in increased H_2O_2 levels that could lead to increased lipid peroxidation[134]. Similar responses may occur during Stage III of pathogenesis. Montalbini et al.[140] monitored increases in xanthine oxidase during incompatible interactions of *Uromyces phaseoli* with bean, but not in susceptible interactions. The inhibition of HR by allopurinol, an inhibitor of xanthine oxidase, lead the authors to suggest xanthine oxidase as a source of AO that leads to HR. This interpretation, however, was questioned by Adam et. al.[27] who did not find similar results in wheat infected with *Perccini reconditi*.

$$\text{Glycollic acid} + O_2 \longrightarrow \text{glyoxylic acid} + H_2O_2 \quad \text{(e)}$$

$$\text{Xanthine} + O_2 \longrightarrow \text{Uric acid} + H_2O_2/O_2^- \quad \text{(f)}$$

$$\text{Uric acid} + O_2 \longrightarrow \text{Allantoin} + H_2O_2/O_2^- \quad \text{(g)}$$

Photosynthetic Decline

Many studies of plant disease have demonstrated a decrease in photosynthesis as disease progresses and symptoms develop.[141–146] Although photosynthetic decline may be triggered in many different ways by a pathogen, ultimately the damage to the photosynthetic machinery is likely due to photooxidation and photoinhibition. Photooxidation and photoinhibition refer to the breakdown of the photosynthetic machinery that occurs to different degrees under normal conditions, but which become more serious during pathogenesis and environmental stress. Photooxidation refers specifically to the oxygen and light-dependent destructive processes such as bleaching of pigments and lipid peroxidation of thylakoid membranes, while photoinhibition refers to the inactivation of the electron carriers in Photosystem II by active oxygen. The two photosystems and intermediate electron carriers are chemically linked together in series to drive electrons from H_2O to $NADP^+$ (Figure 4-11). During periods of stress the plant is often exposed to more light energy than it can accommodate. This results in a temporary period of over-excited pigments and an over-load on the reduced electron carriers leading to AO production and damage to the photosynthetic apparatus.

The production of AO during this period of photosynthetic decline is primarily associated with three processes: 1) singlet oxygen production by pigments, 2) superoxide production by electron carriers, and 3) lipoxy radical production in the thylakoid membrane involved in lipid peroxidation. Fortunately, the chloroplast is able to scavenge most of the AO

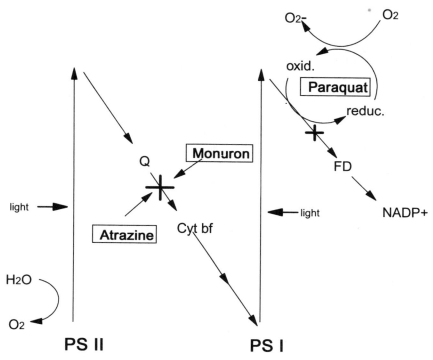

Fig. 4-11. Interruption of photosynthetic electron flow by herbicides. Light excitation of photosystem II (PS II) generates a strong oxidant which oxidizes H_2O to O_2. The reductant formed by this process donates electrons to plastoquinone (Q) which passes them on to the cytochrome *bf* complex (Cyt bf). This complex passes electrons to other electron carriers which link to photosystem I (PS I). Light excitation of PS I generates a strong reductant which transfers electrons through a series of ferrodoxin carriers (FD) and finally to NADP+. Many of the herbicides operate by blocking or altering photosynthetic electron transport producing excessive AO in the chloroplast. Pigments which are now unable to transfer their energy to PS I or II produce singlet oxygen while reduced electron carriers pass electrons on to molecular oxygen producing superoxide and H_2O_2. Severe photooxidation and photoinhibition result and the death of the plant follows. One of the best known herbicides is paraquat, a bipyridylium herbicide, which is able to penetrate chloroplasts. It is able to act as a redox cycling agent by accepting electrons directly from PSI prior to ferrodoxin and then passing them on to oxygen. Two other classes of herbicides, the triazines such as atrazine, and the substituted ureas such as Monuron, block electron transport in the chloroplast between Q and the cytochrome *bf* complex.

produced under normal conditions when it is produced in transient bursts. However, under chronic stressed conditions, AO accumulation exceeds the scavenging capacity in the chloroplast and severe photosynthetic decline and chlorosis occurs.

It is noteworthy to point out here that photosynthesis is pivotal to plant health in that all of its energy is ultimately derived from this process and any damage to the photosynthetic machinery significantly and quickly affects the plant. This is illustrated by the mode of action of many of the commonly used herbicides (Figure 4-11).[147] Herbicides can divert electrons from photosynthesis into AO production and overload the scavenging capacity of the chloroplast. Paraquat, for example, is able to substitute as an electron carrier and pass electrons directly on to molecular oxygen instead of NADP+. Therefore, not only is AO production increased but the production of NADPH, which is needed for AO scavenging by the ascorbate/glutathione cycle, is reduced. It is intriguing to consider that during pathogenesis a similar diversion could occur. The production of an "alternate" electron carrier either by the pathogen or plant could lead to rapid cell death similar to that seen during HR or after paraquat treatment.

Conclusions

In healthy plants, AO is an inevitable side product of many normal processes such as photosynthesis. During normal circumstances, these levels of AO may play a role in normal plant metabolism, but may not accumulate to measurable levels due to the plant's antioxidant mechanisms. However, during pathogenesis and times of stress, AO production increases and may exceed the plant's scavenging capabilities. Considered in isolation, some of the processes involved in AO metabolism (e.g., the use of NAD(P)H to produce as well as to scavenge AO) may appear counterproductive and a futile use of the plants energy. However, as demonstrated in this chapter, such events are generally separated both in time (e.g., stages I–III) and space (e.g., chloroplasts vs. plasma membrane).

In this chapter we have outlined the involvement of AO throughout three stages of pathogenesis. The AO production during Stage I appears to be a deliberate process rather than the inadvertent result of pathogenesis. Early in the plant/pathogen interaction it is conceivable that AO production could function as a signal, triggering subsequent defense-related responses. An interesting comparison could be made between the AO "burst" and the Ca^{++} influx involved in signaling processes. In both cases the plant attempts to keep normal cytoplasmic levels low. Any production of AO in organelles is normally contained within that organelle. Therefore, when a burst of AO is detected, it could serve as a significant signal to the plant, resulting in the triggering other responses.

The role of AO during Stage III tissue degeneration is in need of further investigation. Unlike Stage I, with the immediate and transient production of AO, Stage III is accompanied by chronic low levels of AO which can have devastating effects and would be a constant drain on the NAD(P)H level of the plant cells, especially if a decline in photosynthesis occurred. Studies in fields related to plant pathology, such as stress-related senescence, have provided us with insights into processes that may contribute to cell degeneration in the final stages of pathogenesis. Undoubtedly, processes such as photoinhibition, which have been well-studied in stress physiology, play a role in plant/pathogen interactions as well. As both the methodologies and general understanding of active oxygen chemistry improve, we will be afforded many new and exciting insights into the role of AO in healthy as well as infected or otherwise-stressed plants.

References

1. Soja, G., D. Grill, R. Ebermann, and G. Hendry. 1997. Oxygen, Free Radicals and Environmental Stress in Plants. Vienna, Austria.
2. Baker, C. and E. Orlandi. Sources and Effects of Reactive Oxygen Species in Plants. In: Reactive Oxygen Species in Living Systems: An Interdisciplinary Approach, D. Gilbert, C. Colton, eds., Plenum Publishing, New York. (In Press).
3. Asada, K. and M. Takahashi. 1987. Production and action of active oxygen in photosynthesis. In: Photoinhibition, D. Kyle, C. Osmund, and C. Arntzen, pp. 227-287. CRC Press, Boca Raton, Florida, USA.
4. Robinson, J. 1988. Does O_2 photoreduction occur *in vivo*? Physiol. Plant 72:666.
5. Murphy, T. and C.-K. Auh. 1996. The superoxide synthases of plasma membrane preparations from cultured rose cells. Plant Physiol. 110:621-629.
6. Gross, G. G., C. Janse, and E. F. Elstner. 1977. Involvement of malate, monophenols, and the superoxide radical in hydrogen peroxide formation by isolated cell walls from horseradish (*Armoracia lapathifolia* Gilib.). Planta 136:271-276.
7. Halliwell, B. 1978. Lignin synthesis: The generation of hydrogen peroxide and superoxide by horseradish peroxidase and its stimulation by manganese (II) and phenols. Planta 140:81-88.
8. Chamulitrat, W., M. F. Hughes, T. E. Eling, and R. P. Mason. 1991. Superoxide and peroxil radical generation from the reduction of polyunsaturated fatty acid hydroperoxides by soybean lipoxygenase. Arch. Biochem. Biophys. 290:153-159.
9. Lynch, D. and J. Thompson. 1984. Lipoxygenase-mediated production of superoxide anion in senescing plant tissue. FEBS Lett. 173:251-254.
10. Schoner, S. and G. H. Krause. 1990. Protective systems against active oxygen species in spinach: Response to cold acclimation in excess light. Planta 180:383-389.
11. Asada, K. 1994. Production and action of active oxygen species in photosynthetic tissues. In: Causes of Photooxidative Stress and Amelioration of Defense Systems in Plants, C. Foyer and P. Mullineaux, pp. 77-105. CRC Press, Boca Raton.
12. Daub, M. E. and R. P. Hangarter. 1983. Light-induced production of singlet oxygen and superoxide by the fungal toxin, Cercosporin. Plant Physiol 73:855-857.

13. Salzwedel, J. D. M. H. J. 1989. Effects of singlet oxygen quenchers and pH on the bacterially induced hypersensitive reaction in tobacco suspension cell cultures. Plant Physiol. 90:25-28.
14. Bilger, W. and O. Bjorkman. 1991. Temperature dependence of violaxanthin de-epoxidation and non-photochemical fluorescence quenching in intact leaes of *Gossypium hirsutum* L. and *Malva parviflora* L. Planta 184:226.
15. Baker, C. 1994. Active oxygen metabolism during plant/bacterial recognition. In: Biotechnology and Plant Protection, editors D. Bills and S. Kung, pp. 275-94. World Science, NJ/London.
16. Baker, C. and E. Orlandi. 1995. Active Oxygen in Plant Pathogenesis. Annu. Rev. Phytopathol. 33:299-321.
17. Schroeder, A., G. Martin, and P. Low. 1996. A comparison of methods for the determination of the oxidative burst in whole plants. In: Biology of Plant-Microbe Interactions, G. Stacey, B. Mullin, and P. Gresshoff, pp. 15-20. International Society for Molecular Plant-Microbe Interactions, St. Paul, MN.
18. Low, P. and J. Merida. 1996. The oxidative burst in plant defense: Function and signal transduction. Physiol. Plant. 96:533-542.
19. Low, P. and A. Schroeder. Signal transduction pathways of the plant oxidative burst. pp. 35-52, In: Plant Microbe Interactions, eds. N. Keen and G. Stacey, Chapman and Hall, New York.
20. Schwacke, R. and A. Hager. 1992. Fungal elicitors induce a transient release of active oxygen species from cultured spruce cells that is dependent on Ca^{++} and protein-kinase activity. Planta 187:136-141.
21. Baker, C. J., E. W. Orlandi, and N. M. Mock. 1993. Harpin, an elicitor of the hypersensitive response in tobacco caused by *Erwinia amylovora*, elicits active oxygen production in suspension cells. Plant Physiol. 102:1341-1344.
22. Viard, M.-P., F. Martin, A. Pugin, P. Ricci, and J.-P. Blein. 1994. Protein phosphorylation is induced in tobacco cells by the elicitor cryptogein. Plant Physiol. 104:1245-1249.
23. Chandra, S. and P. S. Low. 1995. Role of phosphorylation in elicitation of the oxidative burst in cultured soybean cells. Proc. Natl. Acad. Sci. USA 92:4120-4123.
24. Kawakita, K. and N. Doke. 1994. Involvement of a GTP-binding protein in signal transduction in potato tubers treated with the fungal elicitor from Phytophthora infestans. Plant Science 96:81-86.
25. Legendre, L., P. F. Heinstein, and P. S. Low. 1992. Evidence for participation of GTP-binding proteins in elicitation of the rapid oxidative burst in cultured soybean cells. J. Biol Chem. 267:20140-20147.
26. Chandra, S., P. F. Heinstein, and P. S. Low. 1996. Activation of phospholipase A by plant defense elicitors. Plant Physiol. 110:979-986.
27. Kiraly, Z., H. El-Zahaby, A. Galal, S. Abdou, A. Adam, B. Barna, and Z. Klement. 1993. Effect of oxy free radicals on plant pathogenic bacteria and fungi and on some plant diseases. In: Oxygen Free Radicals and Scavengers in the Natural Sciences, eds. Mozsik, G, I. Emerit, J. Feher, B. Matkovics, A. Vincze. pp. 9-19. Akademiai Kiado, Budapest, Hungary.
28. Peng, M. and J. Kuc. 1992. Peroxidase-generated hydrogen peroxide as a source of antifungal activity in vitro and on tobacco leaf disks. Phytopathology 82:696-699.
29. Bolwell, G., V. Butt, D. Davies, and A. Zimmerlin. 1995. The origin of the oxidative burst in plants. Free Radical Res. 23:517-532.

30. Doke, N. 1985. NADPH-dependent O_2^- generation in membrane fractions isolated from wounded potato tubers inoculated with *Phytophthora infestans*. Physiol. Plant Pathol. 27:311-322.

31. Auh, C. K. and T. M. Murphy. 1995. Plasma membrane redox enzyme is involved in the synthesis of O_2^- and H_2O_2 by Phytophthora elicitor-stimulated rose cells. Plant Physiol. 107:1241-1247.

32. Low, P. S. and S. C. Dwyer. 1994. Comparison of the oxidative burst signaling pathways of plants and human neutrophils. Proceedings of 1994 Korean Botanical Society, pp. 75-87. Ewha Womans University, Seoul, Korea.

33. Chanock, S. J., J. E. Benna, R. M. Smith, and B. M. Babior. 1994. The Respiratory Burst Oxidase. J. Biol. Chem. 269:24519-24522.

34. Abo, A., E. Pick, A. Hall, N. Totty, C. G. Teahan, and A. W. Segal. 1991. Activation of the NADPH oxidase involves the small GTP-binding protein p21rac1. Nature 353:668-670.

35. Dwyer, S., L. Legendre, P. Heinstein, P. Low, and T. Leto. 1996. Plant and human neutrophil oxidative burst complexes contain immunologically related proteins. Biochim. Biophys. Acta Gen. Subj. 1289:231-237.

36. Desikan, R., J. Hancock, M. Coffey, and S. Neill. 1996. Generation of active oxygen in elicited cells of *Arabidopsis thaliana* is mediated by a NADPH oxidase-like enzyme. FEBS Lett. 382:213-217.

37. Tenhaken, R., A. Levine, L. Brisson, R. Dixon, and C. Lamb. 1995. Function of the oxidative burst in hypersensitive disease resistance. Proc. Natl. Acad. Sci. USA 92:4158-4163.

38. Groom, Q., M. Torres, A. Fordham-Skelton, K. Hammond-Kosack, N. Robinson, and J. Jones. 1996. rbohA, a rice homologue of the mammalian gp91phox respiratory burst oxidase gene. Plant Journal 10:515-522.

39. Xing, T., V. Higgins, and E. Blumwald. 1997. Race-specific elicitors of *Cladosporium fulvum* promote translocation of cytosolic components of NADPH oxidase to the plasma membrane of tomato cells. Plant Cell 9:249-259.

40. May, M., K. Hammond-Kosack, and J. Jones. 1996. Involvement of reactive oxygen species, glutathione metabolism and lipid peroxidation in the Cf-gene dependent defense response of tomato cotyledons induced by race-specific elicitors of *Cladosporium fulvum*. Plant Physiol. 110:1367-1379.

41. Doke, N. 1983. Involvement of superoxide anion generation in the hypersensitive response of potato tuber tissues to infection with an incompatible race of *Phytophthora infestans* and to the hyphal wall components. Physiol. Plant Pathol. 23:345-357.

42. Doke, N. 1983. Generation of superoxide anion by potato tuber protoplasts during the hypersensitive response to hyphal wall components of *Phytophthora infestans* and specific inhibition of the reaction by suppressors of hypersensitivity. Physiol. Plant Pathol. 23:359-367.

43. Doke, N. and K. Tomiyama. 1982. Suppression of the hypersensitive response of potato tuber protoplasts to hyphal wall components by water soluble glucans isolated from *Phytophthora infestans*. Physiol. Plant Pathol. 16:177-186.

44. Chai, H. B. and N. Doke. 1987. Superoxide anion generation: A response of potato leaves to infection with *Phytophthora infestans*. Physiol. Biochem. 77:645-649.

45. Chai, H. B. and N. Doke. 1987. Activation of the potential of potato leaf tissue to react hypersensitively to *Phytophthora infestans* by cytospore germination fluid and the enhancement of this potential by calcium ions. Physiol. Mol. Plant Path. 30:27-37.

46. Sanchez, L. M., N. Doke, and K. Kawakita. 1993. Elicitor-induced chemilumi-nescence in cell suspension cultures of tomato, sweet pepper and tobacco plants and its inhibition by suppressors from *Phytophthora* spp. Plant Science 88:141-148.

47. Sanchez, L. M., Y. Ohno, Y. Miura, K. Kawakita, and N. Doke. 1992. Host selec-tive suppression by water-soluble glucans from *Phytophthora* spp. of hypersen-sitive cell death of suspension-cultured cells from some solanaceous plants caused by hyphal wall elicitors of the fungi. Ann. Phytopathol. Soc. Japan 58:664-670.

48. Doke, N., Y. Miura, H.-B. Chai, and K. Kawakita. 1991. Involvement of active oxygen in induction of plant defense response against infection and injury. In: Active Oxygen/Oxidative Stress and Plant Metabolism, eds. E. Pell, K. Steffen. American Society of Plant Physiologists, Rockville, MD

49. Adam, A., T. Farkas, G. Somlyai, M. Hevesi, and Z. Kiraly. 1989. Consequence of O_2^- generation during a bacterially induced hypersensitive reaction in tobacco: Deterioration of membrane lipids. Physiol. Mol. Plant Path. 34:13-26.

50. Keppler, L. D., C. J. Baker, and M. M. Atkinson. 1989. Active oxygen production during a bacteria-induced hypersensitive reaction in tobacco suspension cells. Phytopathology 79:974-978.

51. Baker, C. J., N. R. O'Neill, L. D. Keppler, and E. W. Orlandi. 1991. Early Re-sponses during plant-bacteria interactions in tobacco cell suspensions. Phytopath-ology 81:1504-1507.

52. Baker, C. and N. Mock. 1994. An improved method for monitoring cell death in cell suspension and leaf disc assays using evans blue. Plant Cell, Tissue Org. Cult. 39:7-12.

53. Orlandi, E. and C. Baker. Unpublished Results.

54. Boller, T. and G. Felix. 1996. Olfaction in plants: Specific perception of common microbial molecules. In: Biology of Plant-Microbe Interactions, eds., G. Stacey, B. Mullin, and P. Gresshoff, pp. 1-8. International Society for Molecular Plant-Microbe Interactions, St. Paul, MN.

55. Orlandi, E. W., S. W. Hutcheson, and C. J. Baker. 1992. Early physiological responses associated with race-specific recognition in soybean leaf tissue and cell suspensions treated with *Pseudomonas syringae* pv. *glycinea*. Physiol. Mol. Plant Pathol. 40:173-180.

56. Glazener, J., E. Orlandi, and C. Baker. 1996. The active oxygen response of cell suspensions to incompatible bacteria is not sufficient to cause hypersensitive cell death. Plant Physiol. 110:759-763.

57. Huang, H., S. W. Hutcheson, and A. Collmer. 1991. Characterization of the *hrp* cluster from *Pseudomonas syringae* pv. *syringae* 61 and TnphoA tagging of genes encoding exported or membrane-spanning hrp proteins. Mol. Plant-Microbe Int. 4:469-476.

58. Wei, Z.-M., R. J. Laby, C. H. Zumoff, D. W. Bauer, S. Y. He, A. Collmer, and S. V. Beer. 1992. Harpin, elicitor of the hypersensitive response produced by the plant pathogen *Erwinia amylovora*. Science 257:85-88.

59. He, S. Y., H.-C. Huang, and A. Collmer. 1993. *Pseudomonas syringae* pv. *syringae* harpin$_{Pss}$: A protein that is secreted via the Hrp pathway and elicits the hypersensitive response in plants. Cell 73:1-20.

60. Alfano, J. and A. Collmer. 1996. Bacterial pathogens in plants: Life up against the wall. Plant Cell 8:1683-1698.

61. Bradley, D. J., P. Kjellbom, and C. J. Lamb. 1992. Elicitor- and wound-induced oxidative cross-linking of a proline-rich plant cell wall protein: A novel, rapid defense response. Cell 70:21-30.

62. Wojtaszek, P., J. Trethowan, and G. P. Bolwell. 1995. Specificity in the immobilization of cell wall proteins in response to different elicitor molecules in suspension-cultured cells of French bean (*Phaseolus vulgaris* L.). Plant Mol. Biol. 28:1075-1087.
63. Brisson, L. F., R. Tenhaken, and C. J. Lamb. 1994. Function of oxidative cross-linking of cell wall structural proteins in plant disease resistance. Plant Cell 6:1703-1712.
64. Storz, G., L. A. Tartaglia, S. B. Farr, and B. N. Ames. 1990. Bacterial defenses against oxidative stress. Trends Genet. 6:363-368.
65. Ahern, H. 1991. Cellular responses to oxidative stress. ASM News 57:627-629.
66. Wu, G., B. Shortt, E. Lawrence, E. Levine, and K. Fitzsimmons. 1995. Disease resistance conferred by expression of a gene encoding H_2O_2-generating glucose oxidase in transgenic potato plants. Plant Cell 7:1357-1368.
67. Baker, C., G. Harmon, J. Glazener, and E. Orlandi. 1995. A non-invasive technique for monitoring peroxidative and hydrogen peroxide-scavenging activities during interactions between bacterial plant pathogens and suspension cells. Plant Physiol. 108:353-359.
68. Baker, C., E. Orlandi, and A. Anderson. 1998. Oxygen metabolism in plant cell culture/bacteria interactions: Role of bacterial concentration and H2O2-scavenging in survival under biological and artificial oxidative stress. Physiol. Mol. Plant Pathol. (in press).
69. Chai, H. B. and N. Doke. 1987. Systemic activation of O_2^- Generating reaction, superoxide dismutase, and peroxidase in potato plants in relation to induction of systemic resistance to *Phytophthora infestans*. Ann. Phytopathol. Soc. Japan 53:585-590.
70. Abdou, E., A. Galal, and B. Barna. 1993. Changes in lipid peroxidation, superoxide dismutase, peroxidase and lipoxygenase enzyme activities in plant/pathogen interactions. Oxygen Free Radicals and Scavengers in the Natural Sciences, eds. Mozsik, G., Emerit, I., Feher, J., Matkovics, B., Vincze A., pp. 29-33. Akademiai Kiado, Budapest, Hungary.
71. Adam, A., C. B. B. Bestwick, and J. Mansfield. 1995. Enzymes regulating the accumulation of active oxygen species during the hypersensitive reaction of bean to *Pseudomonas syringae* pv. *phaseolicola*. Planta 197:240-249.
72. May, M. J. and C. J. Leaver. 1993. Oxidative Stimulation of Glutathione Synthesis in *Arabidopsis thaliana* Suspension Cultures. Plant Physiol 103:621-627.
73. Levine, A. T. R. D. R. L. C. 1994. H_2O_2 from the oxidative burst orchestrates the plant hypersensitive disease resistance response. Cell 79:583-593.
74. Ryals, J., U. Neuenschwander, M. Willits, A. Molina, H.-Y. Steiner, and M. Hunt. 1996. Systemic acquired resistance. Plant Cell 8:1809-1819.
75. Hammerschmidt, R. and J. Kuc. 1982. Lignification as a mechanism for induced systemic resistance in cucumber. Physiol. Plant Path. 20:61-71.
75a. Kuc. J. 1990. A case for self-defense in plants against disease. Phytoparasitica 18:3-8.
76. Chen, Z., H. Silva, and D. F. Klessig. 1993. Active oxygen species in the induction of plant systemic acquired resistance by salicylic acid. Science 262:1883-1886.
77. Rüffer, M., B. Steipe, and M. Zenk. 1995. Evidence against specific binding of salicylic acid to plant catalase. FEBS Lett. 377:175-180.
78. Neuenschwander, U., B. Vernooij, L. Friedrich, S. Uknes, H. Kessmann, and J. Ryals. 1995. Is hydrogen peroxide a second messenger of salicylic acid in systemic acquired resistance? The Plant J. 8:227-233.

79. Klessig, D., J. Durner, Z. Chen, M. Anderson, U. Conrath, H. Du, A. Guo, Y. Liu, J. Shah, H. Silva, H. Takahashi, and Y. Yang. 1996. Studies of the salicyclic acid signal transduction pathway. In: Biology of Plant-Microbe Interactions, eds., G. Stacey, B. Mullin, and P. Gresshoff, pp. 33-38. International Society for Molecular Plant-Microbe Interactions, St. Paul, MN.
80. Apostol, I., P. F. Heinstein, and P. S. Low. 1989. Rapid stimulation of an oxidative burst during elicitation of cultured plant cells. Plant Physiol. 90:109-116.
81. Davis, D., J. Merida, L. Legendre, P. S. Low, and P. Heinstein. 1993. Independent elicitation of the oxidative burst and phytoalexin formation in cultured plant cells. Phytochem. 32:607-611.
82. Montillet, J. L. and N. Degousee. 1991. Hydroperoxydes induce glyceollin accumulation in soybean. Plant Physiol. Biochem. 29:689-694.
83. Degousee, N., C. Triantaphylides, and J.-L. Montillet. 1994. Involvement of oxidative processes in the signaling mechanisms leading to the activation of glyceollin synthesis in soybean. Plant Physiol. 104:945-952.
84. Gomez, L., M. Braga, and S. Dietrich. 1994. Involvement of active oxygen species and peroxidases in phytoalexin production induced in soybean hypocotyls by an elictor from a saprophytic fungus. J. Brazilian Assoc. Adv. Sci. 46:153-156.
85. Rusterucci, C., V. Stallaert, M.-L. Milat, A. Pugin, P. Ricci, and J.-P. Blein. 1996. Relationship between active oxygen species, lipid peroxidation, necrosis, and phytoalexin production induced by elicitins in Nicotiana. Plant Physiol. 111:885-891.
86. Devlin, W. S. and D. L. Gustine. 1992. Involvement of the oxidative burst in phytoalexin accumulation and the hypersensitive reaction. Plant Physiol. 100:1189-1195.
87. Vance, C., T. Kirk, and R. Sherwood. 1980. Lignification as a mechanism of disease resistance. Ann. Rev. Phytopathol. 18:259-288.
88. Beardmore, J. R. J. P. G. W. 1983. Cellular lignification as a factor in the hypersensitive resistance of wheat to stem rust. Physiol. Plant Path. 22:209-220.
89. Moerschbacher, B. M., U. Noll, L. Gorrichon, and H. J. Reisener. 1990. Specific inhibition of lignification breaks hypersensitive resistance of wheat to stem rust. Plant Physiol. 93:465-470.
90. Dean, R. and J. Kuc. 1987. Rapid lignification in response to wounding and infection as a mechanism for induced systemic protection in cucumber. Physiol. Mol. Plant Path. 31:69-81.
91. Freudenberg, K. 1968. The constitution and biosynthesis of lignin. In: Constitution and Biosynthesis of Lignin, eds., K. Freudenberg and A. Neish, pp. 45-122. Springer Verlag, Berlin.
92. Gross, G. 1977. Biosynthesis of lignin and related monomers. Rec. Adv. Phytochem. 11:141-184.
93. Higuchi, T. 1985. Biosynthesis of lignin. In: Biosynthesis and Biodegradation of Wood Components, ed. T. Higuchi, pp. 141-160. Academic Press, Orlando.
94. Sterjiades, R., J. Dean, and K. Eriksson. 1992. Laccase from sycamore maple (Acer psuedoplatanus) polymerizes monolignols. Plant Physiol. 99:1162-1168.
95. Bao, W., D. O'Malley, R. Whetten, and R. Sederoff. 1993. A laccase associated with lignification in Loblolly pine xylem. Science 260:672-674.
96. Harkin, J. and J. Obst. 1973. Lignification in trees: Indication of exclusive peroxidase participation. Science 180:296-298.
97. Bruce, R. J. and C. A. West. 1989. Elicitation of lignin biosynthesis and isoperoxidase activity by pectic fragments in suspension cultures of castor bean. Plant Physiol. 91:889-897.

98. Czaninski, Y., R. M. Sachot, and A. M. Catesson. 1993. Cytochemical Localization of hydrogen peroxide in lignifying cell walls. Ann. Bot. 72:547-550.
99. Olson, P. D. and J. E. Varner. 1993. Hydrogen peroxide and lignification. Plant J. 4:887-892.
100. Theorell, H., R. Holman, and A. Akeson. 1947. Crystalline lipoxidase. Acta Chem. Scand. 1:571-576.
101. Croft, K. P. C., C. R. Voisey, and A. J. Slusarenko. 1990. Mechanism of hypersensitive cell collapse: correlation of increased lipoxygenase activity with membrane damage in leaves of *Phaseolus vulgaris* (L.) inoculated with an avirulent race of *Pseudomonas syringae* pv. *phasicola*. Physiol. Mol. Plant Path. 36:49-62.
102. Keppler, L. D. N. A. 1986. Involvement of membrane lipid peroxidation in the development of a bacterially induced hypersensitive reaction. Phytopathology 76:104-107.
103. Koch, E., B. Meier, H.-G. Eiben, and A. Slusarenko. 1992. A lipoxygenase from leaves of tomato (*Lycopersicon esculentum* Mill.) is induced in response to plant pathogenic pseudomonads. Plant Physiol. 99:571-576.
104. Chamulitrat, W. and R. Mason. 1989. Lipid peroxyl radical intermediates in the peroxidation of polyunsaturated fatty acids by lipoxygenase. J. Biol. Chem. 264:20968-20973.
105. de Groot, J., G. Veldink, J. Vliegenthart, J. Boldingh, R. Wever, and B. van Gelder. 1975. Demonstration by EPR spectroscopy of the functional role of iron in soybean lipoxygenase- 1. Biochim. Biophys. Acta 377:71-79.
106. Kanofsky, J. R. and B. Axelrod. 1986. Singlet oxygen production by soybean lipoxygenase isozymes. J. Biol. Chem. 261:1099-1104.
107. Russell, G. 1957. Deuterium-isotope effects in the autoxidation of aralkyl hydrocarbons. Mechanism of the interaction of peroxy radicals. J. Am. Chem. Soc. 79:3871-3877.
108. Kulkarni, A., J. Chaudhuri, A. Mitra, and I. Richards. 1989. Dioxygenase and peroxidase activities of soybean lipoxygenase: Synergistic interaction between linoleic acid and hydrogen peroxide. Res. Comm. Chem. Path. Pharmacol. 66:287-296.
109. Roy, P., S. Roy, A. Mitra, and A. Kulkarni. 1994. Superoxide generation by lipoxygenase in the presence of NADH and NADPH. Biochim. Biophys. Acta 1214:171-179.
110. Van der Zee, J., T. Eling, and R. Mason. 1989. Formation of free radical metabolites in the reaction between soybean lipoxygenase and its inhibitors. Biochem. 28:8363-8367.
111. Conner, H., V. Fischer, and R. Mason. 1986. A search for oxygen-centered free radicals in the lipoxygenase/linoleic acid system. Biochem. Biophys. Res. Comm. 141:614-621.
112. Chang, D., G. Ringold, and R. Heller. 1992. Cell killing and induction of manganous superoxide dismutase by tumor necrosis factor-alpha is mediated by lipoxygenase metabolites of arachidonic acid. Biochem. Biophys. Res. Comm. 188:538-546.
113. Shimizu, T. and L. S. Wolfe. 1990. Arachidonic Acid Cascade and Signal Transduction. J. Neurochem. 55:1-15.
114. Siedow, J. 1991. Plant lipoxygenase structure and function. Annu. Rev. Plant Physiol. Plant Mol. Biol. 42:145-188.
115. Samuelsson, B., S.-E. L. A. Dahlen, C. Rouzer, and S. Serhan. 1987. Leukotrienes and lipoxins: Structures, biosynthesis, and biological effects. Science 237:1171-1175.

116. Farmer, E. E. and C. A. Ryan. 1992. Octadecanoid precursors of jasmonic acid activate the synthesis of wound-inducible proteinase inhibitors. Plant Cell 4:129-134.
117. Kauss, H., W. Jeblick, J. Ziegler, and W. Krabler. 1994. Pretreatment of parsley (*Pertroselinum crispum* L.) suspension cultures with methyl jasmonate enhances elicitation of activated oxygen species. Plant Physiol. 105:89-94.
118. Buonaurio, R. and N. Umesh Kumar. 1995. Active oxygen-producing and -scavenging enzymes in pepper leaves infected with *Xanthomonas campestris* pv. *vesicatoria*. J. Phytopathol. 143:165-168.
119. Lupu, R., S. Grossman, and Y. Cohen. 1980. The involvement of lipoxygenase and antioxidants in pathogenesis of powdery mildew on tobacco plants. Physiol. Plant Path. 16:241-248.
120. Yamamoto, H. and T. Tani. 1986. Possible involvement of lipoxygenase in the mechanism of resistance of oats to *Puccinia coronata avenae*. J. Phytopathol. 116:329-337.
121. Rojas, M., V. Montes de Gomez, and C. Ocampo. 1993. Stimulation of lipoxygenase activity in cotyledonary leaves of coffee reacting hypersensitively to the coffee leaf rust. Physiol. Mol. Plant Path. 43:209-219.
122. Keppler, L. and Novacky, A. 1987. The initiation of membrane lipid peroxidation during bacteria-induced hypersensitive reaction. Physiol. Mol. Plant Path. 30:233-245.
123. Mittler, R. and E. Lam. 1996. Sacrifice in the face of foes: Pathogen-induced programmed cell death in plants. Trends Microbiol. 4:10-15.
124. Greenberg, J. 1996. Programmed cell death: A way of life for plants. Proc. Natl. Acad. Sci. USA 93:12094-12097.
125. Dangl, J., R. Dietrich, and M. Richberg. 1996. Death don't have no mercy: Cell death programs in plant-microbe interactions. Plant Cell 8:1793-1807.
126. Hockenbery, D., Z. Oltval, X.-M. Yin, C. Milliman, and S. Korsmeyer. 1993. Bcl-2 functions in an antioxidant pathway to prevent apoptosis. Cell 75:241-251.
127. Slater, A., C. Nobel, and S. Orrenius. 1995. The role of intracellular oxidants in apoptosis. Biochim. Biophys. Acta 1271:59-62.
128. Muschel, R., E. Bernhard, L. Garza, W. McKenna, and C. Koch. 1995. Induction of apoptosis at different oxygen tensions: Evidence that oxygen radicals do not mediate apoptotic signaling. Cancer Res. 55:995-998.
129. Jacobson, M. 1996. Reactive oxygen species and programmed cell death. TIBS 21:83-86.
130. Wang, H., J. Li, R. Bostock, and D. Gilchrist. 1996. Apoptosis: A functional paradigm for programmed plant cell death induced by a host-selective phytotoxin and invoked during development. Plant Cell 8:375-391.
131. Ryerson, D. and M. Heath. 1996. Cleavage of nuclear DNA into oligonucleosomal fragments during cell death induced by fungal infection or by abiotic treatments. Plant Cell 8:393-402.
132. Levine, A., R. Pennell, M. Alvarez, R. Palmer, and C. Lamb. 1996. Calcium-mediated apoptosis in a plant hypersensitive disease resistance response. Curr. Biol. 6:427-437.
133. Thompson, J. E., J. H. Brown, P. Gopinadhan, J. F. Todd, and K. Yao. 1991. Membrane phospholipid catabolism primes the production of activated oxygen in senescing tissues. In: Active Oxygen/Oxidative Stress and Plant Metabolism, editors E. Pell and K. Steffen, pp. 57-66. American Society of Plant Physiologists, Rockville, MD.

134. Pastori, G. and L. del Rio. 1994. An activated oxygen-mediated role for peroxisomes in the mechanism of senescence of *Pisum sativum* L. leaves. Planta 193:385-391.
135. Palma, J. M., Garrido M., M. I. Rodriguez-Garcia, and L. A. del Rio. 1991. Peroxisome proliferation and oxidative stress mediated by activated oxygen species in plant peroxisomes. Arch. Biochem. Biophys. 287:68-74.
136. del Rio, L., L. Sandalio, J. Palma, and P. a. C. F. Bueno. 1992. Metabolism of oxygen radicals in peroxisomes and cellular implications. Free Rad. Biol. & Med. 13:557-580.
137. Tolbert, N. 1981. Metabolic pathways in peroxisomes and glyoxysomes. Annu. Rev. Biochem. 50:133-157.
138. Sandalio, L., J. Palma, and L. del Rio. 1987. Localization of manganese superoxide dismutase in peroxidsomes isolated from *Pisum sativum* L. Plant Science 51:1-8.
139. Sandalio, L. M. and L. A. del Rio. 1988. Intraorganellar distribution of superoxide dismutase in plant peroxisomes (Glyoxysomes and leaf peroxisomes). Plant Physiol. 88:1215-1218.
140. Montalbini, P. 1992. Inhibition of hypersensitive response by allopurinol applied to the host in the incompatible relationship between *Phaseolus vulgaris* and *Uromyces phaseoli*. J. Phytopathol. 134:218-228.
141. Magyarosy, A. and B. Buchanan. 1975. Effect of bacteria on photosynthesis of bean leaves. Phytopathology 65:777-780.
142. Mignucci, J. and J. Boyer. 1979. Inhibition of photosynthesis and transpiration in soybean infected by *Microsphaera diffusa*. Phytopathology 69:227-230.
143. Mitchell, R. 1978. Halo blight of beans: Toxin production by several *Pseudomonas phaseolicola* isolates. Physiol Plant Pathol. 13:37-49.
144. Mlodzianowshi, F. and R. Siwecki. 1975. Ultrastructural changes in chloroplasts of *Populus tremula* L. leaves affected by the fungus *Melampsora pintorqua*, Braun. Rostr. Physiol Plant Pathol. 6:1-3.
145. Montalbini, P. and B. Buchanan. 1974. Effect of rust infection on photophosphorylation by isolated chloroplasts. Physiol Plant Pathol. 4:191-196.
146. Spotts, R. and D. Ferree. 1979. Photosynthesis, transpiration, and water potential of apple leaves infected by *Venturia inaequalis*. Phytopathology 69: 717-719.
147. Halliwell, B. 1991. Oxygen radicals: Their formation in plant tissues and their role in herbicide damage. In: Herbicides, editors N. R. Baker and M. P. Percival, pp. 87-129. Elsevier Science Publishers, New York.
148. Schopfer, P. 1994. Histochemical demonstration and localization of H_2O_2 in organs of higher plants by tissue printing on nitrocellulose paper. Plant Physiol. 104:1269-1275.
149. Keppler, L. D. and C. J. Baker. 1989. Superoxide-initiated lipid peroxidation in a bacteria-induced hypersensitive reaction in tobacco cell suspensions. Phytopathology 79:555-562.
150. El-Moshaty, F. I. B., S. M. Pike, A. J. Novacky, and O. P. Sehgal. 1993. Lipid peroxidation and superoxide production in cowpea (*Vigna unguiculata*) leaves infected with tobacco ringspot virus or southern bean mosaic virus. Physiol. Mol. Plant Path. 43:109-119.
151. Moreau, R. A. and S. F. Osman. 1989. The properties of reducing agents released by treatment of *Solanum tuberosum* with elicitors from *Phytophthora infestans*. Physiol. Mol. Plant Path. 35:1-10.

152. Misra, H. P. and I. Fridovich. 1972. The role of superoxide anion in the autoxidation of epinephrine and a simple assay for superoxide dismutase. J. Biol. Chem. 247:3170-3175.

153. Glazener, J., E. W. Orlandi, G. L. Harmon, and C. J. Baker. 1991. An improved method for monitoring active oxygen in bacteria-treated suspension cells using luminol-dependent chemiluminescence. Physiol. Mol. Plant Path. 39:123-133.

154. Auh, C. and T. Murphy. 1995. Plasma membrane redox enzyme is involved in the synthesis of superoxide and hydrogen peroxide by *Phytophthora* elicitor-stimulated rose cells. Plant Physiol. 107:1241-1247.

155. Gönner, M. V., E. Schlösser, and H. Neubacher. 1993. Evidence from electron-spin resonance for the formation of free radicals during infection of *Avena sativa* by *Drechslera* spp. Physiol. Mol. Plant Path. 42:405-412.

156. Kuchitsu, K., H. Kosaka, T. Shiga, and N. Shibuya. 1995. EPR evidence for generation of hydrozxyl radical triggered by N-acetylchitooligosaccharide elicitor and a protein phosphatase inhibitor in suspension-cultured rice cells. Protoplasma 188:138-142.

157. Lindner, W. A., C. Hoffmann, and H. Grisebach. 1988. Rapid elicitor-induced chemiluminescence in soybean cell suspension cultures. Phytochem. 27:2501-2503.

158. Arnott, T. and T. Murphy. 1991. A comparison of the effects of fungal elicitor and ultraviolet radiation on ion transport and hydrogen peroxide synthesis by rose cells. Environ. Exp. Bot. 31:209-216.

159. Anderson, A., K. Rogers, C. Tepper, and K. Blee. 1991. Timing of molecular events following elicitor treatment of plant cells. Physiol. Mol. Plant Path. 38:1-13.

160. Baker, C., N. Mock, J. Glazener, and E. Orlandi. 1993. Recognition responses in pathogen/non-host and race/cultivar interactions involving soybean (*Glycine max*) and *Pseudomonas syringae* pathovars. Physiol. Mol. Plant Path. 43:81-94.

161. Vera-Estrella, R., E. Blumwald, and V. Higgins. 1992. Effect of Specific elicitors of *Cladosporium fulvum* on tomato suspension cells: Evidence for the involvement of active oxygen species. Plant Physiol. 99:1208-1215.

162. Haga, M., Y. Kohno, M. Iwata, and Y. Sekizawa. 1995. Superoxide anion generation in rice blade protoplasts with the blast fungus proteoglucomannan elicitor as determined by CLA-phenyl luminescence and its suppression by treating the elicitor with α-D-Mannosidase. Biosci. Biotech. Biochem. 59:969-973.

163. Mussell, H. W. 1973. Endopolygalacturonase: Evidence for involvement in Verticillium wilt of cotton. Phytopathology 63:62-70.

CHAPTER 5

Suppressor as a Factor Determining Plant-Pathogen Specificity

Tomonori Shiraishi, Tetsuji Yamada, Yuki Ichinose, Akinori Kiba, Kazuhiro Toyoda, Toshiaki Kato, Yasuhiro Murakami, and Hikaru Seki

Plants are, in general, resistant or immune to the vast majority of pathogens. In other words, the number of pathogens that have the potential either to attack or to evoke disease in a given plant species is extremely limited (1). Thus, resistance is the rule and susceptibility is the exception (98). This phenomenon is called "host-parasite specificity", and mechanisms underlying host-parasite specificity contribute one of the most intriguing question that phytopathologists should elucidate.

Physiologically, plants are equipped with both static and active defense mechanisms. The former includes constitutive properties, e.g., the thickness and strength of cell walls and the presence of preformed antibiotics such as saponins and polyphenols. On the other hand, active (induced) resistance involves newly formed inhibitory chemicals and physical barriers, including phytoalexins, active oxygen species, infection inhibitors, pathogenesis-related (PR) proteins, lignin, callose and hydroxyproline-rich glycoproteins (for reviews, see 33, 79, and 106). These factors are thought to protect plants against invasion by pathogens. They are clearly active mechanisms since suppression of active resistance by treatment with certain antibiotics or high temperature results in susceptibility to normally avirulent or incompatible pathogens.

Resistance-inducing metabolites were found in many phytopathogenic bacteria and fungi. A peptide, monilicolin A, was first isolated from the mycelia of *Monilinia fructicola* as a factor inducing phytoalexin production in bean (28). Substances inducing phytoalexin production were named "elicitors" by Keen (65). Later, the term elicitor was used in a broader sense for substances that are able to induce resistance in plants (for review, see 31). Elicitors extracted from the culture filtrates, hyphal cell walls or spore germination fluid of pathogenic fungi include chitin, chitosan, glycoprotein, glucan, lipids, polysaccharide or peptides (30,31,112). Significant roles of elicitors in the induction of plant resistance and their mode of action were reviewed by Lamb *et al.* (79), Dixon and Lamb (33) and Yoshikawa *et al.* (153).

Since elaboration of the "phytoalexin theory" (94), a great deal of information on defense mechanisms has accumulated. However, little is known about mechanisms determining susceptibility or accessibility (susceptibility at the cell level) (109). The occurrence of such mechanisms was indicated by important demonstrations that avirulent pathogens were able to infect plant tissues that had been previously infected by virulent or compatible pathogens (for reviews, see 108,150). Inoculation with a compatible pathogen was reported to predispose potato to infection by an incompatible race of the late blight fungus (137), barley to powdery mildew fungi (76,93, 109,110,141) and oat to an incompatible race of crown rust (135). It was found that 45 out of 51 powdery mildew fungi were able to infect previously mildewed barley leaves and 30 of them formed conidia (141). Ouchi *et al.* (109,110) reported that the inoculation with a compatible race of *Erysiphe graminis hordei* conditioned barley leaves to be accessible within 18 h to infection by incompatible races of the same fungus as well as to wheat and melon powdery mildew fungi normally avirulent to barley. On the other hand, barley leaves pre-inoculated with an incompatible mildew race became inaccessible even to a compatible race within 12 h. Ouchi et al. also demonstrated that leaf cells induced to accessibility or inaccessibility were localized near the originally inoculated site (107). Thus, accessibility and inaccessibility are inducible and irreversible. Later, Kunoh *et al.* (76,77) clarified the timing of establishment of accessibility or inaccessibility by inoculation of barley coleoptiles with a compatible race of *E. graminis* and an avirulent pathogen, *E. pisi*, respectively. *E. pisi* alone was not capable of infecting the barley coleoptile under normal conditions, but 30% of *E. pisi* conidia established infection on cells where *E. graminis* had previously penetrated.

The phenomenon of accessibility was partly involved in formulation of the "phytoalexin theory", since this defense response was delayed in compatible plant-pathogen combinations (10,29,101,102,154,156). These data suggested possible mechanisms as follows: 1) virulent pathogens (compatible races) do not produce elicitors effective on host plant cultivars, at least during an early stage of infection; or 2) pathogens have an aggressive ability to suppress the active resistance mechanisms of their hosts. If the former case was true, the challenging avirulent pathogen (incompatible race) could not establish its infection on the plants which had been pre-inoculated with virulent pathogens (compatible race) since the active resistance, that is rejection reaction, must be induced by effective specific elicitors, which are produced by the challenging incompatible pathogens. However, the fact is that the incompatible challenger is able to establish its infection on the predisposed plants as described above.

Furthermore, as far as we know, there is no pathogen that does not produce an elicitor. It is well known that common compounds such as β-glucans and chitin constitutively present in the hyphal cell walls of many pathogenic fungi act as non-specific elicitors (30). It was also reported that the substances, which were secreted in spore germination fluid of pathogens at the infection sites, induced active resistance even in their own hosts (52,86,127,148,149,158,159). Once the chemical and physical barriers have been established in plant tissues, the penetration, growth and/or reproduction of the pathogens are crucially inhibited (103,126,149). These findings coincide with the previous observations that the challenging compatible pathogens failed to infect on the cells which had been inoculated with incompatible pathogens (76,94,109,110).

These findings strongly suggest that the ability to overcome host resistance is essential for adaptation of the pathogens (47,97). In other words, specificity cannot be explained solely by the activity of elicitors, but is rather determined by substances that are able to circumvent or negate the active resistance of host plants (for reviews, see 54,55,97,108,130). In fact, several phytopathogens have been found to produce non-toxic metabolites, called "suppressors", which block plant active resistance mechanisms otherwise induced by elicitors or avirulent pathogens, in a strict species-specific or race-cultivar-specific manner (Table 5-1).

Suppressors were defined as "determinants for pathogenicity (specificity) without apparent phytotoxicity" (97,100,104): i) they are produced by pathogens at the site of infection; ii) they participate in suppression of active resistance mechanisms initiated by elicitors; iii) some of them induce local susceptibility (accessibility) in host plants; iv) they are host-specific; and v) they are not toxic to plants. The host specific (selective) toxins (HSTs) were first identified as substances which cause necrosis only of host tissues normally attacked by the producing pathogen and were defined as primary determinants of pathogenicity (115). The only fungal genera that produce HSTs, however, are *Alternaria* and *Helminthosporium,* excepting *Phyllosticta maydis*. Therefore, pathogenic fungi belonging to other genera might produce other types of primary determinants of specificity such as suppressors. In fact, the known suppressors do not cause visible damage on host tissues. Thus, suppressors should be distinguished from HSTs. However, it has been reported that the most important role of HSTs in determining specificity may also be the suppression of host defenses, thus

Table 5-1. Suppressors from phytopathogenic fungi [a]

Fungus	Origin	Chemical Nature	Host Plant	Specificity	Defense Suppressed	Induction	Site of Action	Reference
Ascochyta rabei	Culture filtrate	Glycoprotein	Chickpea	Race-cultivar	PA	?	?	67
Botrytis sp.	Germination fluid	Peptide ?	Allium sp.	Genus (species)	General ?	Induced	PM ?	74
Mycosphaerella ligulicola	Germination fluid	Glycopeptide ?	Chrysanthemum	Species	General ?	Induced	?	100
M. melonis	Germination fluid	Glycopeptide ?	Cucumis spp.	Genus (species)	General ?	Induced	?	100
M. pinodes	Germination fluid	Glycopeptide	Pea	Species	General ? I.I., PA, PR proteins	Induced	CW, PM (ATPase, PI)	99
Phytophthora capsici	Mycellia	Glucan	Sweet pepper, tomato	Species	HR, AOS	?	PM ?	114
P. nicotiana	Mycellia	Glycan	Tobacco, tomato	Species	HR, AOS	?	PM ?	114
P. infestans	Zoospore, mycellia	Glycan, phosphoglucan	Potato, tomato	Species, race-cultivar	HR, AOS, PA	?	PM (Ca^{2+}, ADPH-oxidase)	35
P. infestans	Zoospore	Glucan ?	Tomato	Race-cultivar ?	HR	?	?	132
P. glycinea	Culture filtrate	Mannan glyco-protein (invertase)	Soybean	Race-cultivar	PA	?	?	162
Uromyces phaseoli	Infection structure	?	Kidney bean	Species	General ?, silicon deposits	Induced	?	54

[a] AOS = active oxygen species; CW = cell wall; HR = hypersensitive reaction; I.I. = infection-inhibitor; PA = phytoalexin; PI = polyphosphoinositide metabolism; PM = plasma membrane; and PR-proteins = pathogenesis-related proteins such as endochitinase and β-1,3-glucanase.

permitting pathogen infection (23,52,75,105,148,151). In this chapter, the role of suppressors, especially a suppressor from the pea pathogen, *Mycosphaerella pinodes*, is discussed.

Properties of Suppressors

Producers

Ten species of phytopathogenic fungi have been reported to produce substances that suppress the expression of active resistance induced by elicitors or avirulent pathogens (Table 5-1). Doke (35) reported that zoospore constituents of a compatible race of *Phytophthora infestans* blocked or delayed the hypersensitive response of potato tissues induced by inoculation with an incompatible race. Conversely, a corresponding preparation from incompatible races exhibited no or less suppressing activity. Germinating sporangia but not mycelia of *P. infestans* also produced similar suppressing activity toward the hypersensitive reaction of tomato (132). The pathogen of *Mycosphaerella* blight of pea, *M. pinodes*, secretes a low molecular weight glycopeptide suppressor into the spore germination fluid (99,127). Similar substances in the culture filtrates, mycelia or spore germination fluids of several pathogenic fungi were found to suppress NADPH-dependent generation of superoxide (36), the accumulation of phytoalexins (39,67,162), infection inhibitors (149) and the deposition of silicon-containing compounds (53).

In plant-pathogenic bacteria interactions, *Pseudomonas syringae* pathovar *phaseolicola* were reported to suppress active defense of host plants. Preinoculation with a compatible strain of *P. s. phaseolicola* reduced the accumulation of transcripts of defense response genes and phytoalexin in common bean induced by glutathione or *P. s. tabaci* (58). They also demonstrated that heat-killed cells or cells treated with protein synthesis inhibitors lost the suppressor activity, suggesting that *P. s. phaseolicola* has an active mechanism to suppress defense responses (58). Another report showed that non-toxic bacterial metabolite(s) may participate in suppression of plant defense responses. A substance(s) which was able to suppress or delay pea phytoalexin production in a race-cultivar-specific manner was found in culture filtrates and water extracts of *Pseudomonas syringae* pathovar *pisi* (146). In addition, it has recently been reported that a symbiont, *Bradyrhizobium japonicum,* produces cyclic $1,3-1,6-\beta$-glucans that suppress soybean defense reactions induced by $1,3-1,6-\beta$-glucans from *Phytophthora megasperma glycinea* (91). Thus, it is probable that certain bacterial metabolites also play an important role in the suppression of host defense responses.

Chemical Properties

The suppressors so far examined are water-soluble glucan, phospho-glucan, glycopeptide, glycoprotein (such as invertase) or peptide. Suppressors from *P. infestans* were reported to be composed of 17-23 glucose units and contained β-1,3 and β-1,6-glycosidic linkages (39). A mannan-glyco-protein (invertase) in culture medium of *P. megasperma* f. sp. *glycinea* also inhibited the accumulation of a soybean phytoalexin, glyceollin, in a race-cultivar specific manner. The invertase from race 1 of the fungus suppressed glyceollin accumulation in wounded cotyledons of two compatible cultivars induced by a non-specific glucan elicitor from cell walls of the same fungus but those of the incompatible cultivar did not (162). Since heat-denaturation did not affect the race-specific activity of invertase but periodate oxidation completely destroyed its activity to suppress phyto-alexin accumulation induced by the glucan elicitor, it was concluded that the carbohydrate moiety of invertase was essential for its activity. *Asco-chyta rabiei*, the causal agent of brown spot of chickpea, produced a suppressor in culture medium (67). The suppressor was precipitated between 30 to 70% saturation of ammonium sulfate, bound to a concanavalin A-conjugated column and was eluted by α-D-methyl mannoside. A SDS-polyacrylamide gel electrophoresis showed that this fraction contained one major protein band ($Mr > 14.5$ k). From these results, the suppressor was regarded as a glycoprotein.

The structure of suppressors was unknown for a long time, but the structures of two mucin-type suppressors, supprescins A and B from *M. pinodes*, were determined as GalNAc-*O*-Ser-Ser-Gly and Gal(β-1,4) GalNAc-*O*-Ser-Ser-Gly-Asp-Glu-Thr, respectively (128).

Specific Production of Suppressors

The interaction between plants and pathogens initiates when a pathogen contacts the plant surface. Since the majority of phytopathogenic fungi commonly infect through conidiospores in water, the interactions may be mediated by substances secreted into spore germination fluids. Indeed, cystospores of races 4 and 1,2,3,4 of *P. infestans* secreted anionic and non-anionic water-soluble glucans into the germination fluid (40). These glucans were detectable within 3 h and increased up to 6 h after the start of incubation. The glucans suppressed both hypersensitive cell death and production of the phytoalexin, rishitin, in potato tuber tissues challenged by incompatible race 4. Furthermore, it was shown that germination fluids did not elicit cell death or the accumulation of rishitin and that the suppressing activity of the fluid from race 1,2,3,4 was more intense than that from race 4.

A pea pathogen, *M. pinodes* was found to secrete a glycopeptide elicitor (mol. wt. > 70 kDa) and glycopeptide suppressors (mol. wt. < 5 kDa) of the defense responses of pea plants into its pycnospore germination fluid (86,99,126,128,136). The elicitor activity was detectable only in a high molecular weight (>10 kDa) fraction from germination fluid 3 h after spores had been suspended in sterilized water and increased up to 12 h, whereas the elicitor activity produced by *M. pinodes* was equivalent to that of an avirulent fungus *M. ligulicola*, the cause of chrysanthemum ray blight. By contrast, the suppressor activity, which was measured as the reduction of pisatin accumulation induced by the *M. pinodes* elicitor, was found immediately after suspending spores of *M. pinodes* in water. The activity increased up to 9 h. Secretion of the suppressor from hypovirulent strain (OMP-X-76) was lower than that from virulent strain (OMP-1), and avirulent *M. ligulicola* secreted only a negligible amount of suppressor over the course of 24 h. The same results were obtained when spores were placed on pea leaves. These findings can be summarized as follows; 1) the elicitor activity was produced even by a virulent fungus in the spore germination fluid, 2) the secretion of suppressor was more rapid, and, 3) the activity of the suppressor was expressed in the simultaneous presence of elicitor. Thus, it is possible that the suppressor may function in plant cells before penetration by pathogens (129).

Induction of Accessibility

Two infection-inducing factors were isolated from spore germination fluid of *Botrytis* sp., the causal agent of scallion bulb rot (74). Infection hyphae from conidiospores of saprophytic or nonpathogenic *Alternaria alternata* isolates formed at significant levels on plants belonging to the genus *Allium* that had been treated with *Botrytis* spore germination fluid or the purified factors. However, the fluid was unable to induce susceptibility on non-host plants such as strawberry, tomato and Japanese pear. The active substances (mol. wt. < 5 kDa) contained peptide moieties, which may be essential for the activity, but did not induce necrosis in host plants (74).

Treatment of pea tissues with the suppressor from *M. pinodes* allowed infection by many avirulent pathogens such as *Stemphylium sarcinaeforme*, *M. melonis, Alternaria alternata* (strain 15B) and *M. ligulicola*. *A. alternata* and *S. sarcinaeforme* were able to colonize and the former formed conidiospores on suppressor-treated tissues, whereas they were unable to colonize untreated tissues (100). Thus, the suppressor from *M. pinodes* conditioned pea plants to be accessible even to avirulent fungi. *A. alternata* could not infect any of 12 leguminous plants tested, but in the presence of the suppressor, it was able to infect five leguminous species, *Pisum sativum, Lespedeza buergeri, Medicago sativa, Milletia japonica* and *Trifolium pratense* (104). Figure 5-1 shows that *A. alternata* was unable to infect water-treated pea or cowpea tissues but could infect only pea tissues in the presence of the suppressor from *M. pinodes*. Thus, the biological specificity of the suppressor of *M. pinodes* coincided with the host range of the fungus (104).

Similar nontoxic and ninhydrin positive substances were found in the low molecular weight fractions (mol wt. < 10 kDa) from the pycnospore germination fluids of *M. ligulicola* and *M. melonis*, the cause of gummy stem blight of *Cucumis* and *Lagenaria* (100). Table 5-2 shows the relationship between the host range of *M. melonis* and the accessibility-inducing activity of substances from the fungus. Conidiospores of *A. alternata* were unable to infect these plants under normal condition. However, *M. melonis, Cucumis sativus, C. melo* and *Lagenaria leucantha* were infected by *A. alternata* only in the presence of the substance from *M. melonis*. Thus, the substance seems to determine the host range specificity of *M. melonis*.

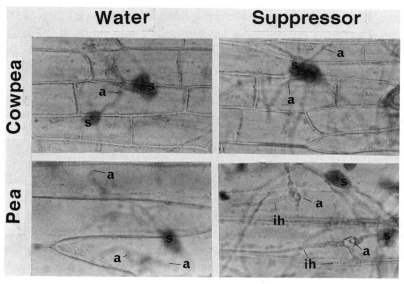

Fig. 5-1. Effect of the suppressor from a pea pathogen, *Mycosphaerella pinodes,* on the establishment of infection by *Alternaria alternata,* strain 15B, on the pea epicotyl and the cowpea hypocotyl tissues. Conidiospores of an avirulent pathogen of pea and cowpea, *A. alternata,* were placed on the surface of these tissues in the absence (water) or presence of 12.5 µg/ml of the *M. pinodes*-suppressor partially purified from the pycnospore germination fluid by the method described by Yoshioka et al. (1990) (160). The germination, appressorial formation and infection were observed 24 h after inoculation. Note that *A. alternata* was able to form its infection hyphae only on the suppressor-treated pea cells. a = appressorium; ih = infection hypha; and s = conidiospore of *A. alternata.*

Table 5-2. Host range of *Mycosphaerella melonis* and effect of the suppressor from *M. melonis* pycnospores on infection by *Alternaria alternata*, strain 15B

Plant Species	*M. melonis*[a]	*A. alternata* 15B[a]	*A. alternata* 15B[a] + Suppressor from *M. melonis*
Chrysanthemum molifolium	−	−	−
Cucumis sativus	+[b]	−	+[b]
C. melo	+[b]	−	+[b]
Lagenaria leucantha	+[b]	−	+[b]
Lycopersicon esculentum	−	−	−
Nicotiana tabacum	−	−	−
Raphanus sativus var. *acanthiformis*	−	−	−
Pisum sativum	−	−	−
Hordeum vulgare	−	−	−

[a] Infection by the fungi was observed 4 days after inoculation.
[b] Sporulation was observed 7 days after inoculation. Infection hyphae were formed (+) or not formed (−)

From these results, the suppressors from *Botrytis* sp., *M. pinodes, M. ligulicola* and *M. melonis* may act as determinants of specificity by suppressing the general resistance of host plants. In other words, host-pathogen compatibility in certain cases seems to be determined by the secretion of specific suppressors.

Interference with the Formation of Chemical Barriers in the Host

As briefly above, active defenses in host plants are inhibited or delayed by suppressors. The hypersensitive reaction in potato tissues or protoplasts, that was induced by an incompatible race or by hyphal wall components (a nonspecific elicitor) of *P. infestans*, was prevented or delayed by pre-treatment with water-soluble glucans from cystospores of the compatible race (35,40,41). It was also shown that the NADPH-dependent superoxide generation, that was induced within a few minutes by the hyphal wall fraction, prior to hypersensitive cell death and production of phytoalexins, was suppressed by the water-soluble glucans in a race-cultivar-specific manner (36,37,38). Furthermore, water-soluble glucans, which were prepared from mycelia of *P. capsici, P. nicotianae* var. *nicotianae* and *P. infestans* also blocked the hypersensitive cell death of suspension-cultured cells of sweet pepper, tobacco and tomato, respectively, caused by hyphal cell wall components (elicitors) from *Phytophthora* spp. These results indicated that the glucan-suppressors acted specifically only on typical host species of the pathogen (114). Thus, the glucan-suppressors seem to participate in determining species-specificity as well as race-cultivar specificity.

As shown in Table 5-3, the glycopeptide elicitor from *M. pinodes* induces defense responses in pea such as the accumulation of a major pea phytoalexin pisatin and an as yet unidentified infection-inhibitor (149), as well as endochitinase and endo-β-1,3-glucanase activities (159). However, these defense responses are markedly suppressed by the concomitant presence of the suppressor with the elicitor. For example, the elicitor induced the accumulation of pisatin within 9 h, activation of phenylalanine ammonia-lyase (PAL) within 3 h and the accumulation of transcripts of PAL and chalcone synthase (CHS) genes within 1 h. However, presence of the suppressor delayed these defense responses for 3 to 6 h (145; Table 5-3). These results indicate that the suppressor temporarily interrupts elicitor action but does not cause any drastic damage to pea cells. In other words, elicitor activity is delayed by the suppressor. These findings may coincide with the delay of defense responses caused by inoculation with compatible races or virulent fungi as discussed above (10,29,101,102,154,156).

The accumulation of pisatin was significantly reduced, even when the suppressor of *M. pinodes* was applied to pea tissues 9-12 h after treatment with elicitor (160). This finding suggests that suppression does not involve competition with the elicitors for binding at receptor sites on plant cells and that the suppressors might affect more fundamental plant functions.

Table 5-3. Chemical properties and biological activities of the suppressor and elicitor from *Mycosphaerella pinodes*

Property	Elicitor	Suppressor
Origin	Pycnospore	Pycnospore
Time of secretion	<3 h	0 h
Composition	Man, Glc, Ser	Gal, GalNAc, Ser, Gly, Asp, Glu, Thr
Biological Activity		
1. Active Resistance		
a. Pisatin	Induce	Delay
b. PR protein	Enhance	Delay
c. Infect. Inhibitor	Induce	Inhibit
d. Specificity	Non-specific	Species-specific
2. Infection	Inhibit	Induce (Species-specific)
3. Toxicity	ND[a]	ND
4. Site of action[b]		
a. PI	Stimulate	Inhibit
b. PM-ATPase	Enhance	Inhibit (non-specific)
c. ATPase (in vivo)	?	Inhibit (Species-specific)
d. H+-pump	ND	Inhibit
e. Na+/K+ efflux	Induce	Inhibit
f. CW ATPase	Enhance (non-specific)	Inhibit (Species-specific)
g. CW peroxidase	Enhance (non-specific)	Inhibit (Species-specific)

[a] Not detected.
[b] PI = polyphosphoinositide metabolism in the plasma membrane; PM-ATPase = plasma membrane ATPase measured in vitro; H+-pump = proton pumping activity measured in vitro; and CW = cell wall-bound activity.

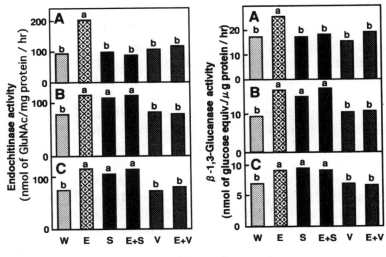

Treatment

Fig. 5-2. Effect of orthovanadate and the suppressor from *Mycosphaerella pinodes* on the activities of endochitinase (left) and endo-β-1,3-glucanase (right) in the seedlings of pea (A), soybean (B) and kidney bean (C) treated with the elicitor from *M. pinodes*. The activities were determined 24 h after the start of treatment with water alone (W), 1 mM orthovanadate alone (V), 500 µg/ml of the elicitor alone, 50 µg/ml of the suppressor alone, the elicitor plus suppressor (E+S) or the elicitor plus orthovanadate (E+V), by the methods described previously (Boller et al. Planta 157, 22-31, 1983; Abels and Forrence, Plant Physiol. 45, 395-400, 1970). Each value represents the mean of results of triplicate experiments. The standard deviations were within 5-6% of the mean values and significant difference (*P* < 0.05) was detected among the different letters at the top of each column. Note that the orthovanadate suppressed non-specifically the increase of both enzyme activities in respective plant species induced by the elicitor, and that the suppressor inhibited those only of pea seedlings but it alone inversely elicited those of nonhosts of *M. pinodes*.

The *M. pinodes* suppressor also blocks accumulation of two red clover phytoalexins, medicarpin and maackiain, but not production of glyceollin in soybean or phaseollin in kidney bean (129). This suppressor also suppresses the activation of pathogenesis-related (PR) proteins, such as endochitinase and β-1,3-glucanase, in pea tissues but not in soybean and kidney bean. In the latter cases, these PR proteins are activated by treatment with the suppressor alone as well as with the *M. pinodes*-elicitor (159; Figure 5-2). These results demonstrate that the suppressor affects defense responses in a species-specific manner.

Effect of Suppressors on Defense Gene Expression

In the potato-*P. infestans* system, northern analysis showed that the transient accumulation of PAL transcripts was induced in aged tuber tissues by inoculation with an incompatible race within 1 h and by treatment with hyphal cell wall components (HWC) within 30 min (156). A suppressor from the compatible race of the fungus suppressed the increase of PAL transcripts induced by HWC, whereas the suppressor from an incompatible race enhanced accumulation of the transcripts (156). These results coincide with *in situ* hybridization data from potato leaf tissues (29). Furthermore, the activity of 3-hydroxy-3-methylglutaryl CoA reductase (HMGR), a key enzyme for sesquiterpenoid phytoalexin biosynthesis, increased after inoculation with an incompatible, but not a compatible race of the fungus. The accumulation of HMGR transcripts was found to be equal in both combinations. However, inoculation with the compatible race of *P. infestans* led to a marked decrease in the amount of HMGR mRNA associated with polysomes (157). Similar data was obtained with glucan-suppressors from the fungus (155). Thus the activity of HMGR in potato might be regulated by the suppressor in a race-cultivar-specific manner, in part at the post-transcriptional level (155,157).

Elicitor prepared from *M. pinodes* induced the accumulation of mRNAs encoding PAL and CHS within 1 h and activated the transcription of *PAL* and *CHS* genes within 5 min in pea (63,145). However, the addition of suppressor to the elicitor resulted in a three hour delay in the accumulation of cellular *PAL* and *CHS*-mRNA and rapid deactivation of the transcription of *PAL* gene(s) within 10 min (144,145). These findings imply that pea defense gene expression is regulated transcriptionally and that a rapid signaling pathway from the cell surface to the nuclei, responding to the elicitor and/or the suppressor, exists in pea cells.

To elucidate this signaling pathway, we investigated the regulation of defense gene expression including the *cis*-regulatory elements and *trans*-acting nuclear factor(s). We previously isolated two elicitor-inducible *PAL* genes, *PSPAL1* and *PSPAL2*, and five elicitor-inducible *CHS* genes, *PSCHS1, 2, 3, 4* and *5,* that contribute to the biosynthesis of pisatin in pea (6,147, Ito *et al.* submitted). In the promoter sequences of these defense genes, we found conserved sequence motifs (Figure 5-3). Putative cis-acting elements, Box-1 (homologous to Box L) and Box-2 (homologous to Box P) were initially reported by Lois *et al.* (81). These elements were AC-rich, and also found in the promoter regions of many defense genes coding for enzymes of the phenylpropanoid biosynthetic pathway such as *PSPAL1* and *2* and *PSCHS1, 2, 3, 4* and *5* (6,147, Ito *et al.* submitted). Although we could not find the third AC-rich conserved sequence motif, Box-3 (96), in the elicitor-inducible *PAL* and *CHS* genes in pea, an AC-rich sequence, CATAACAA, related to Box-3 was found at a similar position in the 5'-transcribed region (6,147). Recently, we found that the 5'-untranslated region of *PSCHS1* was required for maximum induction of the transcript by elicitor (118). This result also suggests that AC-rich sequences in the 5'-transcribed region of the defense genes might be involved in elicitor-mediated activation.

In the distal region of the *PAL* promoter sequences in a variety of plant species, such as pea, bean, parsley and *Arabidopsis thaliana*, AT-rich sequences designated as Box-4 were identified as candidates for the *cis*-regulatory elements (147). We identified another AT-rich sequence in the *PSPAL1* promoter as Box-5 to form a low mobility complex (LMC) with a nuclear extract (63). The LMC formation was markedly increased in a nuclear extract prepared from elicitor-treated tissues rather than from the control tissue (63,117), but was undetectable in nuclear extracts from tissues pretreated with alkaline phosphatase (63). Therefore, phosphoryla-tion of certain nuclear protein(s) seems to be essential for activation of the defense genes. In *PSCHS1*, we showed that the directly repeated AT-rich sequences, TAAAATACT were capable of forming a LMC using a gel mobility shift assay and *in vitro* DNase I-footprinting analysis (117). LMC formation with the AT-rich sequence of the *PSPAL1* promoter was signifi-cantly reduced by addition of heterogeneous AT-rich DNA fragments prepared from other elicitor-inducible genes, such as *PSPAL2*, *PSCHS1* and *PSCHS2*. Since these DNA fragment did not show sequence similarity,

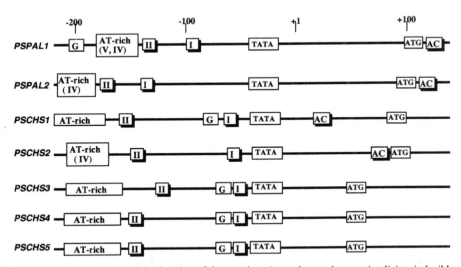

Fig. 5-3. Schematic pattern of the location of the putative *cis*-regulatory elements in elicitor-inducible *PSPAL* and *PSCHS* promoters. TATA boxes and translation initiation codon, ATG, G-box and AT-rich sequence elements including Box-4 and Box-5 were whitely boxed. The AC-rich sequence elements including Box-1, Box-2 and Box-3-like element were represented with the shaded boxes as I, II and AC, respectively.

except for the AT-rich region, these results led us to speculate that putative DNA-binding protein(s) forming LMC could bind to a wide range of AT-rich sequences (63), such as the Box-4 and Box-5 elements in *PSPAL*1 and the TAAAATACT sequence in *PSCHS*1.

The contribution of these conserved sequence elements to elicitor-mediated transcriptional activation was investigated with both transiently transfected pea protoplasts and stably transformed tobacco seedlings. In *PSCHS*1, a loss of function experiment in the transient transfection assay using pea protoplasts revealed that co-existence of Box-1 and the G-Box was indispensable not only for the activation of basal expression but also induction by elicitor. Furthermore, the Box-2 element seems to participate in the enhancement of elicitor-mediated activation, but Box-2 alone is not sufficient (118). In *PSPAL*1, a similar result was also obtained by the transient transfection assay by microprojectile bombardments (Murakami *et al.* submitted) and by analysis of transgenic tobacco plants carrying the various types of chimeric *PSPAL*1 promoters with a reporter gene (our unpublished results). These results together support the importance of the putative *cis*-elements in activation of these defense genes.

Furthermore, a chimeric promoter containing octamerized Box-4/Box-2 units with the Box-1 element of *PSPAL*1 exhibited remarkable deactivation of reporter gene expression in response to suppressor (Murakami *et al.* submitted). This leads us to speculate that the suppressor might affect complex machinery including Box-1, Box-2 and Box-4 required for elicitor-mediated transcriptional activation. In another elicitor-inducible pea *PAL* gene, *PSPAL2*, we observed a prominent change by suppressor-treatment in complex formation of the *PSPAL2* promoter region with a nuclear factor(s) (144). Although nuclear factor(s) related to suppression of defense gene expression have not been determined, treatment of the nuclear extracts with alkaline phosphatase abolished LMC formation. This indicates that phosphorylation of some nuclear protein(s) is necessary for the formation of complex machinery. It may therefore be interesting to examine whether the suppressor of *M. pinodes* interferes with such a process.

Except for the G-Box element, the putative *cis*-regulatory elements can be classified into two groups by sequence similarity. The first is the AC-rich elements such as Box-1, -2 and Box-3-like sequence, that seem to be indispensable for elicitor-responsiveness. The second is the AT-rich elements such as Box-4, -5 and the TAAAATACT elements in *PSCHS*1, that seem to be important in enhancement of elicitor-mediated activation (118, Murakami *et al.* submitted). The arrangements of these elements in the promoter region of the *PAL* and *CHS* genes is Box-3-like sequence, Box-1,

Box-2 and AT-rich elements from the coding region to the 5'-upstream sequence. The G-Box element, often located between Box-1 and Box-2 elements in many CHS genes, is not found in many *PAL* genes. The difference in the components of the putative *cis*-regulatory element might reflect the differential expression patterns and function of the *PAL* and *CHS* gene products.

The suppressor from *M. pinodes* seems to affect different steps of the process of gene expression. For example, expression of the *PSCHS* gene appears to be regulated both by transcriptional activation and at post-transcriptional level (117). This indicates that the suppressor might negatively regulate the expression of defense-related genes at these steps. Post-application of the suppressor into the UV-irradiated pea epicotyls also suppressed significantly the expression of a reporter gene driven from the *PSPAL1* promoter (Murakami et al. submitted). Previous nuclear run-on assays (144) revealed that the suppressor signal is transduced to terminate activated transcription of the *PSPAL* gene within 10 min after the onset of perception from the cell surface to nuclei. Thus, it is proposed that the specific negative signal pathway mediated by suppressor may exist in pea cells independently from the active signal pathway regulating *PSPAL1* gene expression.

Action of Suppressors on Plasma Membrane Functions

Suppression of Plasma Membrane ATPases

There are many reports that fungal elicitors affect ion fluxes in cultured plant cells. As shown in Table 5-3, the suppressor from *M. pinodes* inhibited the changes in pH, Na^+ and K^+ on the surface of pea tissues induced by the elicitor (4,5). These results suggest that the suppressor from *M. pinodes* affects a fundamental function such as a proton pump and/or anti-porter. In fact, the suppressor was found to inhibit the ATPase in pea plasma membranes both *in vitro* and *in vivo* (124,160). However, plasma membrane ATPases of non-host plants, such as kidney bean, soybean, cowpea and barley were also significantly inhibited *in vitro* to varying extents by the suppressor (124). Cytochemical observation by scanning transmission electron microscopy and energy dispersive X-ray analysis, however, showed that, of five plant species tested, the suppressor inhibited the ATPase activity only in pea cells. At the site of infection, the ATPase activity was also inhibited for at least 6 h by inoculation with *M. pinodes* but not with an avirulent pathogen of pea, *M. ligulicola* (124). Such inhibition of ATPase activity by the suppressor *in vivo* perfectly coincides with the specific induction of susceptibility (104) and with the observed delay of defense responses in pea tissues caused by the suppressor (159). In contrast, an effective inhibitor of P-type ATPases, orthovanadate, inhibited *in vivo* ATPase activity in all plant species tested. Orthovanadate non-specifically suppressed the activation of endochitinase and endo-β-1,3-glucanase in the tissues of pea, soybean and kidney bean treated with the elicitor of *M. pinodes* (159; Figure 5-2). Moreover, the reagent also delayed the initiation of pisatin accumulation for 6 h and the mRNAs encoding PAL and CHS for 3 h (158). The data with orthovanadate therefore suggest that plasma membrane ATPases, which carry out many fundamental cellular functions (121), must be closely associated with the regulation of defense responses. In this connection, we found that the suppressor also inhibited proton transport in proteoliposomes reconstituted with pea plasma membrane ATPase (3).

Supprescins from the germination fluid of *M. pinodes* inhibited the accumulation of pisatin in pea tissues, but only supprescin B inhibited the ATPase activity in isolated pea plasma membranes (128). *In vitro* action of the peptide moieties involved in the supprescins on ATPase activity was reported in detail by Kato *et al.* (64). The components of supprescins, such as galactose, *N*-acetylgalactosamine, Ser-Gly, Asp-Glu and respective amino acids did not inhibit the activity of plasma membrane ATPase. However, eight synthesized peptides, Ser-Ser-Gly-Asp-Glu-Thr, Ser-Ser-Gly, Ser-Ser-Gly-Thr-Glu-Asp, Thr-Ser-Gly-Asp-Glu-Thr, Ser-Ser-Gly-Asp, Gly-Asp-Glu-Thr, Asp-Glu-Thr and Gly-Asp-Glu, all inhibited ATPase activity. As shown in Table 5-4, the kinetics of inhibition

Table 5-4. Summary of kinetic data[a] for inhibition of pea plasma membrane ATPase activity by the supprescins and synthesized peptides

Group	Inhibition	Compound[b]	K_i (mM)	K_i' (mM)
1	Competitive	SSGDET	1.6	∞
		SSG	4.6	∞
		SSGTED	8.5	∞
2	Competitive and non-competitive	TSGDET	3.0	6.0
		SSGD	4.0	21.1
		GDET	4.7	17.0
3	Non-competitive	Sup. B	1.0	0.9
		DET	3.9	5.1
		GDE	4.5	6.5
4	(No inhibition)	Sup. A	-	-
		SG	-	-
		GD	-	-
		GGG	-	-

[a] Average Km and Vmax values for the activity of the pea plasma membrane ATPase, in the absence of inhibitors, were 0.32 mM and 7.2 μM Pi · h^{-1} · mg protein^{-1}, BSA equivalent, respectively (64).

[b] The compounds, with the exception of supprescin A (Sup. A) and supprescin B (Sup. B), are given in the single-letter code for amino acids. Thr structures of supprescins A and B are GalNAc-O-Ser-Ser-Gly and Gal-GalNac-O-Ser-Ser.

of the ATPase indicated that the inhibitors could be classified into three groups as 1) competitive; 2) mixed competitive and non-competitive; and 3) non-competitive. The common sequences found in groups 1), 2) and 3) were Ser-Ser-Gly, Gly-Asp and Asp-Glu, respectively. Ser-Ser-Gly-<u>Asp</u>-Glu-Thr had the greatest inhibitory effect among the peptides tested, while Ser-Ser-Gly-<u>Thr</u>-Glu-Asp had only a small effect. Thus, the sequence of Gly-Asp seems to play an important role in the inhibition of ATPase. This result seems to be correlated with the concept that the ATP-binding domain in the P-type ATPase has a Gly-Asp sequence and that the aspartic acid residue in the ATPase binds to the phosphate or adenosine moiety of an ATP molecule (121). Supprescin A did not inhibit ATPase activity but reduced the inhibitory effect of supprescin B. Furthermore, the peptide moiety of supprescin A, Ser-Ser-Gly, inhibited ATPase activity. These findings suggest that supprescin A has the potential to interact with the ATPase.

It was also shown that either supprescin B or the oligopeptides in group 3) and a part of group 2), Gly-Asp-Glu-Thr and Ser-Ser-Gly-Asp, inhibited the activity of acid phosphatase in pea plasma membranes, but supprescin A and other peptides did not. With the exception of Ser-Ser-Gly-Asp, the inhibitors of phosphatases share a common sequence, Asp-Glu. Thus, the peptide moiety of supprescin B, Ser-Ser-Gly-Asp-Glu-Thr, includes at least two active sequences, Ser-Ser-Gly and Asp-Glu. Since the former sequence inhibited ATPase activity in a competitive manner and the latter reduced acid phosphatase activity, it is possible that Ser-Ser-Gly-Asp-Glu-Thr might interact both with the ATP-binding domain and the phosphatase domain in the ATPase molecule (64).

In addition, Ser-Ser-Gly-Asp-Glu-Thr, Ser-Ser-Gly, Gly-Asp-Glu and Asp-Glu-Thr each suppressed the accumulation of pisatin in pea tissues induced by the elicitor from *M. pinodes* (*unpublished*). Our previous report showed that the proteinase-treated suppressor lost the activity of suppression of pisatin production (104). Both observations indicate that the peptide moiety is essential for activity of the *M. pinodes* suppressor.

Suppression of Transmembrane Signaling

There are several reviews on signal transduction regulation by fungal elicitors (34,112). As described above, it appears that the elicitor from *M. pinodes* rapidly (perhaps within 5 min) switches on primary transmembrane signaling leading to pea defense responses, but the suppressor may interrupt certain steps in such a signaling stream. In this section, we will describe the effect of suppressor on polyphosphoinositide (PI) metabolism in plasma membranes.

Polyphosphoinositide metabolism. An earlier report indicated that PI metabolism was not involved in signal transduction leading to phytoalexin biosynthesis in cultured cells of parsley and soybean (133). However, the elicitor and/or suppressor from *M. pinodes* were reported to affect markedly the activity of phosphatidylinositol kinase (PtdIns kinase), phosphatidylinositol monophosphate kinase (PtdInsP kinase) and phospholipase C (PLC) in epicotyl tissues and isolated plasma membranes of pea plants (138,139). In elicitor-treated tissues, rapid accumulation and turnover of phosphatidylinositol bisphosphate (PtdInsP2) was observed within 5 s, but the level of PtdInsP did not increase. From 5 s to 6 min after the start of treatment with the elicitor, a second increase and decrease in levels of ^{32}P-PtdInsP2 were observed. However, in the presence of the suppressor, the first rapid increase in PtdInsP2 was completely blocked, and the second increase was suppressed. The elicitor also induced a biphasic and transient accumulation of IP3 with peak levels at 30 s and 7 min, whereas such an increase was also markedly inhibited by the presence of suppressor. Neomycin, an effective inhibitor of PLC, also inhibited elicitor-induced accumulation of IP3 and pisatin in pea tissues. Interestingly, neomycin also caused pea tissue to be accessible to avirulent fungi (138).

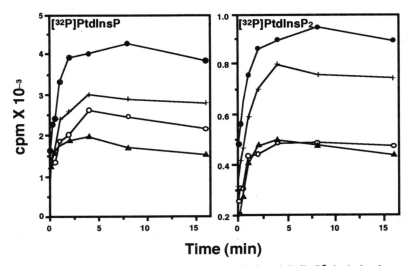

Fig. 5-4. In vitro phosphorylation of endogenous PtdInsP and PtdInsP2 in isolated pea plasma membranes upon the addition of the elicitor (100 μg/ml) from *M. pinodes* alone (●), elicitor plus suppressor (final concentration, 100 μg/ml; ○), suppressor alone (△) or distilled water (+). Isolated plasma membranes were incubated at 25°C with [γ-^{32}P]ATP without exogenous phospholipids. The extent of incorporation of radioactivity from [γ-^{32}P]ATP into both phospholipids was determined by photo-stimulated fluorography with a Bio-imaging analyzer (Bas 2000; Fujix, Tokyo, Japan).

Polyphosphoinositide metabolism in isolated pea plasma membranes was also affected by the elicitor and/or suppressor from *M. pinodes*. A time course study of phosphorylation of PtdIns and PtdInsP showed that the simultaneous incorporation of radioactivity from $[\gamma^{32}P]$ATP into both phospholipids was detectable within 5 s after treatment and increased for up to 4 min (Figure 5-4). The elicitor enhanced phosphorylation of phospholipids within 10 s, a response that was markedly reduced by suppressor. The phosphorylation of PtdIns to PtdInsP, however, was severely inhibited by the suppressor. The phosphorylation of PtdInsP was also inhibited by several other reagents, such as orthovanadate, K-252a (an effective inhibitor of many protein kinases) and neomycin, that suppress the defense response of pea (125,138, 158,159,160). The concomitant presence of these inhibitors markedly inhibited the incorporation of radioactivity into PtdInsP and/or PtdInsP2, as compared with the elicitor-treatment (Figure 5-5). The elicitor and

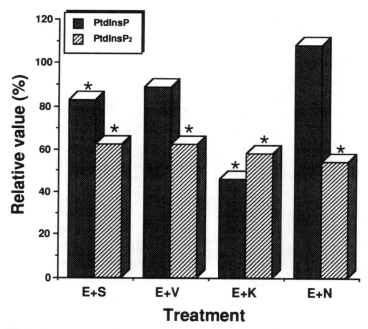

Fig. 5-5. Effects of vanadate, K-252a, neomycin and the suppressor from *Mycosphaerella pinodes* on the accumulation of ^{32}P-labeled PtdInsP and ^{32}P-labeled PtdInsP2 in isolated pea plasma membranes. The assay was carried out at 0°C for 1 min with 100 μg/ml of the elicitor in the presence or in the absence of 1 mM orthovanadate (V), 20 μM K-252a (K), 250 μM neomycin (N) or 100 μg/ml of the suppressor. The relative incorporation of radioactivity was represented as percentage of the result of the elicitor alone (100% = 1,855 ± 52 cpm for PtdInsP and 2,236 ± 82 cpm for PtdInsP2, respectively). * = Significantly different from the control ($P \leq 0.05$).

suppressor also affected the accumulation of IP3 in isolated plasma membranes of pea (138). The elicitor induced a transient increase of IP3 within 30 s. The increase of IP3 in the plasma membrane seems to coincide with the first increase of IP3 observed in pea tissues. No such increase in the level of IP3 was, however, observed in plasma membranes when the suppressor was present. Thus, the elicitor and suppressor from *M. pinodes* regulate rapid changes in PI metabolism in both tissues and isolated plasma membranes of pea plants.

Rapid activation of kinases of inositol phospholipids was induced in plasma membranes from cultured carrot cells that had been pre-treated with driselase and hemicellulase that induce the release of an endogenous elicitor from carrot cell walls (20). The subsequent steps in signal transduction might involve the conversion of PtdInsP2 into IP3 and diacylglycerol (DAG) by PLC (42,78,88,161). However, in tobacco BY-2 cells, it was reported that levels of IP2 but not IP3 increased within 4 min after treatment with the elicitor from *Phytophthora nicotianae* (60). It was assumed that the elicitor activated PtdIns kinase and that its product, PtdInsP, was converted to IP2 by PLC in the cultured cells. Together with these results, our findings suggest that PtdIns kinase, PtdInsP kinase and PLC in plant plasma membranes are closely related to defense responses. In other words, PI metabolism seems to be an indispensable process for the initial elicitation of defense responses of plants, as is the case for responses to light (89,92) or plant hormones (24,43,161). The suppressor from *M. pinodes* disturbs transmembrane signaling causing delayed transcriptional activation of defense genes.

In addition, the activity of phospholipase A (PLA) also seems to be regulated by these fungal signals since increases in lysoPtdInsP and lysoPA were induced by the elicitor from *M. pinodes* in pea plasma membranes within 5 s, while they were suppressed by the presence of the suppressor (Figure 5-6). In this connection, it is of interest that linoleic acid and linolenic acid, which are putative metabolites from phospholipids (PtdInsP or PA) by PLA, elicited the accumulation of pisatin (139). These results suggest that fatty acids might also act as second messengers for defense responses in pea, as described in tomato tissues by Farmer and Ryan (44).

Protein kinases. It has been reported that protein kinases may participate in transmembrane signaling for elicitors. Elicitors rapidly stimulate the phosphorylation of proteins in isolated membranes (45) and in suspension-cultured cells (13,32,46,48,143), and inhibitors of protein kinases affect defense responses (25,61,125). While K-252a was reported to enhance the synthesis of coumarin in cultured parsley cells induced by a fungal elicitor

(25), the inhibitor suppressed the production of pisatin in pea epicotyls (125) and the induction of PAL in tobacco BY-2 cells (61). In pea epicotyls, K-252a inhibited the accumulation of pisatin when K-252a was applied prior to treatment with the elicitor of *M. pinodes*. Application after elicitor treatment inversely stimulated the accumulation of pisatin (125). This finding indicates that the protein phosphorylation essential for elicitation of defense responses occurred immediately after elicitor perception. In addition, the suppressor from *M. pinodes* markedly inhibited, only in the presence of Ca^{2+}, the incorporation of radioactivity from $[\gamma\text{-}^{32}P]ATP$ into specific plasma membrane proteins of 78, 62 and 42 kDa (H. Yoshioka and T. Shiraishi, *unpublished*). Moreover, supprescin B and its peptide moieties, Ser-Ser-Gly-Asp-Glu-Thr, Ser-Ser-Gly-Asp, Glu-Asp-Glu-Thr, Ser-Ser-Gly and Asp-Glu Thr, each suppress the activity of protein kinase C prepared from rat brain (T. Kato and T. Shiraishi; *unpublished*). These results suggest that the suppressor might also be able to interfere with the activity of protein kinase(s) in the signal transduction cascade that leads to defense responses. In this connection, the relationship between resistance genes and protein kinases is extremely interesting. The tomato *Pto* gene, which confers resistance to race 0 of *Pseudomonas syringae* pv. *tomato*, encoded a serine/threonine protein kinase, suggesting its role in a signal transduction pathway (85).

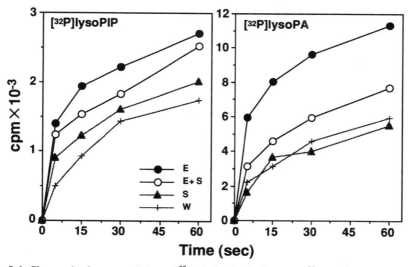

Fig. 5-6. Changes in the accumulation of ^{32}P-labeled lysoPtdInsP and ^{32}P-labeled lysophosphatidic acid in isolated pea plasma membranes. The incorporation of radioactivity from $[\gamma\text{-}^{32}P]ATP$ into both phospholipids was determined at 0°C without exogenous phospholipid in the presence or absence of the elicitor alone (100 μg/ml; ●), elicitor plus suppressor (○), the suppressor alone (100 μg/ml; ▲) or distilled water (+).

Functional Association Between ATPase and PI Metabolism

As mentioned above, the suppressor of *M. pinodes* inhibits both ATPase activity and PI metabolism (124,129,138,139,160). Phospholipids in plasma membranes have been reported to be essential for maintenance and regulation of ATPase activity in higher plants (21,62,89,90,120). Memon and Boss (89) demonstrated that light-induced inhibition of inositol phospholipid kinases in plasma membranes was closely correlated with simultaneous inhibition of P-type ATPase activity, suggesting that surrounding phospholipids regulate the ATPase activity in pea plasma membranes. Only phosphatidylinositol bisphosphate and PtdInsP out of six phospholipids tested stimulated the activity of pea plasma membrane ATPase about 1.82 and 1.13-fold, respectively, at 40 µM. By contrast, a trapper of PtdInsP2, neomycin, inhibited ATPase activity in a non-competitive manner (K. Toyoda and T. Shiraishi; *unpublished*). At least a part of the ATPase activity, therefore, might be regulated by levels of endogenous PtdInsP2 and/or PtdInsP. On the contrary, orthovanadate inhibited PtdIns kinase, PtdInsP kinase and PLC in pea plasma membranes (130, 139). These findings suggest that the inhibition of ATPase activity results in suppression of PI metabolism. In animal cells, ouabain, an inhibitor of Na$^+$/K$^+$-ATPase, was reported to suppress the activation of lymphocytes via its action on phospholipid metabolism (134). Taken together, the results indicate that there exists a functional association (cross-talk) between ATPase and PI metabolism in plant defenses. Recently, it was also found that a physical association also exists between ATPase and PtdIns kinase (140). However, we could not find a strict correlation between species-specificity and effect of the suppressor from *M. pinodes* on these functions in plasma membranes isolated from several plant species (124, K. Toyoda and T. Shiraishi, *unpublished*).

Specific Suppression of Cell Wall Function by Suppressor

It is thought that plants recognize internal and external signals, such as hormones and metabolites of microorganisms, via receptors located on plasma membranes. In fact, binding proteins for elicitors have been found in the microsome fraction or plasma membranes (22,26,95,116,152; for review, see 14). However, their functions and/or effectors are not yet unknown. On the other hand, the cell wall, the exterior and specific organelle of plant cells, is also thought to participate in the recognition and modification of external or internal signals (for reviews, see 112, 131, and 142). For example, plant hormone-binding proteins were reported to exist in cell walls of maize and hyoscyamus (59,82). With respect to microbial stress, the plant cell wall is thought to be the primary site where the secondary signals for defense responses are generated (30,112). Fragments of pectic polysac-charides of plants have been reported to induce the accumulation of phyto-alexins and/or the expression of defense responsive genes (30,50). Several cell wall-bound proteins, such as peroxidase (7,17), proline-rich glyco-protein (17,18), β-fructosidase (11), β-1,3-glucanase (87) and chitinase (12), are also thought to play important roles in plant defense responses. In plant and phytopathogenic bacteria combinations, it was recently found that harpin$_{pss}$ isolated from *Pseudomonas syringae* pv. *syringae* immediately induced extracellular alkalinization in tobacco suspension-cultured cells but harpin-treated protoplasts did not alkalinize the medium (56). The result indicated that harpin may interact with cell walls. However, it is not known if cell wall-bound proteins respond to suppressors and whether receptors exist in cell walls for fungal signals such as suppressors.

Specific Suppression of Cell Wall-Bound ATPase

An early study showed that phosphatase activity, including ATPase, existed in cell walls prepared from corn coleoptiles (72). However, the relationship between these phosphatases and external signals was unknown. As mentioned above, the suppressor from *M. pinodes* as well as ortho-vanadate inhibit the ATPase activities *in vivo* and *in vitro* (64,124,160). However, specific inhibition of the ATPase activity by the suppressor appears at the cell level but not at the isolated plasma membranes (124). These findings suggested that cell walls include certain target molecules for the suppressor and that these putative molecules affect function of the plasma membrane, including ATPase and PI metabolism, and accordingly might play an important role in the determination of specificity.

It was recently shown that certain phosphatases, such as nucleoside triphosphatases, *p*-nitrophenolphosphatase and pyrophosphatases, were tightly bound to cell wall fractions prepared from pea and cowpea seedlings (69,70,71). As shown in Table 5-5, the ATP-hydrolyzing activity in the pea cell wall fraction was considerably different in several properties, such as substrate specificity, dependence both on divalent cations and pH and sensitivity to neomycin, from those in isolated pea plasma membranes. The ATPase activity in cell walls was affected by the elicitor and suppressor from *M. pinodes* (70,71). That is, the elicitor non-specifically enhanced the activity in cell walls of pea, cowpea, kidney bean and soybean (Figure 5-7). On the other hand, the suppressor inhibited the activity only in the pea cell wall but inversely enhanced those of non-host plants of *M. pinodes* (Figure

5-7). Thus, the action of the suppressor on cell wall-bound ATPase showed strict species-specificity even *in vitro*.

In pea plants, even when the tissues were treated with *M. pinodes* elicitor without prior injury, local resistance was induced within 1 h that was accompanied by the production of an as yet unidentified infection-inhibitor. However, the suppressor also blocked this defense response (149) and conditioned pea cells to be susceptible even to avirulent pathogens (104, 127; see Figure 5-1). In this case, the suppressor and elicitor seemed to affect pea cells via cell walls. Orthovanadate, placed on the surface of cells, also inhibited the ATPase activities associated with all membrane systems in the epidermal cells of five plant species tested (124). As mentioned, the suppressor inhibited *in vitro* the ATPase activities in plasma membranes isolated both from host and non-host plants of *M. pinodes*, whereas the ATPase activities of non-host cells were never inhibited *in vivo* by the suppressor as well as isolated cell wall *in vitro*. Together with our previous reports, it is likely that the cell wall (or cell wall-bound ATPases) might affect or regulate the ATPases of other organelles, such as the plasma membrane or vacuole. In other words, inhibition of cell wall-ATPases might decrease the activity of plasma membrane ATPases (129,158,159, 160). If so, cell walls might also participate in acceptance of a virulent pathogen as well as in rejection of an avirulent pathogen. It was reported that tight connections between cell walls and cytosolic microtubles via plasma membranes existed (2,122) and that "Hechtian strands" in plant epidermal cells might be drawn out from the cytoplasmic face of a trans-membrane protein (or protein complex) bound to a wall-to-membrane linker (111). These reports may support the cell wall concept.

The finding that cell wall-bound ATPases are non-specifically stimulated by the elicitor but are inhibited by the suppressor in a species-specific manner also indicates that the putative receptor for fungal signals might

Table 5-5. Properties of ATPases in cell walls and plasma membranes prepared from etiolated pea seedlings

	Plasma Membrane	Cell Wall
Optimal pH	6–7 (6.5–6.7)	5–9 (6 and 8)
Substrate Specificity	ATP>>CTP>GTP >UTP	UTP = CTP>GTP>ATP >PPI = pNPP
Divalent cation requirement	Mn^{2+}, Mg^{2+} (none, 90% loss)	Ca^{2+}, Mn^{2+}, Mg^{2+} (none, 20–40% loss)
Orthovanadate	Inhibited	Inhibited
Neomycin	Inhibited	Not affected
In vitro action of *Mycosphaerella pinodes* suppressor	Inhibited (non-specifically)	Pea-inhibited Non-hosts-activated (species specific)

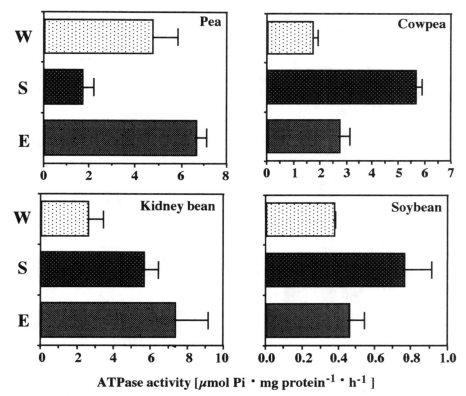

Fig. 5-7. Effects of the suppressor and elicitor from *Mycosphaerella pinodes* on ATPase activities in cell wall fractions prepared from pea, cowpea, kidney bean and soybean. The ATPase assay was carried out at 25°C for 20 min in 30 mM Tris-MES (pH 6.5) containing 3 mM Mg-ATP in the absence (water control; W) or presence of 100 µg/ml of the elicitor alone (E) or 100 µg/ml of the suppressor alone (S) as described by Kiba *et al.*(1995) (71). The protein content of cell wall fractions from pea, cowpea, kidney bean and soybean were 4.0, 14.6, 15.0 and 12.3 mg/g dry wt., respectively. Each value represents the mean with standard deviation (SD) of results from triplicate experiments. Note that the elicitor activated the ATPase activities in cell wall fractions all of species but that the suppressor inhibited the activity only in pea fraction and rather activated those of non-hosts of *M. pinodes* as compared to the water-control.

bind tightly to and affect the cell wall-bound ATPases or that the cell wall-bound ATPases might acts as a receptor and/or a modifier to recognize and change these fungal signals. Alternatively, there remains the possibility that the cell wall-bound ATPase itself might be the receptor for both signals from *M. pinodes*.

Specific Suppression of O_2^- Generation

Significance of oxidative burst in plant defense response was first demonstrated in the interaction between potato and *Phytophthora infestans* (36,37). Thereafter, there have been many reports that active oxygen species such as hydrogen peroxide, hydroxyl radical and superoxide anion may participate in plant-pathogen interactions (8). Generation systems for H_2O_2 and O_2^- exist in plant cell walls (15,49,51,84). However, the role of this system in plant-microbe interactions has been obscure. Recently, we found that O_2^- was generated on the surface of uninjured leaves of pea and cowpea and in fractions solubilized from cell walls of both plants (68, Figure 5-8). The elicitor from *M. pinodes* induced the nitroblue tetrazolium

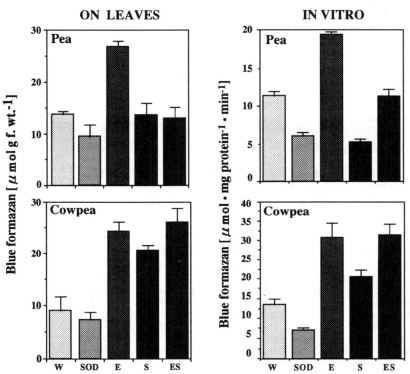

Fig. 5-8. Effect of the elicitor and suppressor from *Mycosphaerella pinodes* on superoxide generation on the surface of pea and cowpea leaves (left) or in the fraction solubilized from cell walls which were prepared from seedlings of both plants (right). In vivo assay was carried out at 25°C for 5 min with 30 mM Tris-Mes (pH 7.6) containing 2.5 µg/ml nitroblue tetrazolium in the absence (W) or presence of 100 µg/ml of the elicitor (E), 100 µg/ml of the suppressor (S), the elicitor plus suppressor (ES) or 300 units of superoxide dismutase (SOD). The reaction mixture was dropped onto the leaf surface. After incubation for 5 min in a moistened chamber, the concentration of blue formazan in each droplet was determined by measuring the absorbance at 560 nm. In vitro assay was carried out in 30 mM Tris-MES (pH. 6.5) containing 20 mM MnCl2, 2.5 µg/ml of nitroblue tetrazolium, 0.5 mM p-coumaric acid in the absence (W) or presence of 100 units of SOD, 100 µg/ml of the elicitor alone (E), 100 µg/ml of the suppressor alone (S), and the elicitor plus suppressor (ES), by the method of Nathan *et al.* (J. Clin. Invest. 48, 1895-1904, 1969). Each value represents the mean with standard deviation from triplicate experiments. Note that the elicitor enhanced superoxide generation in both fraction and that the suppressor significantly inhibited the generation in pea fraction but rather enhanced those in cowpea fraction.

(NBT)-reducing activity on the leaves that was sensitive to superoxide dismutase. The concomitant presence of the *M. pinodes*-suppressor markedly inhibited blue formazan formation on pea leaves but not cowpea leaves (Figure 5-8, left). Moreover, the formation of blue formazan on cowpea leaves was enhanced by the suppressor alone (Figure 5-8,. left-bottom).

A superoxide dismutase-sensitive NBT-reducing activity was also found in the NaCl soluble fraction from cell walls, which were isolated from etiolated seedlings of pea and cowpea (68; Figure 5-8, right). The activity was NAD(P)H-dependent and required manganese ion and *p*-coumaric acid as cofactors but was markedly reduced in the presence of a scavenger of H_2O_2, catalase. The requirement of these co-factors and the strong inhibition by catalase indicated that the O_2^- generation system is sustained by certain cell wall-bound peroxidase(s) as described by Halliwell (51) with horseradish peroxidase. Inhibitors of NADPH oxidase, quinacrine and imidazole, that are bound to flavoprotein (27) and *b*-type cytochrome (57), respectively, did not inhibit or scarcely affected O_2^- generation in the fraction solubilized from cell walls of both plants. On the other hand, salicylhydroxamic acid, an inhibitor of peroxidase, remarkably inhibited O_2^- generation in both fractions. This result also supports the idea that cell wall-bound peroxidase(s) may participate in NADH-dependent O_2^- generation in cell wall fractions but not a NAD(P)H oxidase(s). This hypothesis was already presented by several groups (49,51,83). Rubinstein and Luster (113) doubted the physiological significance of this mechanism since NAD(P)H concentrations required for this reaction were not likely in apoplast. However, it was reported that NAD was contained in cell wall free space (123) and that apoplastic fluid from the elicited kidney bean contained as yet-unidentified reductant (16).

Neomycin did not affect O_2^- generation in both fractions. On the other hand, orthovanadate markedly inhibited activity of the fractions from both plants. Effects of both inhibitors on the formazan-formation seem to coincide with those of cell wall-bound ATPases. That is, the cell wall-bound ATPase was inhibited by orthovanadate but not by neomycin (68; Table 5-5) while the plasma membrane ATPases were inhibited by both inhibitors (160, *unpublished*). These results suggest the possibility that O_2^- generation in the cell wall fractions of pea and cowpea might be regulated together with cell wall-bound ATPases. However, since it was reported that vanadate acted as an inhibitor of peroxidase (119), it is unknown whether the inhibition of O_2^- generation by orthovanadate results from the inhibition of cell wall-bound peroxidase or ATPase.

As shown in Figure 5-8 (right), the elicitor from *M. pinodes* significantly enhanced NBT-reducing activity (blue formazan-formation) in the fractions solubilized from cell walls of pea and cowpea in a non-specific manner. On the other hand, the suppressor inhibited formazan-formation only in the pea fraction. Even the concomitant presence of the suppressor with the elicitor decreased formazan-formation in the pea fraction to the control level. However, in the cowpea fraction, formazan formation was not inhibited by the presence of the suppressor and was inversely enhanced even by the suppressor alone. These results showed that the activity of O_2^- generation in the cell wall fraction is also regulated by both fungal signals and that the suppressor acts on the activity in a species-specific manner. Thus, the NADH-dependent O_2^- generating system in cell wall seems to be tightly correlated with cell wall-bound ATPase. Our recent experiments showed that the cell wall-bound ATPases of pea and cowpea co-purified with the activity of peroxidase(s) by affinity chromatography (A. Kiba and T. Shiraishi, *unpublished*). Further investigations are needed to elucidate not only the relationship between the receptors for these fungal signals and the regulatory systems for O_2^- generation in cell walls but also the role of O_2^- in signal transduction cascade leading to the early defense responses of plants. The results strongly support the idea that the plant cell wall may recognize fungal signals and play a key role in determination of plant-microbe specificity.

Concluding Remarks

The mechanisms involved in determining host-pathogen specificity appear to reside in the ability of the pathogen to overcome host resistance. For the parasitic adaptation, microorganisms must possess the capacities to (a) penetrate plant tissues, (b) overcome the host's resistance and (c) evoke disease (49,97). The second ability seems to be most important in the determination of specificity. From accumulated evidence, the suppressor of *M. pinodes* is thought to be a key factor for overcoming the general resistance of its own host and thereby in determining host specificity. As discussed in this review, data indicate that a suppressor can overcome the action of a non-specific elicitor on host plants but also can often be recognized as a specific elicitor by non-host plants. Therefore, we hypothesize that the suppressor is not only the product of *vir* gene but also that of *avir* gene, dependently upon the combinations.

It is important to understand how a pathogen overcomes defense responses and establishes accessibility in a given plant species. Based on the production of suppressors, a molecular mechanism to explain host specificity is proposed in Figure 5-9. At first, host-pathogen compatibility is established by the production of a species-specific suppressor with common non-specific elicitors. At the next step, a new resistant cultivar, which is capable of degrading the suppressor, of recognizing the suppressor molecule as an elicitor or of neglecting the suppressor, is developed. In the former two cases, resistance may be dominant, but, in the latest case, recessive. At the third step, the rejected pathogen gains the ability to produce a new suppressor overcoming the general resistance in a given species

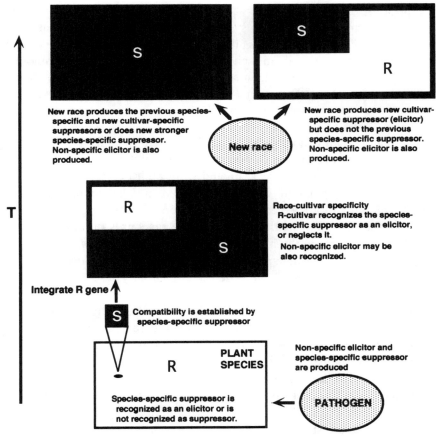

Fig. 5-9. Proposed model of the establishment of specificity between host plants and microbial pathogens based on the production of specific suppressors. The suppressor molecule from a given pathogen may be recognized as i) the suppressor for general resistance by the host or ii) as an elicitor by nonhosts. In both cases, species-specific action of suppressor appears as shown in Fig. 2. In the case that the suppressor is not recognized by plants, defense responses seem also to be triggered by non-specific elicitor.

or cultivars. Both cases may occur; one where a stronger suppressor, effective on all cultivars is produced; another where the activity of the new suppressor is cultivar-specific. The latter may act as a specific elicitor on incompatible cultivars. As mentioned in the Introduction, double inoculation tests indicate that specific elicitation by *avir* gene products alone may fail to explain the specific establishment of infection. Therefore, in this treatise, we did not focus on significance of *avir* gene products, but there have been many excellent works on *avir* genes of fungal and bacterial pathogens (for reviews, see 66,73,80) .

The possible modes of action of suppressors may be to: 1) interfere with perception of elicitor molecules at their receptor site in the host; 2) interfere with the signal-transduction system; 3) block the transcriptional activation of defense genes; 4) suppress the biosynthetic pathways that lead to the establishment of physical and chemical barriers against the pathogens; and/or 5) inhibit fundamental functions, such as ion transport and generation of energy that sustain a series of defense responses. In this review, we demonstrated that the suppressor of *M. pinodes* seems to possess at least properties 2) and 5).

Where and when is host-parasite specificity determined? The suppressor from *M. pinodes* was shown to nonspecifically inhibit *in vitro* the ATPase activities in plasma membranes isolated from both host and nonhost plant tissues. Isolated plant cell walls, however, responded to the suppressor from *M. pinodes* in a species-specific manner. This finding strongly indicates that the system switching plant defense responses on or off exists in the cell wall, at least in the interaction between plants and *M. pinodes*. Our working hypothesis on the specific suppression of defense responses is as follows; A) The suppressor and elicitor are initially bound to respective receptors on the cell wall (in the case of pea-*M. pinodes* system, the suppressor must be recognized earlier than the elicitor); B) On the cell walls of non-host plants, the suppressor of the pathogen is recognized as an elicitor; C) The suppressor inhibits the host cell wall function including ATPase activity and O_2^- generation in a species-specific manner; D) This anesthetized state of the host cell wall system affects function of the plasma membrane including ATPase activity and PI metabolism via an as yet unidentified signal transduction cascade; E) defense responses with defense gene expression are suppressed (delayed) for at least 3-6 h; F) The anesthetized state of cell wall may also reflect on another defense system in cell walls of uninjured host tissues including the production of an infection-inhibitor induced within 1 h after elicitor-treatment.

We believe that infection establishment on a given plant species (or cultivar) by a pathogen is dependent on specific secretion of a substance disturbing fundamental functions, such as signal transduction, ion transport and energy generation in cell wall and membrane systems that are required for defense responses in the host (129). If a compatible plant species (or cultivar) develops the ability to recognize such a suppressor as an elicitor, it will have confounded the pathogen suppressor strategy.

Acknowledgments

We thank Professor N. T. Keen, University of California, Riverside, for helpful suggestions on and critical reading of this manuscript. This work was supported in part by Grants-in-Aid for Scientific Research from the Ministry of Education, Science and Culture of Japan and the Grant for Research for the Future Program from the Japan Society for the Promotion of Sciences (JSPS-RFTF 96L00603). Financial supports from Sankyo Co. Ltd., Tokyo, Japan, and The Life Science Research & Development Institute, Okayama, Japan, are also acknowledged.

References

1. Agrios, G. N. (1978). Plant Pathology (2nd ed.), pp. 703. Academic Press, New York.
2. Akashi, T., and Shibaoka, H. (1991). Involvement of transmembrane proteins in the association of cortical microtubles with the plasma membrane in tobacco BY-2 cell. J. Cell Sci. 98:169-174.
3. Amano, M., Toyoda, K., Ichinose, Y., Yamada, T. and Shiraishi, T. (1995). H+-translocating activity in proteoliposomes reconstituted with pea plasma membrane ATPase and its inhibition by fungal suppressor from Mycosphaerella pinodes. Ann. Phytopath. Soc. Japan 61:369-375.
4. Amano, M., Toyoda, K., Ichinose, Y., Yamada, T. and Shiraishi, T. (1995). Association between defense responses and ion-flux. Plant Cell Physiol. 36:s119.
5. Amano, M., Toyoda, K., Ichinose, Y., Yamada, T. and Shiraishi, T. (1996). Association between defense responses and ion-flux (II). Plant Cell Physiol. 37:s57.
6. An, C. C., Ichinose, Y., Yamada, T., Tanaka, Y., Shiraishi, T. and Oku, H. (1993). Organization of the genes encoding chalcone synthase in *Pisum sativum*. Plant Mol. Biol. 21:789-803.
7. Apostol, I., Heinstein, P. F. and Low, P. S. (1989). Rapid stimulation of an oxidative burst during elicitation of cultured plant cells. Plant Physiol. 90:109-116.
8. Baker, C. J. and Orlandi, E. W. (1995). Active oxygen in plant pathogenesis. Annu. Rev. Phytopathol. 33:299-321.
9. Basse, C. W., Bock, K. and Boller, T. (1992). Elicitor and suppressor of the defense response in tomato cells. Purification and characterization of glycopeptide elicitors and glycan suppressors generated by enzymatic cleavage of yeast invertase. J. Biol. Chem. 267:10258-10265.
10. Bell, J. N., Ryder, T. B., Wingate, V. P. M., Bailey, J. and Lamb, C. J. (1986). Differential accumulation of plant defense gene transcripts in a compatible and an incompatible plant-pathogen interaction. Mol. Cell. Biol. 6:1615-1623.
11. Benhamou, N., Grenier, J. and Chrispeels, M. J. (1991). Accumulation of β-fructosidase in the cell walls of tomato roots following infection by a fungal wilt pathogen. Plant Physiol. 97:739-750.

12. Benhamou, N., Joosten, M. H. A. J. and De Wit, P. J. G. M. (1990). Subcellular localization of chitinase and of its potential substrate in tomato root tissues infected by *Fusarium oxysporum* f. sp. *radicis-lycopersici*. Plant Physiol. 92: 1108-1120.

13. Blower, D. P., Boss, W. F and Trewavas, A. J. (1988). Rapid changes in plasma membrane protein phosphorylation during initiation of cell wall digestion. Plant Physiol. 86:505-509.

14. Boller, T. (1995). Chemoperception of Microbial signals in plant cells. Ann. Rev. Plant Physiol. Plant Mol. Biol. 46:189-214.

14a.Boller,T., Gehri, A., Mauch, F., Vogeli, U. (1983) Chitinase in bean leaves: Induction by ethylene, purification, properties, and possible function *Phaseolus vulgaris*, antibiotic function, defense capacity of plant cells against pathogens. Planta. 157:22-31.

15. Bolwell, G. P., Butt, B. S., Davies, D. and Zimmerlin, A. (1995). The origin of the oxidative burst in plants. Free Rad. Res. 23:517-532.

16. Bolwell, G. P. (1996). The origin of the oxidative burst in plants. Biochem. Soc. Transact. 24:438-441.

17. Bradly, D., Kjellbom, P and Lamb, C. J. (1992). Elicitor- and wound-induced oxidative cross-linking of a proline-rich plant cell wall protein, a novel rapid defense response. Cell 70:21-30.

18. Brisson, L. F., Tenhaken, R. and Lamb, C. (1994). Function of oxidative cross-linking of cell wall structural proteins in plant disease resistance. Plant Cell 6:1703-1712.

19. Bushnell, W. R. and Rowell, J. B. (1981). Suppressors of defense reactions, a model for roles in specificity. Phytopathology 71:1012-1014.

20. Chen, Q. and Boss, W. F. (1990). Short-term treatment with cell wall degrading enzymes increases the activity of the inositol phospholipid kinases and the vanadate-sensitive ATPase of carrot cells. Plant Physiol. 94:1820-1829.

21. Chen, Q. and Boss, W. F. (1991). Neomycin inhibits the phosphatidylinositol mono-phosphate and phosphatidylinositol bisphosphate stimulation of plasma membrane ATPase activity. Plant Physiol. 96:340-343.

22. Cheong, J. J., Alba, R., Cote, F., Enkerli, J. and Hahn, M. G. (1993). Solubilization of functional plasma membrane-localized hepta-b-glucoside elicitor-binding proteins from soybean. Plant Physiol. 103:1173-1182.

23. Comstock, J. C. and Scheffer, R. P. (1973). Role of host-selective toxin in colonization of corn leaves by *Helminthosporium carbonum*. Phytopathol. 63:24-29.

24. Connett, R. J. A. and Hanke, D. E. (1987). Changes in the pattern of phospholipids synthesis during the induction by cytokinin of cell division in soybean suspension cultures. Planta 170:161-167.

25. Conrath, U., Jeblick, W. and Kauss, H. (1991). The protein kinase inhibitor, K-252a, decreases elicitor-induced Ca^{2+} uptake and K^+ release, and increases coumarin synthesis in parsley cells. FEBS Lett. 279:141-144.

26. Cosio, E. G., Feger, M., Miller, C. J., Antelo, L. and Ebel, J. (1996). High-affinity binding of fungal β-glucan elicitors to cell membranes of species of the plant family Fabaceae. Planta 200:92-99.

27. Cross, A. R. and Jones, O. T. G. (1991). Enzymic mechanisms of superoxide production. Biochim. Biophys. Acta 1075:281-291.

28. Cruickshank, I. A. M. and Perrin, D. R. (1968). The isolation and partial characterization of monilicolin A, a polypeptide with phaseollin-inducing activity from *Monilinia fructicola*. Life Science 7:449-458.

29. Cuypers, B., Schmelzer, E. and Hahlbrock, K. (1988). *In situ* localization of rapidly accumulated phenylalanine ammonia-lyase mRNA around penetration sites of *Phytophthora infestans* in potato leaves. Mol. Plant-Microbe Interactions 1:157-160.

30. Darvill, A. G. and Albersheim, P. (1984). Phytoalexins and their elicitors-A defense against microbial infection in plants. Ann. Rev. Plant Physiol. 35:243-275.

31. DeWit, P. J. G. M. (1986). Elicitation on active resistance mechanisms. In: Biology and Molecular Biology of Plant-Pathogen Interactions (J. A. Bailey, ed.), pp.149-169. Springer-Verlag, Berlin.

32. Dietrich, A., Mayer, J. E. and Hahlbrock, K. (1990). Fungal elicitor triggers rapid, and specific protein phosphorylation in parsley cell suspension culture. J. Biol. Chem. 265:6360-6368.

33. Dixon, R. . and Lamb, C. J. (1990). Molecular communication in interactions between plants and microbial pathogens. Annu. Rev. Plant Physiol. Plant. Mol. Biol. 41:339-367.

34. Dixon, R. A., Harrison, M. J. and Lamb, C. J. (1994). Early events in the activation of plant defense responses. Annu. Rev. Phytopathol. 32:479-501.

35. Doke, N. (1975). Prevention of the hypersensitive reaction of potato cells to infection with an incompatible race of *Phytophthora infestans* by constituents of the zoospores. Physiol. Plant Pathol. 7:1-7.

36. Doke, N. (1983). Involvement of superoxide anion generation in the hypersensitive response of potato tuber tissues to infection with an incompatible race of *Phytophthora infestans* and to the hyphal wall components. Physiol. Plant Pathol. 23:345-357.

37. Doke, N. (1983). Generation of superoxide anion by potato tuber protoplasts during the hypersensitive response to hyphal wall components of *Phytophthora infestans* and specific inhibition of the reaction by suppressors of hypersensitivity. Physiol. Plant Pathol. 23:359-367.

38. Doke, N. (1985). NADPH-dependent O_2^- generation in membrane fractions isolated from wounded potato tubers inoculated with *Phytophthora infestans*. Physiol. Plant Pathol. 27:311-322.

39. Doke, N., Garas, N. A. and Kuc, J. (1979). Partial characterization and aspects of the mode of action of a hypersensitivity-inhibiting factor (HIF) isolated from *Phytophthora infestans*. Physiol. Plant Pathol. 15:127-140.

40. Doke, N., Garas, N. A. and Kuc, J. (1980). Effect on host hypersensitivity of suppressors released during the germination of *Phytophthora infestans* cystospores. Phytopathology 70:35-39.

41. Doke, N. and Tomiyama, K. (1977). Effect of high molecular substances released from zoospores of *Phytophthora infestans* on hypersensitive response of potato tubers. Phytopath. Z. 90:236-242.

42. Einspahr, K. J. and Thompson, G. A. Jr. (1990). Transmembrane signaling via phosphatidyl 4,5-bisphosphate hydrolysis in plants. Plant Physiol. 93:361-366.

43. Ettlinger, C. and Lehle, L. (1988). Auxin induces rapid changes in phosphatidylinositol metabolites. Nature 331:176-178.

44. Farmer, E. E. and Ryan, C. A. (1992). Octadecanoid precursors of jasmonic acid activate the synthesis of wound-inducible proteinase inhibitors. Plant Cell 4:129-134.

45. Farmer, E. E., Pearce, G. and Ryan, C. A. (1989). *In vitro* phosphorylation of plant plasma membrane proteins in response to the proteinase inhibitor inducing factor. Proc. Natl. Acad. Sci. USA 86:1539-1542.

46. Felix, G., Grosskopf, D. G., Regenass, M. and Boller, T. (1991). Rapid changes of protein phosphorylation are involved in transduction of the elicitor signal in plant cells. Proc. Natl. Acad. Sci. USA 88:8831-8834.
47. Gaumann, J. (1950). "Principles of Plant Infection," pp. 543. Crosby Lockwood & Son, London.
48. Grab, D., Feger, M. and Ebel, J. (1989). An endogenous factor from soybean (*Glycine max*) cell culture activates phosphorylation of a protein which is dephosphorylated *in vivo* in elicitor-challenged cells. Planta 179:340-348.
49. Gross, G. G., Janse, C. and Elstner, E. F. (1977). Involvement of malate, monophenols, and hydrogen peroxide formation by isolated cell walls from horseradish (*Armoracia lapathifolia* Gilib.). Planta. 136:271-276.
50. Hahn, M. G., Darvill, A. G. and Albersheim, P. (1981). Host-pathogen interactions XIX. The endogenous elicitor, a fragment of a plant cell wall polysaccharide that elicits phytoalexin accumulation in soybeans. Plant Physiol. 68: 1161-1169.
51. Halliwell, B. (1978). Lignin synthesis, the generation of hydrogen peroxide and superoxide by horseradish peroxidase and its stimulation by manganese (II) and phenols. Planta. 140:81-88.
52. Hayami, C., Otani, H., Nishimura, S. and Kohmoto, K. (1982). Induced resistance in pear leaves by spore germination fluids of nonpathogens to *Alternaria alternata*, Japanese pear pathotype and suppression of the induction by AK-toxin. J. Fac. Agric. Tottori Univ. 17:9-18.
53. Heath, M. C. (1980). The suppression of the development of silicon-containing deposits in French bean leaves by exudates of the bean rust fungus and extracts from bean rust-infected tissues. Physiol. Plant Pathol. 18:149-155.
54. Heath, M. C. (1981). A generalized concept of host-parasite specificity. Phytopathology 71:1121-1123.
55. Heath, M. C. (1993). current concepts of the determinants of plant-fungal specificity. In: Host-Specific Toxin: Biosynthesis, Receptor and Molecular Biology (K. Kohmoto and O. C. Yoder ed.) pp.3-21, Tottori Univ. Press, Tottori.
56. Hoyos, M. E., Stanley, C. M., He, S. Y., Pike, S., Pu, X-A. and Novacky, A. (1996). The interaction of harpinPss with plant cell walls. Mol. Plant-Microbe Interactions 9:608-616.
57. Iizuka, T., Kanegasaki, S., Makino, R., Tanaka, T. and Ishimura, Y. (1985). Pyridine and imidazole reversibly inhibit the respiratory burst in porcine and human neutrophils, Evidence for the involvement of cytochrome *b588* in the reaction. Biochem. Biophys. Res. Commun. 130:621-626.
58. Jakobek, J. L., Smith, J. A. and Lindgren, P. B. (1993). Suppression of bean defense responses by *Pseudomonas syringae*. Plant Cell 5:57-83.
59. Jones, A. M. and Herman, E. M. (1993). KDEL-containing auxin binding protein in secreted to the plasma membrane and cell wall. Plant Physiol. 101:595-606.
60. Kamada, Y. and Muto, S. (1994). Stimulation by fungal elicitor of inositol phospholipid turnover in tobacco suspension culture cells. Plant Cell Physiol. 35:397-404.
61. Kamada, Y. and Muto, S. (1994). Protein kinase inhibitors inhibit stimulation of inositol phospholipid turnover and induction of phenylalanine ammonia-lyase in fungal elicitor-treated tobacco suspension culture cells. Plant Cell Physiol. 35:405-409.
62. Kasamo, K. and Nouchi, I. (1987). The role of phospholipids in plasma membrane ATPase activity in *Vigna radiata* L. (mung bean) roots and hypocotyls. Plant Physiol. 83:323-328.

63. Kato, H., Wada, M., Muraya K, Malik, K., Shiraishi, T., Ichinose, Y. and Yamada, T. (1995). Characterization of nuclear factors for elicitor-mediated activation of the promoter of the pea phenylalanine ammonia-lyase gene 1. Plant Physiol. 108:129-139.
64. Kato, T., Shiraishi, T., Toyoda, K., Saitoh, K., Satoh, Y., Tahara, M., Yamada, T. and Oku, H. (1993). Inhibition of ATPase activity in pea plasma membranes by fungal suppressors from Mycosphaerella pinodes and their peptide moieties. Plant Cell Physiol. 34:439-445.
65. Keen, N. T. (1975). Specific elicitors of plant phytoalexin production, determinants of race specificity in pathogens. Science 187:74-75.
66. Keen, N. T. (1990). Gene-for-gene complementarity in plant-pathogen interactions. Annu. Rev. Genet. 24:447-463.
67. Kessmann, H. and Barz, W. (1986). Elicitation and suppression of phytoalexin and isoflavone accumulation in cotyledons of Cicer arietinum L. as caused by wounding and by polymeric components from fungus Ascochyta rabiei. J. Phytopathol. 117:321-335.
68. Kiba, A., Miyake, C., Toyoda, K., Ichinose, Y., Yamada, T. and Shiraishi, T. (1996). Superoxide generation in cell wall during plant-pathogen interaction. Plant Cell Physiol. 37:s127.
69. Kiba, A., Toyoda, K., Ichinose, Y., Yamada, T. and Shiraishi, T. (1994). Specific inhibition by fungal suppressor of ATPase activity tightly binding to cell wall fraction from pea and cowpea. Plant Cell. Physiol. 35:s65.
70. Kiba, A., Toyoda, K., Ichinose, Y., Yamada, T. and Shiraishi, T. (1996). Specific response of partially purified cell wall-bound ATPases to fungal suppressor. Plant Cell Physiol. 37:207-214.
71. Kiba, A., Toyoda, K., Yamada, T., Ichinose, Y. and Shiraishi, T. (1995). Specific inhibition of cell wall-bound ATPases by fungal suppressor from Mycosphaerella pinodes. Plant Cell Physiol. 36:809-817.
72. Kivilaan, A., Beaman, T. C. and Bandurski, R. S. (1961). Enzymatic activities with cell wall preparation from corn coleoptiles. Plant Physiol. 36: 605-610.
73. Knogge, W. (1996). Fungal infection of plants. Plant Cell 8:1711-1722.
74. Kodama, M., Kajiwara, K., Otani, H. and Kohmoto, K. (1989). A host-recognition factor from Botrytis affecting scallion. In: Host-Specific Toxins, Recognition and Specificity Factors in Plant Disease (Kohmoto, K. and Durbin, R. D., eds.), pp. 33-44. Tottori Univ. Press, Tottori.
75. Kohmoto, K., Otani, H. and Nishimura, S. (1987). In: Molecular Determinants of Plant Diseases (Nishimura, S. et al. eds.), pp. 127-143. Japan Sci. Soc. Press, Tokyo.
76. Kunoh, H., Hayashimoto, A., Harui, M. and Ishizaki, H. (1985). Induced accessibility and enhanced resistance at cellular level in barley coleoptiles. I. The significance of timing of fungal invasion. Physiol. Plant Pathol. 27:43-54.
77. Kunoh, H., Katsuragawa, N., Yamaoka, N. and Hayashimoto, A. (1988). Induced accessibility and enhanced inaccessibility at the cellular level in barley coleoptiles. III. Timing and localization of enhanced inaccessibility in a single coleoptile cell and its transfer to an adjacent cell. Physiol. Plant Pathol. 33:81-93.
78. Kurosaki, F., Tsurusawa, Y. and Nishi, A. (1987). Breakdown of phosphatidylinositol during the elicitation of phytoalexin production in cultured carrot cells. Plant Physiol. 85:601-604.
79. Lamb, C. J., Lawton, M. A. and Dron, M. and Dixon, R. A. (1989). Signals and transduction mechanisms for activation of plant defenses against microbial attack. Cell 56:215-224.

80. Leach, J. E. and White, F. F. (1996). Bacterial avirulence genes. Annu. Rev. Phytopathol. 34:153-179.

81. Lois, R., Dietrich, A., Hahlbrock, K. and Schulz, W. (1989). A phenylalanine ammonia-lyase gene from parsley: structure, regulation and identification of elicitor- and light-responsive *cis*-acting elements. EMBO J. 8: 1641-1648.

82. Macdonald, H., Jones, A. M. and King, P. J. (1991). Photoaffinity labeling of soluble auxin-binding proteins. J. Biol. Chem. 266:7393-7399.

83. Mader, M. and Amberg-Ficher, V. (1982). Role of peroxidase in lignification of tobacco cells. Plant Physiol. 70:1128-1131.

84. Mader, M. and Fussl, R. (1982). Role of peroxidase in lignification of tobacco cells. II. Regulation by phenolic compounds. Plant Physiol. 70:1132-1134.

85. Martin, G. B., Brommonschenkel, S. H., Chunwongse, J., Fray, A., Ganal, M. W., Spivey, R., Wu, T., Warle, E. D. and Tanksley, S. D. (1993). Map-based cloning of a protein kinase gene conferring disease resistance in tomato. Science 262:1432-1436.

86. Matsubara, M. and Kuroda, H. (1987). The structure and physiological activity of glycoprotein secreted from conidia of *Mycosphaerella pinodes*. Chem. Pharm. Bull. 35:249-255.

87. Mauch, F. and Staehelin, A. (1989). Functional implications of the subcellular localization of ethylene-induced chitinase and β-1,3 glucanase in bean leaves. Plant Cell 1:447-457.

88. Melin, P. M., Sommarin, M., Sandals, A. S. and Jergil, B. (1987). Identification of Ca^{2+}-stimulated polyphosphoinositide phospholipase C in isolated plant plasma membranes. FEBS Lett. 223:87-91.

89. Memon, A. R. and Boss, W. F. (1990). Rapid light-induced changes in phospho-inositide kinases and H^+-ATPase in plasma membrane of sunflower hypocotyls. J. Biol. Chem. 265:14817-14821.

90. Memon, A. R., Chen, Q. and Boss, W. F. (1989). Inositol phospholipids activate plasma membrane ATPase in plants. Biochem. Biophys. Res. Comm. 162:1295-1301.

91. Mithofer, A., Bhagwat, A. A., Feger, M. and Ebel, J. (1996). Suppression of fungal β-glucan-induced plant defense in soybean (*Glycine max* L.) by cyclic 1,3-1,6-b-glucans from the symbiont *Bradyrhizobium japonicum*. Planta 199:270-275.

92. Morse, M. J., Crain, R. C. and Satter, R. L. (1987). Light-stimulated inositol-phospholipid turnover in *Samanea saman* leaf pulvini. Proc. Natl. Acad. Sci. USA 84:7075-7078.

93. Moseman, J. G. and Greeley, L. W. (1964). Predisposition of wheat by *Erysiphe graminis* f. sp. *tritici* to infection with *Erysiphe graminis hordei*. Phytopathology 54:618.

94. Muller, K. O. and Borger, H. (1940). Experimentelle Untersuchungen uber die Phytophthora-resistenz der Kartoffel. Arb. Biol. Reichsanst. Land Forstwertsch. Berlin 23:189-231.

94a. Nathan,D. G., Baehner,R. L.,Weaver,D. K. (1969). Failure of nitro blue tetrazolium reduction in the phagocytic vacuoles of leukocytes in chronic granulomatous disease. J. Clin. Invest. 48:1895-904

95. Nurnberger, T., Nennstiel, D., Jabs, T., Sacks, W. W.. R., Hahlbrock, K. and Scheel, D. (1994). High affinity binding of a fungal oligopeptide elicitor to parsley plasma membranes triggers multiple defense responses. Cell 78:449-460.

96. Ohl, S., Hedrick, S. A., Choy, J. and Lamb, C. J. (1990). Functional properties of a phenylalanine ammonia-lyase promoter from *Arabidopsis*. Plant Cell 2:837-848.

97. Oku, H. (1980). Determinant for pathogenicity without apparent phytotoxicity in plant disease. Proc. Japan Acad. 56 (Ser. B):367-371.

98. Oku, H. (1994). "Plant Pathogenesis and Disease Control," pp. 193. Lewis Publishers, Boca Raton, FL.

99. Oku. H., Shiraishi, T. and Ouchi, S. (1977). Suppression of induction of phytoalexin, pisatin. Naturwissenschaften 64:643.

100. Oku, H., Shiraishi, T. and Ouchi, S. (1987). Role of specific suppressors in pathogenesis of *Mycosphaerella* species. In: Molecular Determinants of Plant Diseases (S. Nishimura *et al.* eds.), pp.145-156. Japan Sci. Soc. Press/Springer-Verlag, Tokyo.

101. Oku, H., Ouchi, S., Shiraishi, T. and Baba, T. (1975). Pisatin production in powdery mildewed pea seedlings. Phytopathology 65:1263-1267.

102. Oku, H., Ouchi, S., Shiraishi, T., Komoto, Y. and Oki, K. (1975). Phytoalexin activity in barley powdery mildew. Ann. Phytopath. Soc. Japan 41:185-191.

103. Oku, H., Shiraishi, T. and Ouchi, S. (1976). Effect of preliminary administration of pisatin to pea leaf tissues on the subsequent infection by *Erysiphe pisi* DC. Ann. Phytopath. Soc. Japan 42:597-600.

104. Oku, H., Shiraishi, T., Ouchi, S., Ishiura, M. and Matsueda, R. (1980). A new determinant of pathogenicity in plant disease. Naturwissenschaften 67:310.

105. Otani, H., Nishimura, S., Kohmoto, K., Yano, K. and Seno, T. (1975). Nature of specific susceptibility to *Alternaria kikuchiana* in Nijisseiki cultivar among Japanese pears (V). Ann. Phytopath. Soc. Japan 41:467-476.

106. Ouchi, S. (1991). Molecular biology of fungal host-parasite interactions. In: Molecular Strategies of Pathogens and Host Plants (S. S. Patil *et al.* eds.), pp.15-27. Springer-Verlag, New York.

107. Ouchi, S., Hibino, C., Oku, H., Fujiwara, M. and Nakabayashi, H. (1979). The induction of resistance or susceptibility. In: Recognition and Specificity in Plant-Parasite Interactions (J. M. Daly and I. Uritani, eds.), pp. 49-65. Japan Scientific Press, Tokyo.

108. Ouchi, S. and Oku. H. (1981). Susceptibility as a process inducing by pathogens. In: Plant Diseases Control (R. C. Staples and G. H. Toenniessen, eds.), pp. 33-44. John Wiley & Sons, New York.

109. Ouchi, S., Oku, H., Hibino, C. and Akiyama, I. (1974). Induction of accessibility and resistance in leaves of barley by some races of *Erysiphe graminis*. Phytopathol. Z. 79:24-34.

110. Ouchi, S., Oku, H., Hibino, C. and Akiyama, I. (1974). Induction of accessibility to a nonpathogen by preliminary inoculation with a pathogen. Phytopathol. Z. 79:142-154.

111. Pont-Lezica, R. F., McNally, J. G. and Pickard, B. G. (1993). wall-to-membrane linkers in onion epidermis: some hypotheses. Plant, Cell Environ. 16: 111-123.

112. Ralton, J. E., Howlett, B. J. and Clarke, A. E. (1986). Receptor in host-pathogen interactions. In: Hormones, Receptors and Cellular Interactions in Plants (C. M. Chadwick and D. R. Garrod, eds.), pp. 281-318. Cambridge University Press, Cambridge.

113. Rubinstein, B. and Luster, D. G. (1993). Plasma membrane redox activity: components and role in plant processes. Annu. Rev. Plant Physiol. Plant Mol. Biol. 44:131-155.

114. Sanchez, L. M., Ohno, Y., Miura, Y., Kawakita, K. and Doke, N. (1992). Host selective suppression by water-soluble glucans from *Phytophthora* spp. of hypersensitive cell death of suspension-cultured cells from solanaceous plants

caused by hyphal wall elicitors of the fungi. Ann. Phytopath. Soc. Japan 58:664-670.

115. Scheffer, R. and Pringle, R. B. (1976). Pathogen-produced determinants of disease and their effects on host plants. In: The Dynamic Role of Molecular Constituents in Plant Infection (C. J. Mirocha *et al.* eds.) pp. 217-236. American Phytopathological Society, St. Paul, MN.

116. Schmidt, W. E. and Ebel, J. (1987). Specific binding of a fungal glucan phytoalexin elicitor to membrane fractions from soybean *Glycine max*. Proc. Natl. Acad. Sci. USA 84:4117-4121.

117. Seki, H., Ichinose, Y., Kato, H., Shiraishi, T. and Yamada, T. (1996). Analysis of *cis*-regulatory elements involved in the activation of a member of chalcone synthase gene family (*PSCHS1*) in pea. Plant Mol. Biol. 31:479-491.

118. Seki, H., Ichinose, Y., Ito, M., Shiraishi, T. and Yamada, T. (1997). Combined effects of multiple cis-acting elements in elicitor-mediated activation of *PSCHS1* gene. Plant Cell Physiol. 38:96-100.

119. Serra, M. A., Sabbioni, E., Marchesini, A., Pintar, A. and Valott, M. (1990). Vanadate as an inhibitor of plant and mammalian peroxidases. Biol. Trace Ele. Res. 23:151-164.

120. Serrano, R. (1988). Structure and function of proton translocating ATPase in plasma membranes of plants and fungi. Biochim. Biophys. Acta 947:1-28.

121. Serrano, R. (1989). Structure and function of plasma membrane ATPase. Annu. Rev. Plant Physiol. Plant. Mol. Biol. 40:61-94.

122. Shibaoka H. (1993). The use of tobacco BY-2 cells for studies of the plant cytoskeleton. J. Plant Res. Special Issue 3:3-15.

123. Shinkle, J. R., Swoap, S. J. and Jones, R. (1992). Cell wall free space of Cucumis hypocotyls contains NAD and a blue light-regulated peroxidase activity. Plant Physiol. 98:1336-1341.

124. Shiraishi, T., Araki, M., Yoshioka, H., Kobayashi, I., Yamada, T., Ichinose, Y., Kunoh, H. and Oku, H. (1991). Inhibition of ATPase activity in pea plasma membranes in situ by a suppressor from a pea pathogen, *Mycosphaerella pinodes*. Plant Cell Physiol. 32:1067-1075.

125. Shiraishi, T., Hori, N., Yamada, T. and Oku, H. (1990). Suppression of pisatin accumulation by an inhibitor of protein kinase. Ann. Phytopath. Soc. Japan 56:261-264.

126. Shiraishi, T., Oku, H., Tsuji, Y. and Ouchi, S. (1978). Inhibitory effect of pisatin on infection process of *Mycosphaerella pinodes* on pea. Ann. Phytopath. Soc. Japan 44:641-645.

127. Shiraishi, T., Oku, H., Yamashita, M. and Ouchi, S. (1978). Elicitor and suppressor of pisatin induction in spore germination fluid of pea pathogen, *Mycosphaerella pinodes*. Ann. Phytopath. Soc. Japan 44:659-665.

128. Shiraishi, T., Saitoh, K., Kim, H. M., Kato, T., Tahara, M., Oku, H., Yamada, T. and Ichinose, Y. (1992). Two suppressors, Supprescins A and B, secreted by a pea pathogen, *Mycosphaerella pinodes*. Plant Cell Physiol. 33:663-667.

129. Shiraishi, T., Yamada, T., Oku, H., Yoshioka, H. (1991). Suppressor production as a key factor for fungal pathogenesis. In: Molecular Strategies of Pathogens and Host Plants (Patil, S. S. *et al.* eds.), pp. 151-162, Springer-Verlag, New York.

130. Shiraishi, T., Yamada, T., Toyoda, K., Kato, T., Kim, H. M., Ichinose, Y and Oku, H. (1994). Regulation of ATPase and signal transduction for pea defense responses by the suppressor and elicitor from *Mycosphaerella pinodes*. In: Host-Specific Toxin, Biosynthesis, Receptor and Molecular Biology (Kohmoto, K. and Yoder, O. C., eds.), pp. 169-182, Tottori Univ. Press, Tottori.

131. Showalter, A. M. (1993). Structure and function of plant cell wall proteins. Plant Cell 5:9-23.
132. Storti, E., Pelucchini, D., Tegri, S. and Scala, A. (1988). A potential defense mechanism tomato against the late blight disease is suppressed by germinating sporangia-derived substances from *Phytophthora infestans*. J. Phytopathol. 121:275-282.
133. Strasser, H., Hoffmann, C,. Grisebach, H.and Matern, U. (1986). Are polyphosphoinositides involved in signal transduction of elicitor-induced phytoalexin synthesis in cultured plant cells? Z. Naturforsch. 41c:717-724.
134. Szamel, M. and Resch, K. (1981). Inhibition of lymphocyte activation by ouabain. Interference with the early activation of membrane phospholipid metabolism. Biochim. Biophys. Acta 647:297-301.
135. Tani. T., Ouchi, S., Onoe, T. and Naito, N. (1975). Irreversible recognition demonstrated in the hypersensitive response of oat leaves against crown rust fungus. Phytopathology 65:1190-1193.
136. Thanutong, P., Oku, H., Shiraishi, T. and Ouchi, S. (1982). Isolation and partial characterization of an elicitor of pisatin production from spore germination fluid of pea pathogen, *Mycosphaerella pinodes*. Sci. Rep. Fac. Agric. Okayama Univ 59:1-9.
137. Tomiyama, K. (1966). Double infection by an incompatible race of *Phytophthora infestans* of a potato plant cell which has previously been infected by a compatible race. Ann. Phytopath. Soc. Japan 32:181-185.
138. Toyoda, K., Shiraishi, T., Ichinose, Y., Yamada, T. and Oku, H. (1993). Rapid changes in polyphosphoinositide metabolism in pea in response to fungal signals. Plant Cell Physiol. 34:729-735.
139. Toyoda, K., Shiraishi, T., Yoshioka, H., Yamada, T., Ichinose, Y. and Oku, H. (1992). Regulation of polyphosphoinositide metabolism in pea plasma membranes by elicitor and suppressor from a pea pathogen, *Mycosphaerella pinodes*. Plant Cell Physiol 33:445-452.
140. Toyoda, K., Sugimoto, M., Ichinose, Y., Yamada, T. and Shiraishi, T. (1996). Purification and characterization of a binding protein for concanavalin A that elicits phytoalexin production in pea (*Pisum sativum* L.). Plant Cell Physiol. 37:s15.
141. Tsuchiya, K, and Hirata, K. (1973). Growth of various powdery mildew fungi on the barley leaves infected preliminarily with the barley powdery mildew fungus. Ann. Phytopath. Soc. Japan 39:396-403.
142. Varner, J. E. and Lin, L. S. (1989). Plant cell wall architecture. Cell 56:231-239.
143. Viard, M. P., Martin, F., Pugin, A., Ricci, P. and Blein, J. P. (1994). Protein phosphorylation is induced in tobacco cells by the elicitor cryptogein. Plant Physiol. 104:1245-1249.
144. Wada, M., Kato, H., Malik, K., Sriprasertsak, P., Ichinose, Y., Shiraishi, T. and Yamada, T. (1995). A supprescin from a phytopathogenic fungus deactivates transcription of a plant defense gene encoding phenylalanine ammonia-lyase. J. Mol. Biol. 249:513-519.
145. Yamada, T., Hashimoto, H., Shiraishi, T. and Oku, H. (1989). Suppression of pisatin, phenylalanine ammonia-lyase mRNA, and chalcone synthase mRNA by a putative pathogenicity factor from the fungus *Mycosphaerella pinodes*. Mol. Plant-Microbe Interact. 2:256-261.
146. Yamada, T., Hayashi, M., Nakatsuka, S., Muraya, K., Kato, H. and Shiraishi, T. (1994). Suppression of pisatin and phenylalanine ammonia-lyase mRNA in a

compatible interaction between *Pisum sativum* L. cv. Midoriusui and *Pseudomonas syringae* pv. *pisi*. Ann. Phytopath. Soc. Japan 60:66-73.

147. Yamada, T., Tanaka, Y., Sriprasertsak, P., Kato, H., Hashimoto, T., Kawamata, S., Ichinose, Y., Kato, H., Shiraishi, T. and Oku, H. (1992). Phenylalanine ammonia-lyase genes from *Pisum sativum*: Structure, organ-specific expression and regulation by fungal elicitor and suppressor. Plant Cell Physiol. 33:715-725.

148. Yamamoto, M., Nishimura, S., Kohmoto, K. and Otani, H. (1984). Studies on host-specific AF-toxins produced by *Alternaria alternata* strawberry pathotype causing alternaria black spot of strawberry (2) Role of toxins in pathogenesis. Ann. Phytopath. Soc. Japan 50:610-619.

149. Yamamoto, Y., Oku, H., Shiraishi, T., Ouchi, S. and Koshizawa, K. (1986). Non-specific induction of pisatin and local resistance in pea leaves by elicitors from *Mycosphaerella pinodes*, *M. melonis* and *M. ligulicola* and effect of suppressor from *M. pinodes*. J. Phytopath. 117:136-143.

150. Yarwood, C. E. (1959). Predisposition. In: Plant Pathology. Vol. 1, The Diseased Plants (J. G. Horsfall and A. E. Dimond, eds.), pp. 521-562. Academic Press, New York.

151. Yoder, O. C. and Scheffer, R. P. (1969). Role of toxin in early interactions of *Helminthosporium victoriae* with susceptible and resistant oat tissue. Phytopathology 59:1954-1959.

152. Yoshikawa, M., Keen, N. T. and Wang, M. C. (1983). A receptor on soybean membranes for a fungal elicitor of phytoalexin accumulation. Plant Physiol. 73:497-506.

153. Yoshikawa, M., Yamaoka, N. and Takeuchi, Y. (1993). Elicitors, Their significance and primary modes of action in the induction of plant defense reactions. Plant Cell Physiol. 34:1163-1173.

154. Yoshikawa, M., Yamauchi, K. and Masago, H. (1978). Glyceollin, its role in restricting fungal growth in resistant soybean hypocotyls infected with *Phytophthora megasperma* var. *sojae*. Physiol. Plant Pathol. 12:73-82.

155. Yoshioka, H., Hayakawa, Y. and Doke, N. (1993). Transcriptional and post-transcriptional regulation of active defense genes in potato tubers by suppressor glucans from *Phytophthora infestans*. The XV International Botanical Congress, Yokohama, Japan p. 393 (abstr.).

156. Yoshioka, H., Hayakawa, Y. and Doke, N. (1995). Suppression of phenylalanine ammonia-lyase mRNA accumulation by suppressors from *Phytophthora infestans*. Ann. Phytopathol. Soc. Japan 61:7-12.

157. Yoshioka, H., Miyabe, M., Hayakawa, Y. and Doke, N. (1996). Expression of genes for phenylalanine ammonia-lyase and 3-hydroxy-3-methylglutaryl CoA reductase in aged potato tubers infected with *Phytophthora infestans*. Plant Cell Physiol. 37:81-90.

158. Yoshioka, H., Shiraishi, T., Kawamata, S., Nasu, K., Yamada, T., Ichinose, Y. and Oku, H. (1992). Orthovanadate suppresses accumulation of phenylalanine ammonia-lyase mRNA and chalcone synthase mRNA in pea epicotyls induced by elicitor from *Mycosphaerella pinodes*. Plant Cell Physiol. 33:201-204.

159. Yoshioka, H., Shiraishi, T., Nasu, K., Yamada, T., Ichinose, Y. and Oku, H. (1992). Suppression of activation of chitinase and β-1,3-glucanase in pea epicotyls by orthovanadate and suppressor from *Mycosphaerella pinodes*. Ann. Phytopath. Soc. Japan 58:405-410.

160. Yoshioka, H., Shiraishi, T., Yamada, T., Ichinose, Y. and Oku, H. (1990). Suppression of pisatin production and ATPase activity in pea plasma membranes by

orthovanadate, verapamil and a suppressor from *Mycosphaerella pinodes*. Plant Cell Physiol. 31:1139-1146.

161. Zbell, B. and Walter-Back, C. (1988). Signal transduction of auxin on isolated plant cell membranes, Indications for a rapid polyphosphoinositide response stimulated by indoleacetic acid. J. Plant Physiol. 133:353-360.

162. Ziegler, E. and Pontzen, R. (1982). Specific inhibition of glucan-elicited glyceollin accumulation in soybeans by an extracellular mannan-glycoprotein of *Phytophthora megasperma* f. sp. *glycinea*. Physiol. Plant Pathol. 20:321-331.

CHAPTER 6

Identification of Plant Genes Involved in Plant-Microbe Interactions

Peter M. Gresshoff

Biology Leads to Genetics

This chapter focuses on the strategies used to isolate plant genes that control interactive processes such as signaling, response and development. Emphasis is given to results and experiences dealing with the characterization of plant genes involved in the soybean root nodule symbiosis. Positional or map-based cloning strategies are stressed, although this goal has not yet been achieved for a nodulation-related gene.[1,2] However, recent advances in the isolation of disease resistance genes either by insertional[3-5] or map-based procedures [6-9] foreshadow similar successes in the legume symbiosis area.[10-12]

The success of plant breeding in producing productive plant lines resistant to microbial attack demonstrates that plants harbor genes controlling their ability to interact with microorganisms. Plant-microbe interactions involve organisms of different complexities and life styles. Yet plant responses to pathogen attack are more or less similar, involving predominantly the hypersensitive response (HR[13-15]) including the synthesis of defense-related molecules such as phytoalexins and hydrogen peroxide or other oxygen-reactive compounds. Sometimes morphological responses such as galling and encapsulation also occur.

Pathogenic, nonpathogenic or symbiotic plant responses share similarities, despite the involvement of diverse organisms. For example, plants interact symbiotically with microbes such as fungi (as in the case of mycorrhizae to mobilize soil phosphorous), free-living soil bacteria (c.f., *Rhizobium*, *Bradyrhizobium*, to develop nitrogen-fixing nodules capable of converting nitrogen gas to ammonia), pseudomonads (involved in biological control of other invading or pathogenic organisms), and filamentous actinomycetes (e.g., *Frankia*, capable of inducing nodules similarly to *Rhizobium* but with altered morphology and development). Common to several of these plant responses is the fact that the microbe, responding to phenolic signal substances from the plant, causes plant development involving *de novo* cell divisions[16] as well as changes in the phenylpropanoid pathway.[17,18] One concludes that plant responses to microbes, whether against a pathogen or symbiont, appear to follow a basic scheme.[19]

For some pathogenic and symbiotic interactions, the molecular signals between plant and microbe are defined. For example, plants secrete flavonoid substances, such as acetosyringone, to activate *Agrobacterium* virulence genes. Likewise, similar flavone or isoflavone substances induce *Brady(rhizobium)* nodulation genes, which in turn cooperate to produce bacterial nodulation (Nod) factor.[20-23] Figure 6-1 diagrams this interaction for the soybean symbiosis. Nod factors are novel signal molecules belonging to a family of chitin-like oligosaccharides previously shown to be involved in fungal elicitation of wound responses in plants. Nod factors, however, are modified through lipid, sulfate and acetyl decorations which provide specificity and perhaps alter survival of the molecule in the presence of plant-derived chitinase.[22,24,25] Despite the discovery of this essential signal molecule in 1990, the nature of its plant receptor(s?) and the subsequent pathways of signal transfer is still unclear (but compare to advances with phytohormone and light reception pathways; reference 19). However, recent discoveries of a root lectin DB46 from the legume *Dolichos*[26] and a binding factor from cell cultures of *Medicago truncatula*[27] may improve our insight into nod-factor perception. Whether such binding factors are specific and whether they are involved in degradative processes rather than signal perception, requires the further coupling of biochemical and genetic analysis, perhaps using transgenic plants with either overexpression or antisense constructs of either gene.

Plant pathogens also produce signal molecules. Often these are toxins (as for *Cochliobolus carbonum* toxin). The products of several avirulence genes (*avr*) were characterized as polypeptides with variable repeat length units.[28] Again, despite significant advances in the isolation of gene sequences involved in plant disease resistance, there is no evidence that the *avr* gene product receptor has been isolated. However, the nature of several cloned plant disease resistance genes, showing leucine-rich repeats (LRR) and nucleotide binding sites (NBS), suggests that these genes are functionally close to the early reception mechanism of *avr* gene products.[29] For example, the *Xanthomonas oryzae* resistance gene *Xa21* of rice encodes a receptor-type protein kinase, with a leucine-rich region and a transmembrane spanning domain, suggesting that the LRR binds an external signal (possibly the *avr* gene product or a breakdown peptide), transmitting signals across a membrane, where kinase activity phosphorylates another, yet unknown, protein of some signal transduction pathway.[30]

Chemical and developmental plant responses to *Rhizobium* or *Bradyrhizobium* interaction are complex, and it is not certain which response preconditions the next. Some temporal studies have produced a time-line, which means little, as it is conditioned by the ease of detection of a plant response. It will require genetic analysis and the detection of gene interaction in the form of epistasis, transgressive variation, and pleiotropy to evaluate causality and to elucidate the pathways of signal transfer. For

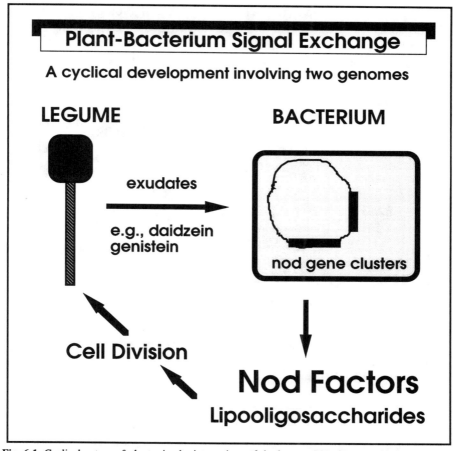

Fig. 6-1. Cyclical nature of plant-microbe interactions of the legume-*Rhizobium* symbiosis. Communication requires the presence of chemical signal molecules, as well as receptors, biosynthetic genes, and response genes, activated by a signal transduction pathway. Although the signal molecules for *Rhizobium* and legumes are broadly understood, the delineation of receptors and components of the signal transduction pathway(s) is the topic of extensive research.

example, it is known that Nod factor application causes both root hair deformation and cortical cell divisions. The former seems to occur first. Yet it is not known whether root hair deformation is needed for cortical cell division. The use of plant mutants that distinguish the two responses will reveal the direction of causality and define the essential nature of processes. At present, being locked into a descriptive mode, we suppose that every observed process is part of the functional cascade leading to symbiosis or, alternatively, plant disease response.

A general paradigm of signal exchange during nodule initiation in legumes without a causal and temporal pattern illustrates the complexity and the need for coupling biochemistry and cell biology with genetics (Figure 6-2). One notes that plant responses occur on many levels. For example, there are changes in host defense enzymes[18,31], initiation of cell division[16], membrane depolarization[24] and calcium efflux.[32]

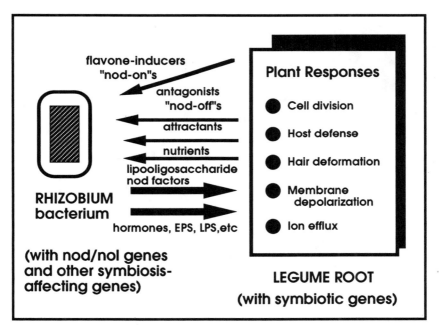

Fig. 6-2. Plant-*Rhizobium* interactions occur in both directions. Plant factors effect chemotaxis, colonization, bacterial attachment and growth, as well as nod gene induction and suppression. Bacterial signals cause both chemical and morphological "defenses," which allow the establishment of a symbiosis in the root nodule. "nod-on" and "nod-off" involves substances which either increase or lower the activity of the nod factor(s). Some of these may be alternate flavones or isoflavones, or simple metabolites such as malate, succinate and riboflavin. Whether these interactions and modulations occur through action at the *nod* gene promoter, or the transport and survival of the nod factor, or the interpretation of the signal by the receptor-signal transduction chain is unknown.

More Mutants Are Needed
for New Gene Identification

Plant genetic analysis is needed for the determination of genetic causality in developmental processes; hence further research should focus on the isolation of additional mutations in the symbiotic or pathogenic processes. Such mutants could be produced by a range of methods, ranging from induced chemical mutagenesis[33] to insertion mutagenesis using either T-DNA[34] or transposable elements, such as AC/Ds of maize.[3,35] Mutants, when characterized through physiology, chemistry and genetics, will provide a causal ranking of observed phenotypes. Some may not be essential for the final symbiotic or pathogenic response. For example, Csanadi and co-workers[36], investigating the early nodulin gene *enod12* in alfalfa nodules used a double null plant, as detected by Southern blotting, to show that its presence is not essential for nodulation, infection and nitrogen fixation. One concluded that ENOD12 functions not on the main pathway of nodule initiation and function and has to be viewed as a response rather than a cause. Similar arguments may be made for other nodulins like Enod2 and even perhaps the most sacrosanct of all nodulins, leghemoglobin.[37] As yet there is no demonstration that these genes and their products have an essential function in nodulation and nitrogen fixation. Clearly the analysis of genetic variants, whether they are of monogenic or polygenic nature, will help reveal causality rather than correlation.

Nodulation and Symbiotic Nitrogen Fixation
as an Example of Plant-Microbe Interactions

. The symbiosis of legumes with soil bacteria belonging to the genus *Bradyrhizobium* or *Rhizobium* is one of the most extensively studied plant-microbe interactions.[24,25,38,39] This symbiosis results in a morphological response by the plant, producing a new organ, the nodule. Underlying such morphological responses are clearly biochemical responses, which are the subject of extensive investigation. Progress in this area is slow, because the biochemical causes of plant morphology in general are poorly understood, comprising mainly descriptive data. In most legumes, nodules develop on roots, although stem and root nodulation exists in some legumes such as *Sesbania rostrata*. Some legumes, such as the Redbud tree in North America or the tropical carob plant fail to nodulate for yet unknown reasons.

Nodules provide a new physiological environment for the inducing bacterium, which itself differentiates to become a nitrogen-fixing endosymbiont (bacteroid). The nitrogen-fixing symbiosis is of substantial agricultural and ecological significance. Nitrogen input into the world's agricultural system relies on the utilization of legumes such as soybean, alfalfa, cowpea, chickpea, pea, beans, and clovers. Additionally, tree legumes such as *Rubinia nigra* or *Acacia* species, contribute to the ecological nitrogen cycle.

Some nonlegume plants also are capable of nitrogen-fixing symbioses. Of largest significance are the *Frankia* (actinomycete)-induced nodules on predominantly woody angiosperms such as *Alnus* or *Casuarina*. These nodules have primitive, branched structures, reminiscent of thickened lateral roots, yet their ability to fix nitrogen is equivalent to that in legumes. The analysis of the actinorhizal symbiosis is severely hampered by experimental problems. The microsymbiont is difficult, if not impossible to culture, most techniques in microbial genetics cannot be applied to the symbiont, and little genetic information exists for the various plant hosts. Moreover, biochemical studies are hampered by the fact that the host and symbiont have long generation times and host tissues are woody and full of phenolics, which interfere with nucleic acid and protein extractions. It is a significant advance that Goetting-Minsky and Mullin[41], Guan et al.[42] and Jacobsen-Lyon et al.[43] isolated nucleic acids from actinorhizal nodules to find nodule-specific gene expression. Of similar impact are the *in situ* hybridization results of Pawlowski and co-workers.[40,44]

Nonlegume nodulation and nitrogen fixation was observed naturally with *Bradyrhizobium* inoculants in the genus *Parasponia*.[45,46] This genus, containing mainly trees and shrubs, grows as a pioneering plant in volcanic areas of South Eastern Asia and the Indonesian archipelago, extending as far as the Fiji Islands. *Parasponia* nodules are induced by *Bradyrhizobium*, which then fixes nitrogen at highly efficient rates, in nodules structurally similar to actinorhizal nodules. Both nodule types are reminiscent of a lateral root developmental program, comprising a persistent meristem and a central vascular bundle.[47]

An additional observation, which underscores the importance of the plant genome during nodulation, are "spontaneous" or NAR (nodulation in the absence of *Rhizobium*) nodules, found in alfalfa (*Medicago sativum*[48,49]) and white clover (*Trifolium repens*[50]). These are detected in a small percentage of plants (about 1–5%). They arise from the identical tissue as normal nodules (the lower cortex close to the endodermis), are subject to autoregulation as well as nitrate inhibition[49] and develop cell types, such as transfer cells, starch grain filled central cells, endodermal and cells associated with the vascular strand, found normally in a *Rhizobium*-induced nodules.[51,52]

Of related interest to the NAR variants are mutants of *Arabidopsis*, which display disease symptoms in the absence of the pathogen[53] or mutants which display systemically acquired resistance (SAR) without a pathogen.[54] Such mutants may be altered in disease response genes. Whether a SAR-like response exists in the control of nodulation is unknown. Related to this point is the observation that supernodulation in many legumes (i.e., the absense of autoregulation of nodulation) is controlled by the shoot of the plant, demonstrating the existence of a systemic signal exchange in the control of nodule number.[55,56]

Several lines of evidence point to the involvement of phytohormones in nodulation.[57-60] This is to be expected as nodules involve plant development and phytohormones, such as auxins, are involved in processes such as rooting, flowering and seed development. Alfalfa seedlings formed nodule-like structures after treatment with auxin-transport inhibitors such as tri-iodo-benzoic acid (TIBA) and naphtyl-phtalamic acid (NPA[60]). More recent studies indicate that the structures may be altered lateral roots, although several nodule-specific genes such as *enod2* and *enod40* are expressed.[61] (NB. ENOD40 is not exclusively nodule-specific as expression was detected in stems[62]). Other studies[63] demonstrated that flavones, commonly increased in plant tissues after microbial interaction, function as natural auxin transport inhibitors, implying that local changes of flavone content may influence local phytohormone concentrations. Inoculation of legume roots leads to alterations of flavone and isoflavone content through activation of phenyl-propanoid enzymes such as PAL (phenylalanine ammonia lyase) and CHS (chalcone synthase[17,18]).

An involvement of phytohormones in nodulation and plant-microbe interactions has been repeatedly demonstrated. Hormone-synthesizing genes, or genes converting an inactive form of a phytohormone to an active one, were found in the T-DNA of either *Agrobacterium tumefaciens* or *Agrobacterium rhizogenes*[64,65] (although recently *rolB* was shown to be a tyrosine phosphatase). Several plant genes involved in early nodulation events, such as *enod2*, a gene for a putative cell wall protein of yet undefined function, and *enod40*, which may function either as a short peptide or a riboregulator[66,67] are regulated by cytokinins, suggesting that alterations in the localized phytohormone levels or sensitivities could lead to mitogenic events. Along this train of thought, NAR alfalfa lines may be altered in a physiological process close to the initial *Rhizobium* signal reception. For example, they may have increased auxin degradation in the root cortex, increased cytokinin levels, or altered flavone levels. They could possess a genetic complement that permits the more rapid transition of early meristematic events, so that root differentiation does not prevent their further development. Observations by E. Kondorosi (Gif-sur-Yvette, France) with NAR plants suggest that NAR alfalfa lines have a higher regeneration potential from cell cultures, supporting the idea that an altered auxin to cytokinin level is present. However, they are clearly not affected in the putative nod-receptor(s), as we were able to isolate NAR phenotypes from a nonnodulating alfalfa line.[68] This suggests that the NN blockage lies causally upstream of NAR, otherwise the nodulation defect would have epistatically blocked nodule development.

Such epistatic interaction was demonstrated for the *nod139* (nonnodulation)[69] and *nts382* (supernodulation) mutations.[70] Nonnodulating double mutants suppressed supernodulation, showing that the *nod139* gene functions in the same developmental cascade prior to the autoregulation control gene (*nts-1*).

Plant Mutations Induced by Mutagenesis

Plant genes control steps of the symbiosis with *Rhizobium* and *Bradyrhizobium*, as evidenced by genetic results based on either chemically and radiation-induced or naturally occurring plant mutants. Both dominant or recessive alleles controlling nodule initiation, nodule number, autoregulation of nodulation, host range, bacterial release, and nitrogen fixation have been described in a large range of legumes. Of great use have been chemically-induced mutations as described for soybean by my group[33,70-72] or pea by LaRue's group[73] and Duc's lab in Dijon.[74] In all these plant species, as well as in *Phaseolus vulgaris*[75], *Melilotus alba* (sweet white clover[76]) and chickpea[77], mutants controlling nodule initiation, nitrogen fixation and nodule number have been isolated (see reference 38 for a review of plant mutants).

A Chemical Mutagenesis Strategy

Insertion mutagenesis is very powerful, as the inserted DNA defines the locus in molecular and genetic terms. However, very few plants possess either efficient transformation systems (needed to introduce a transposable element) or endogenous and mobile transposable elements. Likewise, T-DNA mutagenesis, so successful with *Arabidopsis*, is limited and also requires high transformation frequencies to achieve the required numbers of treated genetic units. Thus, chemical mutagenesis, using predominantly ethyl-methyl-sulphonate (EMS) still is a powerful tool for the generation of genetic conditionality.[38,70,71,75] Below is a brief summary of one mutation strategy used in my laboratory with soybean to induce and select mutants altered in nodulation:

1. Select a soybean cultivar with a stable symbiotic phenotype (or pathogenic response, if mutants in a plant disease response are desired). Presoak 100,000–150,000 seeds in a well-aerated bath for 4–6 h, making sure that the germination fluid is frequently replenished to prevent the buildup of germination-inhibiting exudates from germinating seeds.

2. Expose these seeds to about 0.5% EMS (or similar mutagen) for 6–8 h, ensuring that effective aeration is maintained. (for safety reasons, the exhaust line should be passed through a water trap containing sodium metabisulphite (3%). All manipulations must be done in a well-vented fumehood). Remove the EMS (extreme care needs to be taken here as both the liquid as well as its vapors are toxic and *carcinogenic* to humans) and wash the seeds extensively with fresh and highly aerated buffer.

(N.B.: if irradiation mutagenesis by gamma rays or fast neutrons is used, these "wet" steps are not required. One would suggest exposing dry seeds (about 5–10% seed moisture content) to irradiation levels that restrict germination to about 40 to 60%. Irradiated seeds then would be planted out to allow growth of survivors. Not all survivors will be fertile and pronounced genetic abnormalities may occur along with the desired phenotype, requiring further backcrossing. Accordingly, larger numbers of seeds need to be treated to achieve results equivalent to chemical mutagenesis.

3. Plant the germinating seeds (M1) either in a field, in trays, or in hundreds of individual pots at a high density, as many will fail to germinate (expected loss from mutagenesis and physical damage may be as high as 75%). Scoring for germination frequency as well as phenotypic parameters, such as average plant height after 3 weeks and chlorophyll deficiency sectoring on plant material, may be useful here to evaluate the effectiveness of the mutation treatment at this early M1 stage.

4. Allow M1 plants to mature and harvest M2 seeds from individual plants (as families), or as a bulk harvest. The former is better as it provides additional information on the inheritance and source of putative mutations.

5. Plant M2 seeds either as families or as bulk in a field, trays, agar-filled Petri plates (for small-seeded plants), or pots and select for the desired mutations. When using M2 families, plant about 15 to 20 seeds per family as the genetically effective cell number (GECN) may differ (for soybean it is '2';[78]). The GECN is the number of presumed cells in the resting embryo prior to mutagenesis, which are responsible for the later production of the reproductive tissue of the plant. If the number is '2', and if one cell was mutated in one of its homologs, then the resultant M1 plant will be chimeric for a mutation-carrying, heterozygous cell lineage as well as a wild-type cell lineage, resulting in an M2 segregation of 7 wild type to 1 mutant.

6. Select mutant phenotypes and self-fertilize these to obtain M3 seeds needed for the determination of genetic stability, and further phenotypic and genetic evaluation. Alternatively, putative M2 mutants should be crossed to the wild type to generate backcross material, which will be needed for further genetic analysis and germplasm conversion.

Such mutagenesis under optimized conditions is time-, space-, and labor-intensive. The M2 family screen may require about 3,000–5,000 families to be screened with 15–20 plants per family. This involves 45,000–100,000 plants. An equivalent number of plants would need to be screened in a bulk selection. It requires three growth seasons to get confirmed data. If one uses a small-seeded plant with a fast generation time (such as *Lotus japonicus* or *Medicago truncatula*), mutant selection can be carried out in the growth chamber and considerable time and space can be saved.

Unfortunately, there is no guarantee for success. For example, we used EMS mutagenized soybean seed of cultivar Bragg, which previously had yielded the supernodulation and nonnodulation[70] as well as nitrate reductase deficient mutants[79] to select glyphosate (Round-Up) resistant material. Despite the screening of nearly 150,000 M2 seeds, no economically useful trait was discovered. In other words, EMS mutagenesis is not always the way to a specific phenotype.

The utility of genetics in developmental studies is well understood in prokaryotes (*Escherichia coli, Rhizobium, Agrobacterium*, etc.) and in eukaryotic model systems like mouse, fruit fly, nematode, and yeast. Only recently has the importance of genetic analysis in plant development reached the foreground, mainly due to the establishment of the model plant *Arabidopsis thaliana* (see Koncz et al.[80] for a useful description of the history, genetics, and molecular biology of the plant).

Symbiotic Gene Identification in the Legume-*Rhizobium* Symbiosis

Several mutants altering plant-microbe interactions during nodulation were isolated in soybean and other legumes; highlighted here are the findings for soybean, which are indicative of the spectrum of mutations.

The first mutant class fails to produce nodules. Locus *rj1* was originally discovered as a naturally occurring mutant. In 1986 Carroll et al[33] used EMS mutagenesis to isolate three nonnodulation mutants, nod49, nod772 and nod139. The former two were found to be allelic to *rj₁*.[81] Their phenotype is controlled by the root as was found for nod139.[55,82] Mutant nod139 is a single recessive Mendelian mutation showing 3:1 F2 segregation with parent cultivar Bragg. However, in crosses with *Glycine soja* (the wild progenitor of commercial soybean) it showed 15:1 segregation.[83]

This suggested inheritance of two loci, consistent with the ancestral tetra-ploid nature of soybean[84] and supported by a parallel study on an independent nonnodulation mutant of soybean isolated by Jim Harper's group.[85] The two new gene loci were labeled rj_5 and rj_6. It appears as if nod139 represents an allele at the rj_6 locus and that cultivar Bragg already has a natural deletion or mutation at the rj_5 locus, which is functional in *G. soja*. Other soybean nodulation mutations are called *Rj2*, *Rj3* and *Rj4*, and control nodulation often in a bacterial inoculum-specific manner.[38]

Mutant nod139 is of specific interest as it fails to respond to bacterial inoculation, showing neither root hair curling nor cortical cell induction.[86] Furthermore, although nonnodulating, it exhibits normal production of isoflavone nod gene inducers, root adhesion and colonization, and root lectin levels.[87,88] Mutant nod139, altered in rj_5 and rj_6 appears to be altered close to the initial stages of nodulation signal reception or transduction. However, the mutant is capable of entering a symbiosis with mycorrhizae, which in turn can be stimulated by the co-inoculation with *B. japonicum*.[89] Thus, the mutant must recognize *B. japonicum* somehow, and therefore cannot be affected in the nod factor receptor itself. Alternatively, the recognition of *B. japonicum* could occur through several receptor systems, one leading to systemic responses, such as increased phytoalexin production caused through activation of the phenylpropanoid pathway, another leading to cell division, and yet another to root hair deformation.

A further class of plant mutations found in soybean are those showing the supernodulation phenotype.[71,72] Supernodulation appears to be caused by the absence or lowering of the endogenous autoregulation of nodulation feedback system, in which early nodulation stages induce a systemic cascade involving the shoot of the plant to block further growth of initiated nodule meristems. This limits the number of functional nodules per plant. The mutants are recessive, unlinked to rj_1 and rj_6, and are capable of nodulation in the presence of otherwise inhibitory levels of soil nitrate.

Interestingly, the nodulation phenotype of these mutants, exemplifed by thousands of nodules per plant, is controlled by the genotype of the shoot.[55,56] The simplest model for this control proposes that the root of the plant perceives the onset of cell divisions caused by *B. japonicum* inoculation, resulting in a signal to the plant shoot. In the shoot, the signal is interpreted resulting in the translocation of a shoot-derived-inhibitor (SDI), which blocks further meristematic development of young nodule primordia. Cell division events which occurred first, are proposed to have advanced sufficiently to develop autonomy, so that they are 'resistant' to SDI. This process causes the characteristic nodulation in the upper part of root systems, when inoculated with sufficient numbers of competent rhizobia.

The process is systemic[90] and is mutated in the supernodulation mutants. Supernodulation mutants also develop nodules in the presence of otherwise inhibitory levels of nitrate, that is, they show a nitrate tolerant symbiosis (hence 'nts'). This suggests that the internal mechanism of autoregulation shares at least one step with the inhibition pathway stemming from nitrate exposure. Interestingly, nitrate exposure has a localized, i.e., nonsystemic effect on nodulation. It is possible that nitrate alters the sensitivity or translocation of SDI to the root. Hence the absence of SDI (i.e., *nts* mutation) is pleiotropically resulting in nitrate tolerance.

In pea (*Pisum sativum*) the range of symbiotic mutants is more extensive. There are at least 20 loci conditioning nonnodulation or reduced nodulation.[38,73] Supernodulation mutants were discovered to be either controlled through the root (as in nod3 plants[91]) or the shoot.[74] Several mutants showing a nonfixation phenotype have been isolated, as were mutants lacking both nodulation and mycorrhizal interactions. Similar Myc⁻ and Nod⁻ mutants were found in *Medicago truncatula*[93] suggesting that the pleiotropic phenotype may be common in some legume species. For soybean, Nod⁻ plants (nod49 and nod139) still interact with mycorrhizae, as was found for Nod⁻ mutants of *Lotus japonicus* (M. Parniske, *personal communication*). Perhaps determinate and indeterminate nodulation-types of legumes differ in this aspect.

Detection of Molecular Polymorphisms

Although chromosome walking (the isolation of overlapping genomic clones to cover the unknown stretch from a molecular marker to a gene of interest) was successful in prokaryotes, lower eukaryotes, and humans (e.g., for the identification of the cystic fibrosis gene), it also offers some disadvantages, as repeated DNA regions lead the walk into different, unlinked genome regions. The alternative strategy is called "chromosome landing," in which molecular markers are isolated, closely linked to a gene of interest, that the optimally flanking markers are contained on one single high molecular weight DNA clone (Figure 6-3).

To achieve this, high marker densities are required, which are now achievable because of several technological developments in the last 5 years. These developments stem mainly from improved DNA testing techniques using Polymerase Chain Reaction (PCR)-based methods and their coupling to classical plant breeding methods and genetic stocks. A major genetic tool uses introgressed near-isogenic lines (NILs). In such material, lines differ by only one phenotype, hopefully controlled by a single genomic region. The remainder of the genome of the NILs is assumed to be identical, because of repeated backcrossing to one parent. This means that the region of the investigated gene, together with some neighboring sequences, will be the only assumed difference between two DNA samples taken from such NILs. This process is also called "backcross conversion" or "gene introgression". Larger backcross generation numbers clearly will increase the degree of genomic isogenicity. For example, a backcross conversion of a soybean line for a resistance gene taken to BC6 contains still about 0.5% of the introgressed genome. With a soybean genome size of about 1,070 megabases (Mb), this still represents a genomic contribution from the donor parent of 5 Mb.

Marker-assisted selection represents a marriage of classical plant breeding and modern molecular biology. It allows the tracking of a trait-related region through a breeding program, permitting detection in the heterozygous state. The strategy has many advantages because the molecular phenotype is independent of the environment, and can be obtained from the youngest plant part (Figure 6-4). Moreover, multiple regions can be followed without fear of epistasis or other gene interactions. However, the approach can be cost-ineffective. At present marker technologies are still at best semi-automated, often relying on gel-based separation. Data acquisition is costly as highly trained staff is needed and expensive supplies (like

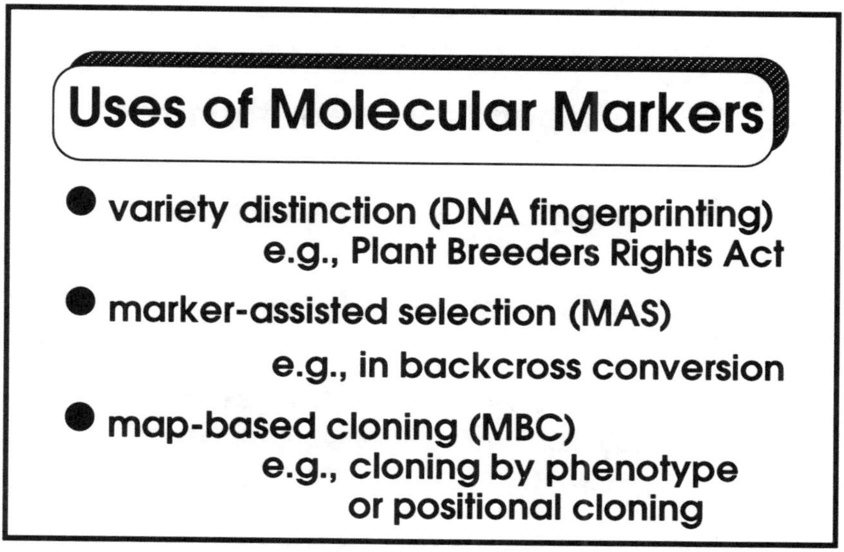

Uses of Molecular Markers

- variety distinction (DNA fingerprinting)
  e.g., Plant Breeders Rights Act
- marker-assisted selection (MAS)
  e.g., in backcross conversion
- map-based cloning (MBC)
  e.g., cloning by phenotype
  or positional cloning

Fig. 6-3. Some uses of molecular markers in plant genetics. Markers may be RFLPs,[103] PCR markers,[104] arbitrary primed markers,[95-98] simple sequence repeats (SSR or microsatellites),[105] AFLPs (also known as selected restriction fragment amplification),[106] or isozymes.

enzymes and kits) as well as instrumentation (like thermocyclers, scanners, etc.) are needed. There also exists a paradox, which is seldom discussed. It is most easy to obtain a linked molecular marker if the phenotype of the desired trait is clear-cut. In contrast, traits which suffer from variable penetrance or gene interaction, are mapped with relatively low accuracy. The paradox is derived from the fact that the former trait in many cases should suffice without the molecular marker. One hopes that further development of DNA analysis technology, including automated DNA extraction, can solve these problems. For example, the use of silicon-wafer affixed oligonucleotide arrays in the form of genosensors may be a future solution for broad scale agricultural application of marker technology.

The second genetic tool uses bulked segregant analysis (BSA) for plant gene discovery.[94] Two plant lines differing in a trait are crossed to produce an F1 and subsequent F2. Optimally the F2 is advanced to an F3 to distinguish dominant and recessive homozygotes. About 7–15 homozygotes of each phenotypic class are pooled in bulks of equal DNA content. Because of segregation and recombination at meiosis of the F1, most genomic

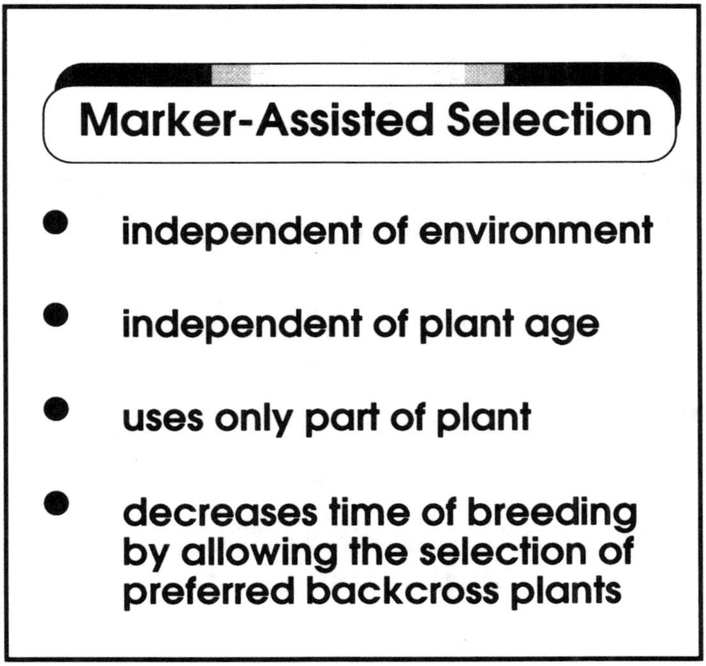

Marker-Assisted Selection

- **independent of environment**

- **independent of plant age**

- **uses only part of plant**

- **decreases time of breeding by allowing the selection of preferred backcross plants**

Fig. 6-4. The advantages of marker-assisted selection (MAS). Traits need to be clearly scoreable to achieve good linkage information. MAS has special application for traits that take long plant life, such as wood quality, wine making ability, sweetness of fruit, etc.

regions are equally present in both pools, except for the region flanking the gene controlling the distinguishing trait. The two DNA bulks are screened for molecular polymorphisms similar to near-isogenic lines. In many ways, bulks create a similar genomic condition as NILs, without the need for long backcross conversion.

The major technical advance stemmed from the introduction of genome screening techniques, which are more efficient than restriction fragment length polymorphisms (RFLP) in detecting molecular polymorphisms. These approaches are based on PCR, but involve significant modifications. An essential factor is the nature of the primer used to target the amplified region.

PCR is based on prior knowledge of the target DNA sequence and uses two different and specific primers of about 15 to 25 nucleotides in length. The anticipated result is a unique amplification product, terminated on each end by the primer sequence. The new DNA scanning techniques work on a different paradigm. Prior sequence information is not needed; hence smaller primer recognition sequences based on arbitrary sequence selection, are used. In arbitrary primer technology, the primer is usually used in singularity.[94,95] This targets and scans multiple regions of the genome concurrently. Three arbitrary primed methods were developed concurrently and independently. Most commonly used is the RAPD (randomly amplified polymorphic DNA) method developed by the DuPont Company.[97] Similar is the AP-PCR (arbitrarily primed-PCR) method.[98] DAF (DNA amplification fingerprinting) was developed with the shortest primers and produces a larger number of amplification products than RAPD and is commonly used together with thin polyacrylamide gels (5 or 10%) which are silver stained to allow high resolution DNA separation.[95,99,100] The similarities and differences between the methods have been reviewed.[101,102] Silver staining kits are available from several companies and provide a useful alternative to the autoradiography used for the detection of DNA bands in DNA sequencing, AP-PCR, as well as differential display, microsatellites, and AFLP (amplification fragment length polymorphism (see later details).

DAF analysis allowed the distinction of plant genotypes for pedigree analysis[107] and phylogenetic comparisons.[108] For example, 10 soybean lines were supplied for a determination of their relatedness. Many of them were breeding lines of known pedigree. It took seven DAF primers to distinguish nine of these. Indeed, with prior knowledge, only three primers were required. The DAF-based dendrogram of relatedness closely matches that from the pedigree and that produced by RFLP analysis. However, the RFLP dendrogram required 53 selected probes to obtain the same resolution. Clearly, arbitrary primer technology is robust and can provide data equivalent to RFLPs. Parallel scanning of the genome with short arbitrary primers (either linear or in mini-hairpin configuration) and high resolution PCR product separation on sequencing gels results in clear molecular markers useful for mapping.[109]

Recently another multiplex technology was developed by KeyGene Inc. in the Netherlands.[106] The technology, called selective restriction fragment amplification (SRFA), or commercialized under the acronym AFLP, generates characteristic DNA profiles from restriction fragments, that were ligated to two distinct adapters (N.B.: the term AFLP was originally coined and published by my laboratory[95] in 1991 to describe polymorphisms generated by DAF; however, because of broad usage of the term in relation to SRFA, and in an attempt to avoid confusion, its connection to DAF is discouraged). PCR primers, specific to the adapter sequences, but extended at their 3 prime end by three nucleotides, amplify a subset of restriction fragments. About 100–120 products are usually separated by PAGE (polyacrylamide gel electrophoresis) and visualized by automated DNA sequencing or autoradiography. Optimally separation works with 70–80 bands, numbers that can also be achieved with DAF using minihairpin primers and 60 cm long PAGE gels (S. Abbitt, *personal communication*). Markers produced by AFLP and the *Eco*RI and *Mse*I restriction kit are predominantly (85–90%) dominant; hybridization of cloned polymorphic AFLP products to genomic blots suggests that most represent repeated DNA. This may lead to clustering of AFLP markers on maps and lowers their potential value as tools for map-based cloning and marker-assisted selection. The potential of using alternative restriction enzymes such as *Pst*I and *Hind*III may altered the nature of SRFA produced products. However, even clustered AFLPs are highly valuable for variety distinction, taking advantage of the increased genetic plasticity and structural tolerance of repeated DNA regions.

Other forms of molecular markers exist. Of these, simple sequence repeats (also called SSRs or microsatellites) are of great value. These usually are di- or trinucleotide repeats. Akkaya et al.[105] and Weising et al.[110,111] demonstrated their application to plant breeding. Because they are based on a PCR strategy, small amounts of starting material are needed. Furthermore, they tend to exist as highly variable loci with as many as 20 different alleles being seen in some soybean SSRs, giving a greater resolution power. Additionally they are codominant. However, they have one major limitation, because SSRs require substantial prior knowledge. This is made easily available for elite species such as soybean, humans, *Arabidopsis*, and maize. However, fringe species will not be accessible in the near future because of the lack of sufficient numbers of known SSR loci, caused by limited financial support and research interest. Additionally, SSRs may be hypervariable, and any bulk tissue may contain multiple forms of the locus. Furthermore it was shown that some SSRs cause phenotypic effects, especially when they are with in coding or regulatory regions of expressed genes. Lack of genetic neutrality would interfere with random transmission. Recently, we mapped the early nodulin gene *enod2b* using both a RFLP as well as an SSR marker to linkage group 1 of the USDA soybean map.[112] The SSR was contained in a putative regulatory region in the 5 prime region of the gene. The use of the SSR allowed the distinction from the duplicate gene, which lacked the SSR.

DAF markers, like RAPD and AP-PCR polymorphisms, are predominantly inherited in a dominant Mendelian fashion.[109,113] In soybean about 15% of all DAF markers were inherited in a maternal way, suggesting cytoplasmic origins.[113] A small fraction of polymorphisms behave in an unexplainable fashion. Some apparently codominant markers are also found. Nonparental bands may arise from unusual PCR events such as jumping PCR or hybrid band formation during annealing. Both these processes are aided by the high content of repeated DNA sequences in eukaryotic DNA. Recently we have improved DAF by increasing the annealing temperature to 55°C from 30°C (F. Ghassemi and P. Gresshoff, *unpublished data*). Also a 10% PAGE is used.[96] This has increased even more the robustness of DAF markers.

DAF markers have been mapped onto the soybean recombinant inbred (RIL) map[113] as well as the map of the model legume *Lotus japonicus*.[109] Polymorphic amplification products have been isolated from silver stained PAGE gels, cloned by T-tailing and blunt end methods, sequenced and converted into sequence characterized amplified regions (SCARs).[104,114] DAF generated molecular markers derived from inserts into yeast artificial chromosomes (YAC), providing extra tags for the potential construction of overlapping YAC clones (called contigs).[115,116]

Amplification methods are susceptible to variation stemming from thermocycler parameters, DNA concentration and quality. However, proper experimental care and elevated (55°C) annealing temperature provide excellent reproducibility. For example, we have detected identical amplification profiles from the same sample DNA over a one year period. Only small variation in staining intensity, but not band presence was detected during a controlled study of soybean DAF profiles.[117] It is presumed that this robustness stems from the optimized DAF conditions pertaining to high primer:low template DNA concentration, the use of Stoffel DNA polymerase (from Perkin-Elmer Inc.), and the application of PAGE and silver staining.[95,101,102,116]

If many markers distinguishing either NILs or BSA DNA pools are detected and shown to be repeatable, the level of recombination in the individuals and the segregating F2 population needs to be obtained. This provides the data for genetic distance as expressed in centimorgans (cM; one centimorgan is defined as being 1% recombination). Molecular markers are arranged relative to the trait locus of interest using computer calculations based on maximum likelihood, such as MAPMAKER.[118] Such calculations provide marker order and marker distances, but these are only statistical, and may be distorted by short distances and the absence of recombination events between certain markers. In that case, it is advantageous to screen larger F2 populations (per-

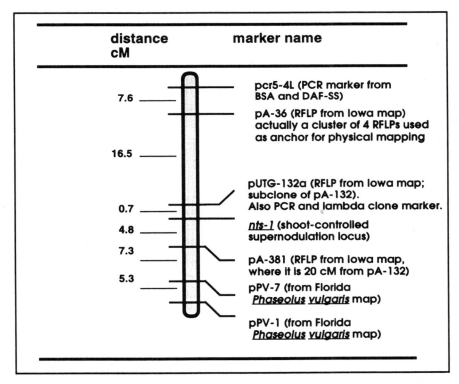

Fig. 6-5. Terminal region of MLG-H of soybean. Markers were generated from genomic *Pst*I clones of soybean (pA-36, pA-381) or common bean (Bng67 [=pPV1] and Bng112 [=pPV7]). Interestingly Bng67 and Bng112 cannot be separated in common bean presumably because of interference in recombination caused by using distantly related parents for the mapping population.[132] Marker pcr5-4L was detected by bulked segregant analysis and DAF screening (104). Marker pUTG-132a is a *Pst*I subclone of the genomic pA-132 marker on the USDA RFLP map. It was converted to a PCR marker, which detects a mappable polymorphism in the same position.[104] Its sequence suggests that the region is AT-rich, unique in the genome, strongly conserved among several soybean cultivars, and not transcribed. *nts-1* is present in both soybean mutant nts382 and nts1007.

haps 1,000–2,000 plants) for recombination in the critical interval, using flanking morphological markers. Here again the repertoire of morphological traits provided by classical plant breeding and genetics is of value, and favors plants with a strong genetical database, such as tomato, barley, wheat, corn, rice and Arabidopsis. If this is not feasible, as it was for the region of the supernodulation gene of soybean on linkage group H[12,104] (Figure 6-5), it is necessary to convert flanking RFLP markers (such as pA-36 and pA-381) to PCR markers, so that large populations can be scored efficiently for recombinants by a PCR screen.

Mapping Symbiotic Genes

Classical genetic maps of legumes are sparsely decorated by morphological and isoenzyme markers. The major genetic activity in most crop legumes focused on agronomic traits, many of which are controlled by interacting genes or quantitative trait loci (QTLs). The advent of molecular markers has enriched maps in many plant species including legumes. However, the number of mapped loci controlling plant-microbe interactions during pathogenesis or symbiosis is relatively small for most legumes. For example, the soybean nonnodulation locus *rj1* was mapped 40 cM from the stem-faciated gene (too large a distance for either plant breeder or molecular biologist!). The restricted nodulation gene *Rj2* was mapped onto Molecular Linkage Group (MLG) J[119] in a cluster of other pathogen resistance genes and possible quantitative trait loci (QTLs) for cyst nematode resistance. The supernodulation *nts-1* gene was mapped to MLG-H of the USDA-ARS map.[12] The early nodulin gene *enod2*, encoding a hydroxyproline-rich cell wall protein was mapped to MLG-A (112). Leghemoglobin RFLPs segregate independently of *nts-1* and *enod2* (F. Ghassemi, *unpublished data*). nod139 (*rj6*) and nod49 (*rj1*), although controlling the same nonnodulation phenotype, segregate independently from each other and from *nts-1*. These findings suggest that genes controlling or involved in the soybean root nodule symbiosis are scattered throughout the genome.

The most extensive mapping in legumes has been achieved in pea, taking advantage of an extensive map and the availability of tester lines. While many *sym* genes and nodulin ESTs (expressed sequence tags) map across the genome (as in soybean), a cluster of several genes is noted on chromosome 1. This includes the *nod3* gene (controlling root-controlled supernodulation and nitrate tolerant symbiosis in pea), leghemoglobin, the early nodulin *enod40*, and several *sym* mutant loci affecting nodulation efficiency. This clustering in pea may be coincidental. As yet no EST has co-mapped with a known nodulation or nitrogen fixation mutation. Likewise no mutated gene with a known phenotype has been explained by a cloned gene sequence. This state of the art is therefore much more primitive than that seen with plant genes involved in plant disease resistance and pathogen interactions[120], and represents a major challenge as well as stumbling block in the pursuit of molecular mechanisms controlling symbioses.

Mapping of symbiosis-related genes helps resolve evolutionary relationships between legume and provides data for chromosome and genome evolution. Additionally, markers close to nodulation genes serve in marker-assisted selection and backcross conversion (Figure 6-4). However, the most challenging application lies in the area of gene discovery and isolation using map-based cloning or positional cloning. As yet no symbiotic gene has been positionally cloned in any legume. However, recent developments of the necessary tools, including recombinant inbred lines[109,121-123], segregating populations[12], YAC and BAC libraries[124-127], the application of multiplex marker technologies[95,97,106,113,128], more complete maps[103,121,129], gene transfer, and use of model legumes[109,130] and application of both chromosome walking and landing approaches as well as insertional mutagenesis[35], increase the chance for success in the near future.

The analysis of the first discovered supernodulation gene *nts-1* in soybean [12] is perhaps the most advanced and is used here to illustrate the present status of map-based cloning in legumes. Multiple independent mutants were isolated, all showing strong nodulation ability. All mutations behaved as recessives, and failed to complement in hybrid plants, suggesting alteration at the same locus (*nts-1*).

In the absence of a detectable altered gene product for these mutants[131] steps were taken to initiate map-based cloning of *nts-1*. Alleles from mutant nts382 and nts1007 were mapped within 0.7 cM of the pUTG-132a marker (part of pA-132) and about 4 cM from pA-381 marker on linkage group H of the USDA map[12,104] (Figure 6-5). This genomic region appears to be diploid condition, which possibly aided the isolation of the recessive mutations. Diploidy was confirmed by the molecular analysis of the pUTG-132a polymorphism[104], which accounted for all polymorphic bands in a Southern by DNA sequence variation, and by *in situ* hybridization using pUTG-132a as a probe (P. Keim, University of Northern Arizona, Flagstaff, *unpublished data*). Using additional RFLP probes derived from *Phaseolus vulgaris*, the map region was extended.[132] At present it is essential to find other markers close to the supernodulation gene; this may be achieved by application of DAF and AFLP approaches using DNA pools derived from RFLP verified homozygotes in the *nts-1* region.

To obtain a physical measure of the genetic distances in linkage group H, an analysis of comigration and cohybridization of closely linked molecular markers on high molecular weight DNA, separated by pulsed field gel electrophoresis[133](PFGE) was done.[11] High molecular weight DNA was isolated from protoplasts embedded in low melting point agarose. *In situ* lysis and digestion with rare cutting endonucleases such as *Sfi*I and *Not*I produced DNA fragments as large as 2 megabases (Mb). These were separated by PFGE and detected by hybridization with linked RFLP markers. The pA-36 RFLP cluster, containing three markers within 1.1 cM, about 16 cM from *nts-1* was chosen. Two of these failed to show recombination in 66 F2 plants in the Iowa mapping population.[103] The probes detected more than one fragment, presumably from their homeologous chromosomes caused by the ancestral tetraploidization. To assign the mapped marker to a specific PFGE fragment, PFGE gel lanes were excised, then restriction-digested *in situ* using the restriction nuclease that originally generated the RFLP in the parents. The digested DNA, still maintained in the PFGE gel lane piece, was electrophoresed in a perpendicular dimension. This revealed which PFGE fragment harbored the mapped RFLP marker. The analysis concluded that in this particular region of the soybean genome, one centimorgan is equivalent to less than 500 kb of DNA. The two markers, pA-9d and pK-89, which were not separated by recombination in the original 66 plants, were on the same 240 kb fragment. These estimations are consistent with the overall genomic calculation; 1,000 markers are spread over about 3,000 cM representing 1,050 megabases of DNA in a haploid genome.[134] This gives an average marker density of 3 cM per marker and about 360 kb per cM. While the estimate of about 500 kb per cM is valid for the pA-36 region, it is expected to differ in other map regions due to recombination interference, heterogeneity, and recombinational hotspots. The future mapping of polymorphic probe pairs (PPPs) derived from endclones of known-sized and map-anchored YAC inserts will test this hypothesis.

The molecular genetic analysis of symbiotic mutants has been slow and hampered by the small number of laboratories working in common with one species. Most focused work was with pea and soybean, but recently activity has increased with model legumes such as *Lotus japonicus*[109] and *Medicago truncatula*.[93] Some mutant loci have been mapped and associated with cloned molecular markers, providing the hope for eventual gene isolation through map-based cloning.[1] The classical crop legumes have been difficult organisms for this purpose. They all possess large haploid genomes (soybean: 1,050 Mb; pea: 4,000 Mb) and often are characterized by tetraploidy, and large repeated DNA content. But the great hindrance towards progress with the large-seeded legumes has been their recalcitrance to genetic transformation. Although gene transfer has been achieved, using either *Agrobacterium* or biolistics, the frequencies are low and the culture time from gene transfer to fertile plant is long. The unavailability of high frequency gene transfer prevents the complementation of recessive mutants by cloned wild-type sequences, as well as the alternative approach for gene discovery through insertional mutagenesis using either T-DNA or transposable elements(*Ac/Ds* launched from T-DNA inserts).[34,35]

Established Gene Isolation Methods

Oligonucleotide Probes

The first genes were isolated from plants by laborious technologies. Proteins that were expressed at a high level were purified, then analyzed for their terminal amino acid sequence. Taking codon usage preferences into account, synthetic oligonucleotides could be synthesized permitting the screening of cDNA libraries for partial or complete clones. Clones were sequenced, and genomic clones were found by homology. Examples of this strategy were seed storage proteins in cereals and leghemoglobin in legumes. This approach is very "straight-forward", but runs into difficulty when the gene product is not available or if the cDNA does not contain the clone. These situations arise when low abundance transcripts, often showing high cell specificity are causative for delicate developmental alterations. It is such targets that we can expect to be altered in mutants affecting plant-microbe interactions.

cDNA Cloning

A more efficient method of gene isolation is the isolation of cDNA clones using differential hybridization to organ- or tissue -specific RNA preparations. An example of the cDNA approach was the initial isolation of nodulin clones. Nodulins that are detectable after inoculation with a virulent *Rhizobium* or *Bradyrhizobium* before the onset of nitrogen fixation (about day 12 to 14 for soybeans) are considered as 'early nodulins' whereas those only expressed afterwards are late nodulins. Nodulins were defined as proteins which are specifically found in developing or mature nodules.[37] This definition may be too limiting, as many nodulins have now been found to be expressed in other organs. For example, nodulin 35 (the enzyme uricase) is also present in young cotyledons. Likewise the early nodulin *enod40*, an "orphan" clone with an uncertain coding potential[61,135], is also expressed in the stem of uninoculated soybean plants. Interestingly, the *enod40* transcript is nearly 1,400 basepairs (bp) in length, contains no long open-reading frame, but shows interspecific conservation for a 36 bp region that codes for a putative oligopeptide of yet unverified existence and function. Likewise the early nodulins ENOD5 and ENOD12, initially considered to be infection-related, were found in flowers of pea plants. It is possible, however, that some of the secondary expression stems from the presence of duplicated members of a multigene family, each member possessing organ-specificity. The nodulin concept, despite the lessened specificity to the nodule organ, is still useful as these proteins are clearly preferentially expressed in the nodule or during early nodule development.

Many cDNA clones have unknown function, but some can be assigned through sequence similarity. The increased resource of bioinformatics taking advantage of completed genomes for yeast and increasing genome saturation for nematode, human, *Arabidopsis*, rice and tomato, as well as several completely sequenced prokaryotic genomes provides a major boost to achieve new insights into putative function of "orphan" DNA clones. Despite this huge data base, it is only suggestive. Function needs to be confirmed by biological, genetical, and biochemical means. An example is nodulin 26 (GmN26), a protein located in the symbiosome membrane surrounding the nitrogen-fixing bacteroid.[20,136] Sequence analysis predicted a protein with several transmembrane domains, consistent with the immunogold localization. Antisense mutagenesis resulted in a nitrogen fixation loss, suggesting a function in transport across the symbiosome membrane. Chimeric gene constructions with the β-glucuronidase (*gus*) gene showed that the promoter is active even in a non-legume plant like tobacco. Biochemical analysis by Dan Roberts' group in Knoxville[137] demonstrates a transport capability of nodulin 26, as the *in vitro* expressed gene product could be inserted into artificial membranes.

Frequently cDNA clones define polymorphic site, thus becoming an expressed sequence tag (EST). ESTs can be detected by the clone itself or the set of specific PCR primers, which detect a polymorphic product. The potential exists for co-mapping an EST to a mutational site, previously defined and mapped only through its phenotype. This requires saturated maps and accurate breeding populations. Co-mapped disease resistance or symbiotic loci provide powerful starting points for further analysis as candidate clones can be transferred, tested for complementation and tissue-specific expression (N.B., again we see the need for gene transfer as part of the verification process during gene discovery and isolation). Most optimally, an EST will show no segregation with a phenotypic marker, making the EST a potential candidate gene associated with this phenotype. At present nearly 60,000 ESTs are available in the human genome, representing perhaps 50% of the entire coding genome. In *Arabidopsis* the number of ESTs is approaching 20,000, more than half the suspected genes of the model plant.

Clone-by-Phone or Synteny

This gene discovery procedure demands that a gene sequence or clone is already available from another organism and that sufficient DNA sequence homology exists between it and the presumed gene in one's organism (or library) of interest. Accordingly, the term "gene discovery" is perhaps incorrect. Clearly, smaller evolutionary gaps increase the chance for success. The gene hunting may use screening by Southern hybridization or PCR. The PCR approach assumes that sufficient sequence similarity exists to design functional primers; alternatively less stringent PCR annealing conditions such as Touch-down PCR, or hybridizations with lower washing temperatures may provide positive signals.

Several examples are available for the clone-by-phone approach. Chalcone synthetase (CHS) was first isolated from parsley and other plants without focus on plant-microbe symbiosis. Estabrook and Sengupta-Gopalan[18] used a CHS and PAL (phenylalanine ammonia lyase) probe to find homologous clones in soybean. Multigene families were discovered for both enzymes, each possessing specific untranslated regions that allowed symbiosis-specific expression studies. These showed that only one CHS clone was symbiotically important.

Most nodulin genes are discovered only once. Thereafter, clone-by-phone approaches pull out the same clone from different legumes. This allows sequence comparisons and evaluation of putatively functional domains, as judged by the distribution of sequence variation and conservation. Regions of assumed common function are called "boxes", such as the NICE box or "destruction box" (N.B., The use of this term stems from the early days of prokaryotic genetics, when specific regions controlling gene expression were drawn with boxes around them to highlight their presence).

The concept of synteny has increased the power of information transfer from one organism to another. While cloning-by-phone relies on sequence conservation, synteny involves the maintenance of marker (gene) order along a chromosomal segment. Again, the closer the two organism, the greater the synteny. For example, soybean and mungbean share many regions for molecular marker order, while soybean and common bean, presumably being more evolutionary distant, share only some regions. We have looked for synteny between common bean and soybean linkage group H, containing supernodulation locus nts-1. In common bean several RFLP markers cluster at the end of a linkage group.[132] These markers detected polymorphisms in soybean lines Minsoy and Noir I, that we used to construct RILs. It was possible to map these markers apart, but in the same map order as in common bean. When placed onto a map for the supernodulation locus (made from *G. soja* and *G. max* nts382), it was possible to show that they were terminal markers, as in common bean. Such data provide evidence for the putative position of the supernodulation locus in common bean, pointing to other RFLPs that may be closer starting points for future gene isolation. Synteny has been a powerful new tool in cereals, where the genomes of rice, corn, wheat and barley are closely related in terms of marker sequence as well as map order.

Map Distortion and Recombinant Inbred Line (RIL) Maps

Many maps are generated from parent lines that show high degree of molecular polymorphism. For example, initial map construction for soybean used a cross of *Glycine max* (the domesticated soybean) with *G. soja* (the ancestral, weedy soybean).[139] Although their karyotypes are similar as judged by pachytene chromosome analysis[84] and their progeny are fully fertile giving Mendelian segregation of markers in the F2, it is possible that regional inversions or deletions distort recombination frequencies. Since then, other *Glycine max* × *G. max* maps were constructed to increase map information and utility. A set of RILs derived from a cross of soybean line Minsoy and Noir 1, has proven useful for mapping.[121,122] If RILs are sufficiently advanced through single seed descent from an original F2 population, the degree of homozygosity should increase. Heterozygotes are terminal as they will "decay" into homozygotes, which in turn are stable. Additional recombination events during further generations only serve to increase the degree of homozygous chromosomal regions. Whereas F2 populations of most sexually maintained plants are terminal, because the DNA or leaf material in the freezer will eventually run out, just as restricted DNA on Nylon membranes can no longer be stripped for further reprobing, RILs have the advantage of being literally immortal. This allows the sharing of seed material, in turn facilitating multi-laboratory mapping on the same population. RILs also open the possibility for the isolation of large DNA pieces, e.g., in the form of yeast artificial chromosomes (YACs) or PFGE separated fragments, necessary for some mapping operations (e.g., telomere mapping). Nearly 300 Minsoy-Noir 1 RILs were characterized for a number of agronomic characteristics (such as oil and protein content, lodging ability, flower time, etc.), which permits the detection of molecular associations to qualitative characteristics. For example, we mapped the early nodulin gene *enod2* to linkage group 3 of the Minsoy-Noir I map[112], close to the region of a soybean cyst nematode race 3 resistance gene (*Rhg4*;[140]), the *"I"* locus (controlling seed coat color presumably through chalcone synthase)[141] and a QTL for seed hardness.[142] The *enod2* region interacts strongly with linkage group W1 explaining genetic variability for pod-filling period (time between flowering and seed maturity). Linkage group 3 of the Minsoy-Noir I map is equivalent to linkage group A of the *G. max* × *G. soja* map, sharing many markers.

RIL-derived maps have another advantage (Figure 6-6). Many polymorphisms detected by arbitrary primer technology (such as RAPD or DAF) are dominant; this means that heterozygotes are not distinguishable, leading to a loss of data points. RILs, by definition, become progressively homozygous for genomic regions. The soybean RILs produced are an F6:5 stage (meaning that single seed descent was carried out for 6 generations, then 5 generations of bulk harvests occurred), so most of the genome is expected to contain homozygous or fixed regions, as was confirmed with *enod2*. Dominant DAF markers can therefore be mapped using every plant. Some chromosomal regions in RILs remained heterozygous at frequencies higher than statistically expected. Mapping of some members of the *cdc2* protein kinase gene family found about 30% heterozygotes.[143] Some regions of the genome may be preferentially maintained in a heterozygous

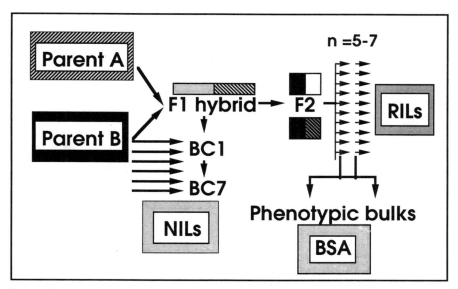

Fig. 6-6. Genetic tools available for gene discovery in plants. A variety of genetic populations are used to map and eventually isolate plant genes involved in plant-microbe interactions. All these areas require classical genetics and plant breeding and provide an excellent contact point between "new" and "old" genetics.

condition suggesting the presence of either lethal genes or heterosis. Both aspects are of interest as they provide new insights into plant growth and reproductive fitness.

Because of the need for maximum polymorphic content, necessitated by RFLP and EST clones, but possibly less needed in view of DAF and AFLP as well as SSR technology, genetic maps in most organisms are based on wide crosses. For example, most mapping in rice is done involving *indica* and *japonica* rice, as is tomato, alfalfa, pea, and common bean. However, large genetic distance may also lead to regional suppression of recombination, resulting in molecular markers unable to be separated by recombination. Likewise some maps, such as in alfalfa, give a small total genome size in terms of recombination distance (around 600 cM compared to 3,000 cM for soybean). This may be caused by incomplete saturation or negative interference caused by large genetic distances between parent and resultant failure to pair chromosomes during meiosis.

Model Legumes for High Transformation and Insertional Mutagenesis

The need for gene verification by functional complementation as well as the desire to identify genes by insertional mutagenesis has increased the emphasis on plants with smaller genomes, excellent genetics, and efficient transformation systems. Spurred by the success with *Arabidopsis thaliana*[80], several legumes have been developed as "model legumes". Most actively researched are *Lotus japonicus* (model for determinate nodulation as in soybean, bean and siratro;[35,109,126,144,145] and *Medicago truncatula* (Australian barrel medic, model for indeterminate nodulation as in pea and alfalfa).[93,130,146,147]

Chemical mutagenesis, as described above, is excellent for producing allelic variation, which can be detected as nonlethal phenotypes. It was used successfully in these model legumes as well. However, the small nature of the presumed DNA alteration makes its detection of such mutation difficult, requiring positional cloning techniques. More optimal may be an insertional mutagenesis approach, similar to the one used successfully for symbiotic genes of microbial partners in various plant-microbe interactions. Insertional mutagenesis in plants occurs naturally in species that harbor transposable elements, such as *Ac/DS/spm* in maize or *Tam* in *Antirrhinum*. If not, transformation can provide such elements [148] or T-DNA derived from *Agrobacterium*.[149,150] The gene introduction approach is not feasible in many plants because of a low incidence of transformation and inefficient plant regeneration.

As yet there is no reported evidence for insertion mutagenesis in legumes leading to symbiotic mutations. For this reason, model legumes have been introduced to help with new gene isolation approaches. *Lotus japonicus* is self-fertile, produces many seeds (2,000 per plant), has numerous and large flowers (10 mm in length), grows fast (10–12 weeks generation time), but most importantly, it has a small genome (about 400 Mb). This decreases the experimental difficulties associated with positional cloning stemming from the distribution of repeated DNA and the length of unique DNA sequences within the genome. Additionally, *L. japonicus* is easily transformed by *Agrobacterium tumefaciens* and *Agrobacterium rhizogenes*, allowing the introduction of novel genes, but also opening up the potential of insertion T-DNA mutagenesis, which has been so successful with *Arabidopsis thaliana*.[80] In our laboratory we optimized transformation and regeneration protocols, permitting the testing of primary transformants of *L. japonicus* 4 to 6 months after gene transfer.[151] Hundreds of fertile, transgenic lines of genotype Gifu have been produced using different gene constructs. Plant regeneration can occur from hypocotyl segments or transgenic roots produced by *A. rhizogenes*. About 70% of plants show co-transfer of the binary plasmid and the T-DNA. The T-DNA of *A. rhizogenes* did not interfere with nodulation, permitting the nodulation analysis in primary transformant roots. It may be possible soon to produce symbiotic mutants of *L. japonicus* through T-DNA insertions, verify the linkage of the phenotype and the T-DNA, and isolate flanking sequences as candidate genes for symbiotic plant functions. As an alternative to T-DNA insertions, foreign transposable elements (the maize Ac/Ds) were introduced into *L. japonicus*.[152] Whether one procedure targets genes better than the other, remains to be determined experimentally.

An efficient genetic system has been developed for two *L. japonicus* genotypes (Gifu and Funakura).[109] Sexual hybridization with a 50% success is possible generating segregating F2 populations. The original F2 was advanced to the F6 as RILs and is available for public usage. Morphological traits and molecular markers generated by DAF segregated in a Mendelian fashion and were mapped. Polymorphisms were detected for nodule-specific and symbiosis-related genes sequences, although the frequency of genetic variation was rather low. This may reflect the genetic distance between Gifu and Funakura, which on the one side provides good karyotype pairing and low recombination interference, but on the other side may increase the effort to detect molecular polymorphisms based on RFLPs. The availability of multiplex procedures like DAF and AFLP may remove this consideration altogether. It is only a matter of effort to produce sufficient markers to cover the small genome, allowing the mapping of newly induced symbiotic mutations. to facilitate the development of map-based strategies, complementing insertional mutagenesis approaches, high molecular weight DNA of *L. japonicus* was cloned into bacterial artificial chromosome (BAC) and yeast artificial chromosome (YAC) vectors.[126]

Similar potential exists for *Medicago truncatula*. This plant also has the advantage of a small genome (600 Mb) and a large data base for its microsymbiont *Rhizobium meliloti*. However, there are some disadvantages which need to be overcome. Seed release is extremely difficult requiring special tools and rollers; flower size is small and requires special emasculation skills. This slows research, when thousands of progeny are required. Transformation and regeneration protocols are still in their early stages and need optimization to facilitate large scale application in strategies such as insertion mutagenesis. The genetic map of *M. truncatula*, however, is growing fast due to the efforts of the French research group headed by Th. Huguet.

Cloning of Mapped Regions into YACs and BACs

A high density of markers and the ordering of markers using recombination events between them, permits the isolation of candidates YAC[153] or BACs.[154] Figure 6-7 cartoons a YAC and a BAC construct carrying plant DNA. Vector pYAC4 is used most frequently taking advantage of its *Eco*RI restriction site inside the suppressor gene. For BACs vector pBeLoBACII and its *Hind*III site is most commonly used. To clone legume DNA, nuclei are isolated from young, preferably etiolated leaf protoplasts, is partially digested in microbeads with appropriate restriction nuclease to produce large DNA fragments. Partial digestion needs to be monitored by PFGE and varies with species as well as time. Once genomic DNA is partially digested to produce a smear of high molecular weight DNA, it is separated by PFGE to make a compression zone, allowing smaller pieces to separate from the desired DNA (above 250 kb). After melting the agarose of the PFGE gel, high molecular weight DNA is obtained ready for ligation into the vector. Prior to transformation into yeast or bacterial spheroplasts a further size selection may be used to separate unligated DNA; however, this step was found to lower the efficiency of transformation.[126,127]

The working paradigm assumes that large pieces of cloned DNA, by virtue of containing flanking markers must also contain the gene of interest between them. This again underscores the need for large segregating populations, that allow the proper ordering of markers around a gene of interest. If only one closely linked marker is available (as in the case of nts-1 in soybean; see above), it is necessary to isolate an extremely large piece of DNA, or consider the construction of a contiguous region (a contig). These large pieces often have the problem of being chimeric, or being the product of rearrangements in the highly recombinogenic yeast. Yeast has the ability to take up more than one YAC clone.[126] Recently, the use of a recombination-deficient yeast (*rad52*, or a double mutant with *rad51*) is hoped to decrease the frequency of chimerism and internal rearrangements. However, these host strains are not very fit to permit efficient regeneration from spheroplasts after electroporation/transformation treatment.

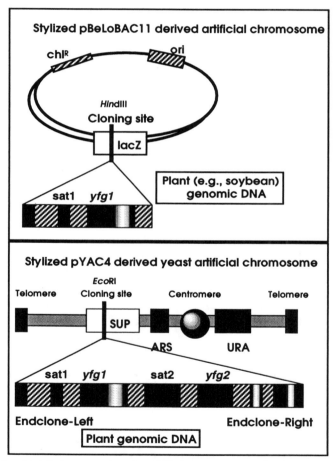

Fig. 6-7. YAC and BAC clones carrying plant DNA insert. *yfg* = your favorite gene; sat = repeated DNA unit; EC = endclone (right or left). Hatched regions may harbor additional molecular markers possibly identified by arbitrary single primer technology (used as DAF or RAPD) or microsatellites. Such markers, especially if they represent unique sequences, allow the detection of an overlapping YAC or BAC clone.

Chimerism of YACs can be determined by two methods. Endclones of the insert can be isolated by ligation of a known vectorette fragment to restriction-digested YAC DNA. Specifically designed PCR primers permit the preferential amplification of the endclones. These can be used as probes in mapping experiments. To map or anchor a YAC, two polymorphic endclones (polymorphic probe pairs, PPP) are needed. For soybean, we found that endclones were detected in the size range of 300 to 700 bp. About half of them hybridized to unique total DNA fragments, allowing the detection of a polymorphism. The other half detected repeated DNA, consistent with the expected composition of the soybean genome based on classical reassociation kinetics. This means that only 25% of all YACs in soybean are flanked by PPPs. If endclones are polymorphic between parents and they map together, no chimerism is inferred. However, internal deletions and inversions still may have occurred and cannot be detected by mapping of endclones. To evaluate internal stability of a YAC clone, a detailed restriction or amplification markers map is necessary for the insert and the corresponding region in the genomic DNA. Mapped PPPs do not only provide information concerning chimerism, but also allow a regional evaluation of the conversion factor for genetic to physical distance.

This genetic approach can be shortened by a visual method. For example, soybean YACs with differing content of repeated DNA (as determined by Southern blotting with total genomic DNA) were isolated by pulsed field gel electrophoresis, eluted from gel slices, then amplified with degenerate short PCR primers, and fluorescently labeled.[127] The amplification products were hybridized onto metaphase chromosomes and interphase nuclei. Bright signals were obtained, exceeding resolutions found with fluorescent *in situ* hybridization (FISH) using cosmid clones. YACs containing predominantly unique DNA produced two foci, while YACs with putatively high repeated DNA content gave multiple foci or even general chromosome painting. Some YACs produced four or six foci, suggesting either YAC chimerism, duplicated genome regions, or multigene families.

This approach is quick and, if yielding only two foci, gives quite accurate analysis of chimerism. However, like endclone mapping it does not detect internal rearrangements. An additional benefit of the chromosome identification technique is that it could be used to place YACs onto soybean chromosomes. All 20 soybean chromosomes have been defined by their length and heterochromatin distribution in pachytene analysis.[84] This cytological approach is laborious and requires considerable skill, maintained by only few researchers. However, Ted Hymowitz (Urbana, Ill., *personal communication*) obtained all 20 trisomics, permitting the use of FISH to different lines. If a YAC is known to harbor a mapped RFLP marker, then the position of the YAC and the marker on a cytologically defined chromosome is possible. Again, one sees an example of technology merging; molecular genetics and cytogenetics overlap to produce new insights, unattainable by either one alone.

To overcome the problems of stability in YACs, alternative cloning vectors have been introduced. The most practical one is based on the F plasmid of *Escherichia coli*. Into this single copy plasmid, high molecular weight DNA of several plants has been cloned.[126,155-157] Insert sizes tend to be smaller (about 100–150 kb per insert) than those found in YACs (150–300 kb per insert), but stability should compensate for this. Moreover, once efficient cloning procedures have been worked out for the DNA of interest (NB., protocols are not directly applicable from species to species), one or two transformations, carried out by electroporation into *E. coli* spheroplasts, suffice to generate a BAC library.

The first isolation of a plant disease resistance gene by positional or map-based cloning was achieved through the use of a YAC clone in tomato.[8] They isolated a *Pseudomonas* (*Pto*) resistance gene from tomato using NILs and a closely linked RFLP marker. This achievement was quickly followed by others, who isolated plant disease resistance genes from several plants such as Arabidopsis, tobacco, rice, flax, and potato (see reference 120 for a recent overview). The essential feature of the newly discovered genes is that most seem to be membrane-associated, possessing the ability to interact with other proteins as well as nucleotides, and that plant disease resistances involving different pathogen types are basically similar. Who would have guessed that a virus resistance gene shares structural similarities with a fungal resistance gene? Further details on these genes and their isolation are given in Volume 1 of this series.[158]

Genetics Leads to Structure

The underlying paradigm of gene isolation is that once the structure, i.e., the primary DNA sequence is known, the derived amino acid sequence will permit the construction of a predicted gene product. This product then can be analyzed to predict function. Primary DNA sequence can also be analyzed for inherent features such as the distribution of untranslated regions, DNA regulatory regions ("boxes") as well as compared to obtain glimpses of evolutionary relationships.

However, determining function is the foremost goal; often this is not achieved. In the electronic data bases one finds a growing number of "orphan" genes, being defined as DNA sequences known by the coding sequence, but with unknown function. Often computer searches find relatedness to other proteins that may not necessarily indicate the correct function. An example of an orphan gene in the plant-microbe interaction area is the *enod2* (early nodulin 2) gene of legumes[159], found in cDNA libraries of young nodule tissue. *In situ* hybridization places its gene product in the developing nodule parenchyma; the derived amino acid sequence shows a repeat heptamer, rich in proline. Accordingly, the protein was labeled a hydroxy-proline rich glycoprotein, similar to those existing in plant cell walls. Its function was postulated to be related to the variable oxygen barrier in legume nodules, a conclusion based entirely on the fact that ENOD2 existed in cell walls, and that the oxygen barrier may function in the nodule parenchyma. No other evidence was available as no mutant lacking ENOD2 has been found. However, recent experiments failed to show any effect of antisense expression of ENOD2.[160]

Similar examples of complete sequence information, but lack of functional information exists for DNA satellites as well as telomeric repeat regions. A 92-bp DNA satellite of soybean was discovered.[161] It appears to exist as a few large blocks of greater than a million basepairs, an observation confirmed by *in situ* hybridization.[162] Despite the knowledge of the sequence of 10 copies of the SB92 satellite we have no idea of its function. Likewise, several telomere-associated regions of soybean were isolated, mapped, and sequenced (A. Kolchinsky, *personal communication*), without knowing there function. In other words, structure is fundamental to understanding, but it does not always lead to function.

At times one wonders whether Albert Einstein's idea, namely that "Not everything that's counted, counts" is relevant here, as could be the thought of determinism and probabilistic events in biology expressed by the Nobel laureate Jacques Monod in his book entitled "Chance and Necessity".[163]

New Genetic Methods

The previous sections listed the genetic and molecular tools used for gene discovery and gene identification. Essentially, there are three avenues to be pursued. First, *insertion mutagenesis*; this is feasible in plants with resident transposable elements or plants with high transformation frequencies. Duplicated genes will not be found, because they most likely lack a phenotype. Second, *map-based cloning*; this has been achieved but only in a limited number of plants for which the necessary "genetic tools" were available. Limiting here is the need for a closely linked molecular marker as well as a YAC or BAC library. Transformation needed for functional complementation also would be a plus. Third, *differential display*; this is a biochemical technique that detects mRNA sequences that are specifically expressed in different tissues or genotypes. The approach may falter because of undefined amplification parameters leading to many false positives as well as the fact that the procedure will favor RNA sequences that are more abundant. Below is a more detailed analysis of these three approaches as they relate to molecular plant-microbe interactions.[26]

Insertional Mutagenesis

The recognition of transposable elements as a mutagenic tool was first used for gene isolation in bacteria. Bacterial transposons defined new genes governing plant-microbe interactions by combining several practical advantages. The insertion was a singular event (as compared to chemical mutagenesis which will alter DNA in several places), generally created a mutant phenotype, such as nonnodulation, combined with the definition of the mutant site through a molecular insertion. Because transposons like Tn5 are not restricted by *EcoRI*, it was possible to isolate clones with flanking DNA from cosmid or phage libraries.[165] Such isolated clones were verified for function by reinsertion into the bacterium to attain the original symbiotic phenotype.

Plants also contain transposable elements. First recognized by Barbara McClintock in *Zea mays*[166,167], they can be used for insertional mutagenesis. Using *Dissociator* (Ds) or *Activator* (Ac) elements several maize genes were first identified, then isolated. The first gene conferring disease resistance, the HM1 gene of maize, was cloned using the *Mutator* insertional element.[168] It encodes a toxin reductase, which inactivates the elicitor toxin produced by the pathogen *Cochliobolus carbonum* race 1. As such it does not fit into the classical gene-for-gene relationship of plant-microbe interaction as proposed by Flor.[169,170]

However, no other plant is known to possess an efficient internal transposable element system such as corn. Although transposable elements were found in *Arabidopsis* (Tag1), and *Antirrhinum* (TamI), these offer a more limited spectrum for insertion mutagenesis. Some other plants show evidence for transposable elements. For example, in soybean there are DNA sequences most likely part of transposable elements; some are found within genes such as one of the seed lectin genes of soybean; some give variegation of flower color.[171] However, there is little genetic evidence for their mobility in soybean. Similar reports exist for other plants such as rice and petunia. For this reason transgenic plants were generated harboring heterologous transposable elements. The most commonly used system uses the maize *Ac/Ds* elements. Transposition of these elements was demonstrated early in transgenic tobacco.[172] A petunia flower color gene which controls the cellular pH, and therefore anthocyanin pigmentation, was tagged and cloned.[173] Recently, plant disease resistance genes were cloned by the same strategy.[4,29,120,158] Most exciting is the fact that the cloned resistance genes contain conserved domains. This permitted the search of resistance gene analogs used degenerate PCR primers leading to the isolation of multiple genes that appear to map close to known resistance genes in soybean.[174]

The *Ac* element transferred into tobacco (*Nicotiana tabacum*) resulted in a mutation conferring resistance to tobacco mosaic virus (TMV)[4] The unstable mutation cosegregated with the element, allowing the isolation of the gene region. The *N* gene of tobacco was isolated and characterized, showing leucine-rich repeats (LRRs) and a nucleotide binding site. Similarity was detected for interleukin receptors. In a parallel approach the L^6 gene of flax, known for its classical involvement in the gene-for-gene relationship between flax and the fungus *Melampsora lini*[170] was cloned. The L^6 gene also codes for two LRR domains as well as a putative P-loop.[5] However, no membrane-spanning domains were recognized.

A large number of tomato transformants carrying the *Ac* element in an expression vector provided the launching sites around the genome to increase the chance for a "jump" to a plant disease resistance gene.[3] This was successfully achieved for the *Cf9* gene conferring resistance to the fungus *Cladosporium fulvum*.[175,176] *Cf9* also contains domains of LRRs and may be a membrane-anchored, extra-cytoplasmic protein. *Cf9* controls resistance to a race-specific elicitor, that has been isolated. The elicitor is a secreted 28 amino acid peptide rich in cysteines. The existence of LRRs is consistent with a binding activity to this elicitor peptide. However, biochemical binding studies with mutant and wild-type membranes failed to show a difference.[175]

Sequence analysis of the *N*, *Rps*, and L^6 genes suggested interesting patterns of similarity, although a viral, bacterial, and fungal disease resistance phenotypes are controlled by them. All have P-loops, known to bind nucleotide phosphates like ATP or GTP[177], and LRRs, known to be involved in protein-protein interactions. The RPS and N proteins have 49% similar sequences, of which half are identical.[7] With these initial successes demonstrating the feasibility of insertion mutagenesis for the isolation of resistance genes in plants, one can be certain that more examples will be achieved in the near future.

Noticeably absent in this initial list of successful gene tagging are genes involved in nitrogen-fixing symbioses or mycorrhizal associations. A major reason may be the inability of *Arabidopsis* to enter these symbioses[178], and the fact that plants capable of abundantly researched symbiotic interactions usually are inefficiently transformable by *Agrobacterium* or biolistics (microprojectile bombardment). This motivated the increased research effort with model legumes that fulfill these requirements. The transposition of the maize *Ac/Ds* elements in *L. japonicus* was already achieved[35], as has been the high efficiency transformation of the same plant.[151]

The T-DNA sequences of *A. tumefaciens* can function as an insertional mutagenic element. First discovered by Kenneth Feldman[179,180] in *Arabidopsis*, the phenomenon was used extensively to discover many new genes. For example, many genes such as *agamous* and *apetela*, affecting flower morphogenesis, were isolated after T-DNA insertion. To confirm the causality of the T-DNA insertion and the mutation event, sufficiently large F2 populations must be screened for cosegregation. In parallel allelism tests of the T-DNA insertion mutant and a confirmed mutant, if available, are useful to confirm gene assignment. This was possible for some mutant, but may not be for others. In general, nearly half of the mutations are not caused by the insertion; at times the insertion is in the proximity of the gene but not within the gene, suggesting that chromatin-based or conformational changes control gene expression.

Once cosegregation is established, a genomic fragment flanking the T-DNA is isolated.[181] Using such flanking DNA as a new probe, larger DNA fragments either as cosmids, YACs, BACs, or cDNA clones homologous to the flanking DNA may be isolated. Putative gene sequences are detected by homology to known cDNA sequences; however, the final confirmation for gene assignment requires the reinsertion of the candidate clone into a recessive mutant plant, thereby correcting the mutant phenotype.

Another elegant form of gene tagging utilizes T-DNA as an insertional mutagen, but is extended as the T-DNA harbors a promoter-less reporter gene.[149-151,182] Such insertions do not only disrupt the gene, but also possibly tag it for expression studies as the promoter-less reporter gene is activated. Both β-glucuronidase and luciferase are good potential reporter genes as could be green fluorescent protein (GFP). The former two were shown to function well in transgenic *L. japonicus* plants.[151] The original insertion may also lead to a mutant phenotype in the next generation. Assuming that the tagged gene exists in a single copy, and that there are no alternative pathways, the insertion will also result in a phenotype. This approach is feasible with plants of high transformation frequencies.

Positional Cloning

The most successful strategy for unknown gene isolation in the plant-microbe interaction area is called positional or map-based cloning.[1,70] The approach requires knowledge of the relevant trait and its inheritance. Then, using pedigree analysis, an association is found between the inheritance of the phenotype and a molecular marker. Usually these markers are RFLP, arbitrary primed amplification markers, or microsatellites. The linked marker should be as close as possible, usually less than 1 cM. Furthermore, flanking molecular markers are advantageous, but not necessary, to bracket the target gene. The molecular tag acts as a means to isolate genomic subclones; these may be YACs, BACs, or cosmids. The latter approach is feasible in organisms with a developed genetic and molecular map (such as yeast, humans, or *Arabidopsis*). A candidate YAC/BAC or contig needs to be subdivided to detect the candidate region of the gene. This is done through the study of rare regional recombinants, which are detected from flanking markers, the transformation and functional complementation of the dominant gene[30], or the screening of cDNA libraries with the candidate YAC/BAC, or contig clones.[6]

Once a candidate cDNA or subclone is found, functional complementation is required to confirm gene assignment. This may be achieved by the production of stable transformants or transient gene expression studies.[7] Figure 6-8 summarizes the general steps used in map-based cloning, although it must be stressed here that various alternative avenues are available and need to be taken for specific genetic conditions.

The map-based cloning approach was demonstrated for several plant genes during 1993. At first, genes involved in plant development and hormone sensitivity were isolated. Researchers at Cornell University[8] were first to isolate using map-based cloning a plant disease resistance gene directly involved in plant-microbe interaction. Their study represented years of painstaking research developing molecular tools in tomato to achieve the cloning. Molecular markers were first associated with the *Pto* (*Pseudomonas* resistance in tomato) gene; YACs were generated and physical mapping was accomplished, providing a measure of physical distance between markers. Because recombination of eukaryotes is not equal in all genomic regions, genetic distances may be misleading (see above).

Pto was mapped with only one RFLP marker as a reference point to find a YAC clone. A cDNA homologous to the candidate YAC was able to correct the mutant phenotype. Furthermore, DNA sequencing of the region in several mutants detected sequence variation in open reading frames consistent with the gene assignment. The DNA sequence of the *pto* gene suggests that it is a protein kinase, consistent with its function in a signal transduction pathway. Of interest is the fact that the gene is neighbored by other, seemingly similar genes, suggesting that multiple functions may be found in this region dealing with specificities in plant-microbe interaction or external signal recognition. One of these genes is *Fen*, coding for resistance against the insecticide Fenthion.[175] This gene has 80% DNA sequence similarity to *Pto* and shows protein kinase activity when expressed in *E. coli*.

Gene interaction possibilities are significant as another gene, *Prf*, which lies close to *Pto*, is required for Fen and Pto activities. *Prf* does not contain regions suggestive of protein kinase activity, but contains regions of LRRs,

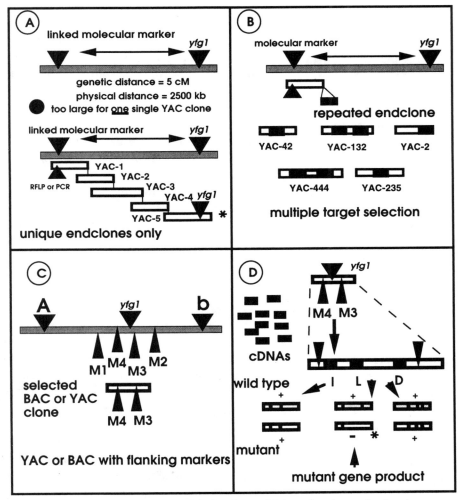

Fig. 6-8. Steps in map-based cloning of the imaginary *yfg1* (your favorite gene) using chromosome walking or landing strategies. Once a candidate YAC or BAC is isolated, several strategies can be pursued. Some mutational alterations cannot be determined through analysis of cDNA clones and may be difficult to isolate. In this case transformation of subclones from the dominant gene region into the recessive mutant is valuable, but requires effective gene transfer systems.

P-loops (nucleotide phosphate binding domains), and several membrane spanning domains. It is feasible that the PRF protein is a membrane receptor, which interacts with the insecticide as well as the *avrPto* elicitor from the bacterium. This reception then 'hands' on the signal to the FEN or PTO proteins.[120]

One further plant disease resistance gene, the *RPS2* gene of *Arabidopsis*, which controls resistance to *Pseudomonas syringae*, was cloned by map-based cloning in 1994 in two separate laboratories.[6,7] Cloning strategies were similar. The study of Bent et al.[6] used a YAC homologous to a flanking marker. This allowed the detection of a candidate cDNA clone, which in turn was sequenced and confirmed to be related to the mutations in several alleles at the *Rps* locus. The Ausubel group[7] took advantage of the fact that the *RPS2* gene maps closely to the *abi* (abscisic acid insensitivity) gene. A contig was made available, taking advantage of the international *Arabidopsis* community, and searched for candidate cDNA clones. Stable transformation of plants was not used to verify their gene assignment; instead transient expression after biolistic bombardment was used. This approach may work in several systems and may provide a time-saving when requiring an otherwise lengthy transformation process. Thus cloning was more rapid than if the task was started from the bare minimum, which can be expected in many other legume situations.

DNA sequencing of the *RPS2* gene revealed extensive motifs similar to other resistance genes. Two major features were the presence of several LRRs as well as a nucleotide phosphate binding domain. Both these features suggest roles in signal recognition and transduction similar to the N, L^6, and *Xa21* disease resistance genes (see above).

The *Xa21* gene was also detected through map-based cloning involving closely linked molecular markers, followed by the isolation of a candidate YAC and BAC, which then was screened for the gene itself using transformation of subclones. It features a transmembrane LLRs, which may function as a receptor to the bacteria-derived ligand.[120] It also contains a cytoplasmic region that appears to have protein kinase activity, suggesting a role in the signal transduction pathway. The efficient transformation technology of rice came to use, as regenerated plants were produced fast and *en masse*, to be easily screened for the phenotype of the dominant resistance gene.

Recently, the first positional cloning in legumes was announced as researchers in France[183] found close linkage of an RAPD marker to the anthracnose (*Colletotrichum lindemuthianum*) resistance gene *Co2* (a.k.a. *ARE*) in *Phaseolus vulgaris*. The marker was sequenced to reveal 23 LLRs and other domains (D2, D3) also found in fully cloned and confirmed resistance genes. Although further verification is needed, this discovery represents a further advance in the field for legumes, and an absolute stroke of luck, finding a polymorphism right inside the gene of interest.

Differential Display or RNA Profiling to Isolate New Genes

The development of arbitrary primer technology in the early 1990s resulted in significant new approaches for gene detection and isolation (see above). Most pronounced are differential display (DD-RT-PCR)[164,184], cDAF,[95] and RAP-PCR.[185] The procedures rely on the fact that mixed populations of mRNA, isolated as polyA$^+$ material, will share many sequences, but may differ in the presence or abundance of some that are differentially expressed depending on the physiological or developmental state of development. Such RNAs are copied into cDNA by reverse transcriptase using a polyT primer extended at its 3' end with either A, G, or C. This provides selectivity of the RNA population. Alternatively, a two-nucleotide extension could be used. For RAP-PCR this first synthesis occurs with a short arbitrary primer, making it more suitable for pro-karyotic analysis. The cDNA population is then amplified using a short arbitrary primer as used by DAF or RAPD. Repeated PCR with the arbitrary primer and the polyT (or arbitrary primer) increases the amount of DNA, so that the mixture can be resolved by PAGE sequencing gel technology using either autoradiography, silver staining, fluorescent detection, or capillary electrophoresis. Although most cDNA bands will be identical, some will differ representing changed mRNA molecules in the original tissue (Figure 6-9). The technique tends to generate several false differences, which necessitates the need for retesting and confirmation of differences by Northern blots, dot blots, and/or RT-PCR. It is feasible that an increased annealing temperature may help the arbitrary primer to anneal more precisely and under less variation induced by thermocycler gradients, DNA pipetting errors, and variable ramping times. for example, an increase from 30°C to 55°C in arbitrarily primed DAF reactions led to an improved DNA profile and greater robustness. Such improvements seem to depend on higher primer (3 μM) concentrations than normally used in PCR or differential display.

The use of arbitrary-primed RNA profiling resulted in the discovery of several known and some novel ESTs from late nodules of *Lotus japonicus*[184] as well as from stem nodules of *Sesbania*[186]. New genes found in mature nitrogen-fixing nodules of *L. japonicus* include protein phosphatase 2C, a peptide transporter protein, and a nodule-specific form of cytochrome P450. A total of 88 unique cDNAs were discovered, of which 22 shared significant homology to DNA or protein sequences in respective data bases.

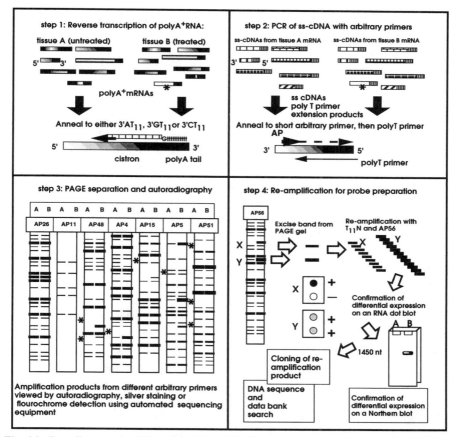

Fig. 6-9. Gene discovery by differential display PCR. These approaches generate new knowledge of genes involved in plant-microbe interaction. Alternative variations exist, optimizing the application to different purposes and experimental preferences.

The application of this elegant technique promises to enrich the spectrum of known genes involved in symbiotic and pathogenic processes.

Functional Complementation in Yeast and Bacteria

It has recently been possible to clone plant genes by functional complementation of yeast or bacterial mutations. By necessity these involve basic and preserved metabolic functions such as those of nitrogen assimilation[187], cell cycle,[188,189] or peptide transport.[190] Genes directly involved in plant development and plant-microbe interaction are not assessable by this procedure, although serendipity may yield an unexpected result when a metabolic function is reflected in a symbiotic or pathogenic phenotype.

Structure Leads to Function Sometimes

The ultimate paradigm underlying the search for genes is that their isolation will lead to the determination of their nucleotide structure. This structure in turn is presumed to suggest a molecular and biochemical function. With a growing data base of known proteins and genes, this determination is progressively more feasible. Isolated genes usually provide access to regulatory sequences such as promoters. These have been fused to reporter genes such as β-glucuronidase to find physiological answers about their expression in transgenic plants. Nucleotide sequences of isolated genes are used to provide primers to RT-PCR giving high resolution detection of expression in various tissues. Alternatively the genes can be used as probes for *in situ* hybridization to tissue section, providing spatial and temporal information about their expression.

However, when there is no corresponding sequence in the data base, the isolated gene, despite a large set of expression data and nucleotide sequence, becomes an "orphan gene", for which no function has yet been determined. Present analysis of the symbiotic interaction between legumes and rhizobia has produced a large number of such orphan genes. Indeed, most early nodulins and several symbiosome membrane proteins are examples. At present the only approach is to disrupt the function of the orphan gene by antisense mutagenesis. For example, an antisense *rab* gene construct was introduced into soybean roots with the *A. rhizogenes* strain K599 to interrupt the expression of that gene during nodulation of soybean. This was a Baconian-type of experiment, based on the philosophy of: "What happens, if we . . .". The symbiosis became nonfunctional, suggesting that *rab* is involved in symbiosome trafficking and function. However, great care needs to be taken when analyzing results based on *A. rhizogenes* transformations as resident bacterial genes controlling phytohormone susceptibility may produce their own phenotype. An example for this effect comes from the study of Bond et al.[192], who found that some soybean roots, transformed with the same *A. rhizogenes* strain K599, developed nodules with persistent meristematic clusters. Such nodules are not normally seen on soybean, which develops determinate, nonmeristematic nodules. The developmental alteration leads to a loss of nitrogen fixation and deterioration of symbiosome. It is possible that positional or copy number effects altered the expression of transferred *rol* genes leading to altered symbiotic properties. In contrast, nodules formed on transgenic *L. japonicus*, using different *Agrobacterium* strains, developed normally, suggesting plant species and bacterial plasmid dependence.[151]

Conclusions

Discovery, isolation and assignment of function for genes involved in and essential for plant-microbe interactions has made significant advances in the last decade.[193] A large number of bacterial and fungal genes has been characterized. The parallel development in plants required the establishment of new techniques such as map-based cloning, chemical and insertional mutagenesis, and differential display. The future is filled with promise, as focus is given to control genes, rather than controlled genes. Clearly, biochemical (or molecular biological) methods preferentially isolated gene products of greater abundance. One anticipates the impact made by insertion mutagenesis in model legume species, that will provide gene sequences useful for gene hunting in other, less amenable legumes. Eventually more orphan genes will be assigned function, as some will be shown to correlate through detailed mapping to existing symbiosis-affecting mutations. More complex maps will permit the recognition of related map orders in different legumes (synteny), pointing the way to new genes in new legume species. Hopefully, one will understand more the influence of the nuclear environment on gene expression and differentiation. The structure and regulation of multigene families will be better understood, as will the interaction with other genetic components (epistasis). Polygenic traits will be better understood and predicted through marker-based selection and breeding. The potential of characterizing RILs for symbiotic parameters and the detection of chromosomal regions that influence any potential variation is most promising as new mechanisms can be detected. In short, there is much to be done, but this can occur only if support is given to this direction of science, so important for agriculture, food security, and the environment.

Acknowledgments

Research from the author's laboratory, as reported here, was supported by the Ivan Racheff Endowment, the Tennessee Agricultural Experiment Station, the Human Frontiers Science Programme, the NSF-Pakistan project, the IAEA (Vienna), the United Soybean Board, the Tennessee Soybean Promotion Board, and a NATO linkage grant. The author thanks staff of the Racheff Chair of Plant Molecular Genetics for providing data.

References

1. Wicking, C. and B. Williamson. 1991. From linked marker to gene. Trends in Genetics 7:288-290.
2. Young, N. D. and R. L. Phillips. 1994. Cloning plant genes known only by phenotype. Plant Cell 6:1193-1195.

3. Jones, D. A., C. M. Thomas, K. E. Hammond-Kosack, P. J. Balint-Kurti, and J. D. G. Jones. 1994. Isolation of the tomato *Cf-9* gene for resistance to *Cladosporium fulvum* by transposon tagging. Science 266:789-793.

4. Whitham, S., S. P. Dinesh-Kumar, D. Choi, R. Hehl, C. Corr, and B. Baker. 1994. The product of the tobacco mosaic virus resistance gene *N*: Similarity to Toll and the interleukin-1 receptor. Cell 78:1101-1115.

5. Lawrence, G. J., E. J. Finnegan, M. A. Ayliffe, and J. G. Ellis. 1995. The *L6* gene for flax rust resistance is related to the *Arabidopsis* bacterial resistance gene *RPS2* and the viral resistance gene *N*. Plant Cell 7:1195-1206.

6. Bent, A. F., B. N. Kunkel, D. Dahlbeck, K. L. Brown, R. Schmidt, J. Giraudat, J. Leung, and B. J. Staskawicz. 1994. *RPS2* of *Arabidopsis thaliana*: A leucine-rich repeat class of plant disease resistance genes. Science 265:1856-1860.

7. Mindrinos, M., F. Katagiri, G.-L. Yu, and F. M. Ausubel. 1994. The *A. thaliana* disease resistance gene *RPS2* encodes a protein containing a nucleotide-binding site and leucine-rich repeats. Cell 78:1089-1099.

8. Martin, G. B., S. Brommonschenkel, J. Chunwogse, A. Frary, M. W. Ganal, R. Spivey, T. Wu, E. D Earle, and S. D. Tanksley. 1993. Map-based cloning of a protein kinase gene conferring disease resistance in tomato. Science 262:1432-1436.

9. Grant, M. R., L. Godiard, E. Straube, T. Ashfield, J. Lewald, A. Sattler, R. W. Innes, and J. L. Dangl. 1995. Structure of the *Arabidopsis RPM1* gene enabling dual specificity disease resistance. Science 269:843-846.

10. Gresshoff, P.M. 1995. Moving closer to the positional cloning of legume nodulation genes. In Nitrogen Fixation: Fundamentals and Applications. eds. I. A. Tikhonovich, V. I. Romanov, N. A. Provorov and W. E. Newton. Kluwer Academic Publishers, Dortrecht, Netherlands, pp. 416-420.

11. Funke, R. P., A. M. Kolchinsky, and P. M. Gresshoff. 1993. Physical mapping of a region in the soybean (*Glycine max*) genome containing duplicated sequences. Plant Mol. Biol. 22:437-446.

12. Landau-Ellis, D, S. Angermüller, R. Shoemaker, and P. M. Gresshoff. 1991. The genetic locus controlling supernodulation in soybean (*Glycine max* L.) cosegregates tightly with a cloned molecular marker. Mol. Gen. Genet. 228:221-226.

13. Lamb, C. J. 1994. Plant disease resistance genes in signal perception and transduction. Cell 76:419-422.

14. Keen, N. T. 1990. Gene-for-gene complementarity in plant-pathogen interactions. Ann. Rev. Genet. 24:447-453.

15. Keen, N. T. 1992. The molecular biology of disease resistance. Plant Molec. Biol. 19:109-122.

16. Calvert, H. E., M. K. Pence, M. Pierce, N. S. A. Malik, and W. D. Bauer. 1984. Anatomical analysis of the development and distribution of *Rhizobium* infections in soybean roots. Can. J. Bot. 62:2375-2383.

17. Recourt, K. A., J. van Tunen, L. A. Mur, A. N. N. van Brussel, B. J. J. Lugtenberg, and J. W. Kijne. 1992. Activation of flavonoid biosynthesis in roots of *Vicia sativa* subsp. *nigra* plants by inoculation with *Rhizobium leguminosarum* biovar. *viciae*. Plant Mol. Biol. 19:411-420.

18. Estabrook, E. M. and C. Sengupta Gopalan. 1991. Differential expression of phenylalanine ammonia lyase (PAL) and chalcone synthase (CHS) during soybean nodule development. Plant Cell 3:299-308.

19. Bowler, C. and N-H. Chua. 1995. Emerging themes of plant signal transduction. Plant Cell 6:1529-1541.

20. Verma, D. P. S. 1992. Signals in root nodule organogenesis and endocytosis of *Rhizobium*. Plant Cell 4:373-382.
21. Fisher, R. F., and S. R. Long. 1992. *Rhizobium*-plant signal exchange. Nature 357:655-660.
22. Lerouge, P., P. Roche, C. Faucher, F. Maillet, G. Truchet, J. C. Promé, and J. Dénarié. 1990. Symbiotic host specificity of *Rhizobium meliloti* is determined by a sulphated and acylated glucosamine oligosaccharide signal. Nature 344: 781-784.
23. Carlson, R. W., N. P. J. Price and G. Stacey. 1994. The biosynthesis of rhizobial lipo-oligosaccharide nodulation signal molecules. Mol. Plant-Microbe Interact. 7:684-695.
24. Long, S. 1996. *Rhizobium* symbiosis: Nod factors in perspectives. Plant Cell 8:1885-1898.
25. Geurts, R. and H. Franssen. 1996. Signal transduction in *Rhizobium*-induced nodule formation. Plant Physiol. 112:447-453.
26. Etzler, M. E. and J. B. Murphy. 1996. Do legume vegetative tissue lectins play roles in plant-microbial interactions? In The Biology of Plant -Microbe Interaction. eds. G. Stacey, B. Mullin and P. M. Gresshoff. pp. 105-110. publ. ISMPMI, St. Paul, MN.
27. Niebel. A., J-J. Bono, R. Ranjeva, and J. V. Cullimore. 1997. Identification of a high affinity binding site for lipooligosaccharidic NodRm factors in the micro-somal fraction of *Medicago* cell suspension cultures. Mol. Plant Microbe Interact. 10:132-134.
28. Keen, N. T. 1996. Bacterial determinants of pathogenicity and avirulence-an overview. In The Biology of Plant -Microbe Interaction. eds. G. Stacey, B. Mullin and P. M. Gresshoff. publ. ISMPMI, St. Paul, MN. pp. 145-152.
29. Jones, J. D. G. 1994. Paranoid plants have their genes examined. Curr. Biol. 4:749-751.
30. Song, W. Y., G. L. Wang, L. Chen, K. S. Kim, T. Holsten, B. Wang, Z. Zhai, L. H. Zhu, C. Fauceut, and P. C. Ronald. 1995. The rice disease resistance gene, *Xa21*, encoded receptor kinase-like protein. Science 270:1804-1806.
31. Sanchez, F., J. E. Padilla, H. Pérez, and M. Lara. 1991. Control of nodulin genes in root-nodule development and metabolism. Ann. Rev. Plant Physiol. Plant Mol. Biol. 42:507-528.
32. Yang, W.-C., C. deBlank, I. Meskiene, H. Hirt, J. Bakker, A. van Kammen, H. Franssen, and T. Bisseling. 1994. *Rhizobium* Nod factors reactivate the cell cycle during infection and nodule primordium formation, but the cycle is only completed in primordium formation. Plant Cell 6:1415-1426.
33. Carroll, B. J., D. L. McNeil and P. M. Gresshoff. 1986. Mutagenesis of soybean (*Glycine max* (L.) Merr.) and the isolation of non-nodulating mutants. Plant Sci. 47, 109-114.
34. Koncz, C., N. Martini, R. Mayerhofer, Z. Koncz-Kalman, H. Körber, G. P. Rédei, and J. Schell. 1989. High frequency T-DNA mediated gene tagging in plants. Proc. Nat. Acad. Sci. USA 86:8467-8471.
35. Thykjær, T., J. Stiller, K. Handberg, J. Jones, and J. Stougaard. 1995. The maize transposable element *Ac* is mobile in the legume *Lotus japonicus*. Plant Mol. Biol. 27:981-993.
36. Csanadi, G., J. Szécsi, P. Kalo, G. Endre, A. Kondorosi, E. Kondorosi, and G. Kiss. 1994. *enod12*, an early nodulin gene is not required for nodule formation and nitrogen fixation in alfalfa. Plant Cell 6:201-213.

37. Legocki, R. and D. P. S. Verma. 1979. A nodule-specific plant protein (Nodulin-36) from soybean. Science 205:190-193.

38. Caetano-Anollés, G. and P. M. Gresshoff. 1991. Plant genetic control of nodulation. Ann. Rev. Microbiology 45:345-382.

39. Hirsch, A. M. 1992. Developmental biology of legume nodulation. New Phytologist 122:211-237.

40. Pawlowski, K,. and T. Bisseling. 1996. Rhizobial and actinorhizal symbioses: What are the shared features? Plant Cell 8:1899-1913,

41. Goetting-Minsky, M. P. and B. C. Mullin. 1994. Differential gene expression in an actinorhizal symbiosis: evidence for a nodule-specific cysteine proteinase. Proc. Nat. Acad. Sci. USA 91:9891-9895.

42. Guan, C., K. Pawlowski, and T. Bisseling. 1995. Nodulation in legumes and actinorhizal plants. In Nitrogen Fixation: Horizons and Application, eds. I. Tikhonovich, N. Provorov, V. Romanov, and W. E. Newton, pp. 49-59. Kluwer Academic Publishers, Dordrecht, Netherlands.

43. Jacobsen-Lyon, K., E. Østergaard Jensen, J.-E. Jørgensen, K. A. Marcker, W. J. Peacock, and E. S. Dennis. 1995. Symbiotic and nonsymbiotic hemoglobin genes of *Casuarina glauca*. Plant Cell 7:213- 223.

44. Pawlowski, K., A. Ribeiro, C. Guan, A. M. Berry, and T. Bisseling. Actinorhizal nodules. In Biological Fixation of Nitrogen for Ecology and Sustainable Agriculture. eds. A. Legocki, H. Bothe, and A. Pühler. pp 267-270. Springer Verlag Berlin Heidelberg.

45. Trinick, M. J. 1979. Structure of nitrogen-fixing nodules formed by *Rhizobium* on roots of *Parasponia andersonii* Planch. Can. J. Microbiol. 25:565-578.

46. Webster, G., M. R. Davey, and E. C. Cocking. 1995. *Parasponia* with rhizobia: a neglected non-legume nitrogen-fixing symbiosis. AgBiotech News and Information 7:119-124.

47. Newcomb, W. 1983. Nodule morphogenesis and differentiation. Int. Rev. Cytol. 13:(suppl) S247-S297.

48. Truchet, G., D. G. Barker, S. Camut, F. deBilly, J. Vasse, and T. Huguet. 1989. Alfalfa nodulation in the absence of *Rhizobium*. Mol. Gen. Genet. 219:65-68.

49. Caetano-Anollés, G., P. A. Joshi, and P. M. Gresshoff. 1991. Spontaneous nodules induce feedback suppression of nodulation in alfalfa. Planta 183:77-82.

50. Blauenfeldt, J., P. A. Joshi, P. M. Gresshoff, and G. Caetano-Anollés. 1994. Nodulation of white clover (*Trifolium repens*) in the absence of *Rhizobium*. Protoplasma 179:106-110.

51. Joshi, P.A., G. Caetano-Anollés, E. T. Graham, and P. M. Gresshoff. 1991. Ontogeny and ultrastructure of spontaneous nodules in alfalfa (*Medicago sativum*). Protoplasma. 162:1-11.

52. Joshi, P. A., G. Caetano-Anollés, E. T. Graham, and P. M. Gresshoff. 1993. Ultrastructure of transfer cells in spontaneous nodules in alfalfa (*Medicago sativum*). Protoplasma 172:64-76.

53. Dietrich, R. A., T. P. Delaney, S. J. Uknes, E. R. Ward, J. A. Ryals, and J. L. Dengl. 1994. *Arabidopsis* mutants simulating disease resistance responses. Cell 77:565-577.

54. Bowling, S. A., A. Guo, H. Cao, S. Gordon, D. F. Klessig, and X. Dong. 1994. A mutation in *Arabidopsis* that leads to constitutive expression of systemic acquired resistance. Plant Cell 6:1845-1857.

55. Delves, A. C., A. Mathews, D. A. Day, A. S. Carter, B. J. Carroll, and P. M. Gresshoff. 1986. Regulation of the soybean-*Rhizobium* symbiosis by shoot and root factors. Plant Physiol. 82:588-590.

56. Gresshoff, P. M., A. J. Krotzky, A. Mathews, D. A. Day, K. A. Schuller, J. Olsson, A. C. Delves, and B. J. Carroll. 1988. Suppression of the symbiotic supernodulation symptoms of soybean. J. Plant Physiol.132:417-423.

57. Thimann, K. V. 1936. On the physiology of the formation of nodule on legume roots. Proc. Natl. Acad. Sci. USA 22:511-514.

58. Gresshoff, P. M. and A. C. Delves. 1986. Plant genetic approaches to symbiotic nodulation and nitrogen fixation in legumes. In A. D. Blonstein and P. J. King (eds.): Plant Gene Research III. A Genetical Approach to Plant Biochemistry, pp. 159-206. Springer Verlag, Wien.

59. Gresshoff, P. M. and G. Caetano-Anollés. 1992. Systemic regulation of nodulation in legumes. In: Gresshoff, P. M. (ed.) Current Topics in Plant Molecular Biology. pp. 87-100, CRC Press, Publ. Boca Raton, Fl.

60. Hirsch, A. M., T. V. Bhuvaneswari, J. E. Torrey, and T. Bisseling. 1989. Early nodulins are induced in alfalfa root outgrowths elicited by auxin transport inhibitors. Proc. Nat. Acad. Sci. USA 86:1244-1248.

61. Hirsch, A. M. and Y. Fang. 1994. Plant hormones and nodulation: what's the connection? Plant Mol. Biol. 26:5-9.

62. Kouchi, H. and S. Hata. 1993. Isolation and characterization of novel nodulin cDNAs representing genes expressed at the early stages of soybean nodule development. Mol. Gen. Genet. 238:106-119.

63. Jacobs, M. and P. H. Rubery. 1988. Naturally-occurring auxin transport regulators. Science 241:346-349.

64. Estruch, J. J., D. Chriqui, K. Grossman, J. Schell, and A. Spena. 1991. The plant oncogene *rolC* is responsible for the release of cytokinins from glucoside conjugates. EMBO J. 10:2889-2895.

65. Estruch, J. J., J. Schell and A. Spena. 1991. The protein encoded by the *rolB* plant oncogene hydrolyses indole glucosides. EMBO J. 10:3125-3128.

66. Asad, S., Y. Fang, K. Wycoff, and A. M. Hirsch. 1994. Isolation and characterization of cDNA and genomic clones of *Ms*ENOD40; transcripts are detected in meristematic cells. Protoplasma 183:10-23.

67. Crespi, M. D., E. Jurkevitch, M. Pioret, Y. d'Aubenton-Carafa, G. Petrovics, E. Kondorosi, and A. Kondorosi. 1994. enod40, a gene expressed during nodule organogenesis, codes for a non-translatable RNA involved in plant growth. EMBO J. 13:5099-5112.

68. Dudley, M. E. and S. R. Long. 1989. A non-nodulating alfalfa mutant displays neither root hair curling nor early cell division in response to *Rhizobium meliloti*. Plant Cell 1:65-72.

69. Mathews, A., B. J. Carroll, and P. M. Gresshoff. 1990. The genetic interaction between non-nodulation and supernodulation in soybean: an example of developmental epistasis. Theor. Appl. Genet. 79:125-130.

70. Gresshoff, P. M. 1993. Molecular genetic analysis of nodulation genes in soybean. Plant Breeding Rev. 11:275-318.

71. Carroll, B. J., D. L. McNeil and P. M. Gresshoff. 1985. Isolation and properties of soybean mutants which nodulate in the presence of high nitrate concentrations. Proc. Nat. Acad. Sci. USA 82:4162-4166.

72. Carroll, B. J., D. L. McNeil and P. M. Gresshoff. 1985. A supernodulation and nitrate tolerant symbiotic (*nts*) soybean mutant. Plant Physiol. 78:34-40.

73. Kneen, B. E. and T. A. LaRue. 1984. Nodulation resistant mutants of *Pisum sativum* (L.). J. Hered. 75:238-240.

74. Kneen, B. E. and T. A. LaRue. 1988. Induced symbiosis mutants of pea (*Pisum sativum*) and sweet clover (*Melilotus alba annua*). Plant Sci. 58:177-182.
75. Park, S. J. and B. R. Buttery. 1988. Nodulation mutants of white bean (*Phaseolus vulgaris* L.) induced by ethyl-methane sulphonate. Can. J. Plant Sci. 68:199-202.
76. Holl, F. B. 1983. Plant genetics: manipulation of the host. Can. J. Microbiol. 29:945-953.
77. Davies, T. M., K. W. Foster, and D. A. Phillips. 1985. Non-nodulation mutants of chickpea. Crop Sci. 25:345-348.
78. Carroll, B.J., P. M. Gresshoff, and A. C. Delves. 1988. Inheritance of super-nodulation in soybean and estimation of the genetically effective cell number. Theor. Appl. Genet. 76:54-58.
79. Carroll, B. J. and P. M. Gresshoff. 1986. Isolation and initial characterisation of soybean mutants with an altered constitutive nitrate reductase mutants NR328 and NR345 of soybean (*Glycine max*). Plant Physiol. 81:572-576.
80. Koncz, C., N.-H. Chua, and J. Schell. 1992. Methods in Arabidopsis Research. World Scientific, publ., Singapore.
81. Mathews, A., B. J. Carroll, and P. M. Gresshoff. 1989. A new recessive gene conditioning non-nodulation in soybean. J. Hered. 80:357-360.
82. Mathews, A., B. J. Carroll, and P. M. Gresshoff. 1992. Studies on the root control of non-nodulation and plant growth of non-nodulating mutants and a supernodulating mutant of soybean (*Glycine max* (L.) Merr.) Plant Science 83:35-43.
83. Gresshoff, P. M. and G. Caetano-Anollés. 1992. Systemic regulation of nodulation in legumes. In Plant Biotechnology and Development. ed. P. M. Gresshoff. pp 87-100. CRC Press, Boca Raton, FL.
84. Singh, R. and T. Hymowitz. 1988. The genomic relationship between *Glycine max* (L.) Merr. and *Glycine soja* (Sieb. and Zucc.) as revealed by pachytene chromosome analysis. Theor. Appl. Genet. 76:705-711.
85. Pracht, J. E., Nickell, D., and J. E. Harper. 1993. Genes controlling nodulation in soybean: Rj_5 and Rj_6. Crop Sci. 33:711-713.
86. Mathews, A., B. J. Carroll, and P. M. Gresshoff. 1989. Development of *Bradyrhizobium* infections in supernodulating and non-nodulating mutants of soybean (*Glycine max* (L.) Merrill). Protoplasma 150:40-47.
87. Mathews, A., R. M. Kosslak, C. Sengupta-Gopalan, E. R. Appelbaum, B. J. Carroll, and P. M. Gresshoff. 1989. Biological characterization of root exudates and extracts from nonnodulating and supernodulating soybean mutants. Mol. Plant-Microbe Interact. 2:283-290.
88. Sutherland, T. D., B. J. Bassam, L.J. Schuller, P. M. Gresshoff. 1990. Early nodulation signals in wild type and symbiotic mutants of soybean (*Glycine max* (L.) Merr. cv. Bragg). Mol. Plant-Microbe Interact. 3:122-128.
89. Xie, Z.-P., C. Staehelin, H. Vier-Heilig, A. Wiemken, S. Jabbouri, W. J . Broughton, R. Vogeli-Lang, and T. Boller. 1995. Nodulation factors differ in their potential to promote mycorrhizal colonization of soybeans. In Nitrogen Fixation: Fundamentals and applications, eds. I. Tikhanovich, N. A. Provorov, V. I. Romanov and W. E. Newton. p. 320. Kluwer Academic Publishers, Dortrecht, Netherlands,
90. Olsson, J. E., P. Nakao, B. B. Bohlool, and P. M. Gresshoff. 1989. Lack of systemic suppression of nodulation in split root systems of supernodulating soybean (*Glycine max* (L.) Merr.) mutants. Plant Physiol. 90:1347-1352.
91. Jacobsen, E. and W. J. Feenstra. 1984. A new pea mutant with efficient nodulation in the presence of nitrate. Plant Sci. Lett. 33:337-344.

92. Duc, G. and A. Messager. 1989. Mutagenesis of pea (*Pisum sativum* L.) and the isolation of mutants for nodulation and nitrogen fixation. Plant Sci. 60:207-213.
93. Sagan, M., D. Morandi, E. Tarengui, and G. Duc. 1995. Selection of nodulation and mycorrhizal mutants of the model legume *Medicago truncatula* (Gaertn.) after gamma ray mutagenesis. Plant Science 111:63-71.
94. Michelmore, R. W., I. Paran, and R. V. Kesseli. 1991. Identification of markers linked to disease resistance genes by bulked segregant analysis: a rapid method to detect markers in specific genomic regions using segregating populations. Proc. Nat. Acad. Sci. USA 88:9828-9832.
95. Caetano-Anollés, G., B. J. Bassam, and P. M. Gresshoff. 1991. DNA amplification fingerprinting using very short arbitrary oligonucleotide primers. BioTechnology 9:553-557.
96. Bassam, B. J. and S. Bentley. 1995. Arbitrary primer technology: basic theory and practice. In 10th Biennial Australasian Plant Pathology Society Conference Workshop Manual, CRC Univ. of Queensland, Brisbane, publ. pp. 16-25.
97. Williams, J. G. K., A. R. Kubelik, K. I. Livak, J. A. Rafalski, and S. V. Tingey. 1990. DNA polymorphisms amplified by arbitrary primers are useful as genetic markers. Nucl. Acids Res. 18:6531-6535.
98. Welsh, J. and M. McClelland. 1990. Fingerprinting genomes using PCR with arbitrary primers. Nucl. Acids Res. 18:7213-7218.
99. Caetano-Anollés, G., B. J. Bassam, and P. M. Gresshoff. 1991. DNA amplification fingerprinting: a strategy for genome analysis. Plant Molecular Biology Reporter 9:292-305.
100. Bassam, B. J., G. Caetano-Anollés, and P. M. Gresshoff. 1991. A fast and sensitive silver-staining for DNA in polyacrylamide gels. Analytical Biochemistry 196:80-83.
101. Gresshoff, P. M. 1994. Plant genome analysis by single arbitrary primer amplification. Probe 4:32-36.
102. Caetano-Anollés, G. and P. M. Gresshoff. 1994. Staining nucleic acids with silver: an alternative to radioisotopic and fluorescent labeling. Promega Notes 45:13-18.
103. Shoemaker, R. C., and T. C. Olson. 1993. Molecular linkage map of soybean. In Genetics Maps; Locus Maps of Complex Genomes, ed. S. J. O'Brien. pp. 6.131-6.138. Cold Springs Harbor Press, New York.
104. Kolchinsky, A.M., D. Landau-Ellis, and P. M. Gresshoff. 1997. Map order and linkage distances of molecular markers close to the supernodulation (*nts-1*) locus of soybean. Mol. Gen. Genet. 254:29-36.
105. Akkaya, M. S., A. A. Bhagwat, and P. B. Cregan. 1992. Length polymorphism of simple sequence repeat DNA in soybean. Genetics 132:1131-1139.
106. Vos, P., R. Hogers, M. Bleeker, M. Reijans, T. van de Lee, M. Hornes, A. Frijters, J. Pot, J. Peleman, M. Kuiper, and M. Zabeau. 1995. AFLP: a new technique for DNA fingerprinting. Nucl. Acid Res. 23:4407-4414.
107. Prabhu, R. R., H. Jessen, D., Webb, S. Luk, S. Smith, and P. M. Gresshoff. 1997. Genetic relatedness among soybean (*Glycine max* L.) lines revealed by DNA Amplification fingerprinting, RFLP and pedigree data. Crop Science 37:1590-1595.
108. Caetano-Anollés, G., L. M. Callahan, and P. M. Gresshoff. 1997. Inferring the origin of bermudagrass (*Cynodon*) off-types by DNA amplification fingerprinting in phyto-forensic applications. Crop Science 37:81-87.
109. Jiang, Q. and P. M. Gresshoff. 1997. Classical and molecular genetics of the model legume *Lotus japonicus*. Mol. Plant-Microbe Interact. 10:59-68.

110. Weising, K., F. Weigand, A. J. Driesel, G. Kahl, H. Zischler, and J. T. Epplen. 1989. Polymorphic simple GATA/GACA repeats in plant genomes. Nucl. Acids Res. 17:10128-10131.
111. Weising, K., J. Kaemmer, F. Weigand, J. T. Epplen, and G. Kahl. 1992. Oligonucleotide fingerprinting reveals various probe-dependent levels of informativeness in chickpea (*Cicer aruentinum*). Genome 35:436-442.
112. Ghassemi, F. and P. M. Gresshoff. 1998. The early *enod2* and the leghemoglobin (*lbc3*) genes segregate independently from other known soybean symbiotic genes. Mol. Plant-Microbe Int. 11:6-13.
113. Prabhu, R. R. and P. M. Gresshoff. 1994. Inheritance of polymorphic markers generated by short single oligonucleotides using DNA amplification fingerprinting in soybean. Plant Mol. Biol. 26:105-116.
114. Paran, I. and R. W. Michelmore. 1993. Development of reliable PCR markers linked to downy mildew resistance genes in lettuce. Theor. Appl. Genet. 85: 985-993
115. Kolchinsky, A., R. P. Funke, and P. M. Gresshoff. 1993. DAF-amplified fragments can be used as markers for DNA from pulse field gels. Biotechniques 14:400-403.
116. Caetano-Anollés, G. and P. M. Gresshoff. 1994. DNA amplification fingerprinting using arbitrary mini-hairpin oligonucleotide primers. BioTechnology 12:619-623.
117. Gresshoff, P. M. and A. MacKenzie. 1994. Low experimental variability of DNA profiles generated by arbitrary primer based amplification (DAF) of soybean. Chin. J. Bot. 6:1-6.
118. Lander, E. S., P. Green, J. Abrahamson, A. Barlow, M. J., Daly, S. E. Lincoln, and L. Newburg. 1987. MAPMAKER: an interactive computer package for constructing primary genetic linkage maps of experimental and natural populations. Genomics 1:174-181.
119. Polzin, K. W., D. G Lohnes, C. D. Nickell, and R. C. Shoemaker. 1994. Integration of *Rps2*, *Rmd*, and *Rj2* into linkage group J of the soybean molecular map. J. Hered. 85:300-303.
120. Martin, G., X. Tang, J. Zhou, R. Frederick, Y. Jia and Y.-T. Loh. 1996. Signal recognition and transduction in bacterial speck disease resistance of tomato. In The Biology of Plant -Microbe Interaction. eds. G. Stacey, B. Mullin and P. M. Gresshoff. pp. 9 -14. publ. ISMPMI, St. Paul, MN.
121. Lark, K. G., J. M. Weisemann, B. F. Mathews, R. Palmer, K. Chase, and T. Macalma. 1993. A genetic map of soybean (*Glycine max* L.) using an interspecific cross of two cultivars 'Minsoy' and 'Noir 1'.Theor. Appl. Genet. 86:901-906.
122. Mansur, L. M., K. G. Lark, H. Kross, and A. Oliverira. 1993. Interval mapping of quantitative trait loci for reproductive, morphological, and seed traits in soybean (*Glycine max* L.). Theor. Appl. Genet. 86:907-913.
123. Concibido, V. C., R. L. Denny, S. R. Boutin, R. Hautea, J. H. Orf, and N. D. Young. 1994. DNA marker analysis of loci underlying resistance to soybean cyst nematode (*Heterodera glycines* Ichinohe). Crop Sci. 34:240-246.
124. Funke, R. P., A. M. Kolchinsky, and P. M. Gresshoff. 1994. High EDTA concentrations cause entrapment of small DNA molecules in the compression zone of pulsed field gels, resulting in smaller than expected insert sizes in YACs prepared from size selected DNA. Nucleic Acids Res. 22:2708-2709.
125. Gresshoff, P. M., R. P. Funke, G. Caetano-Anollés, R. R. Prabhu, S. Pillai, A. M. Kolchinsky, and D. Landau-Ellis. 1995. Application of high molecular weight DNA cloning in legume nodulation gene analysis. In The Use of Induced

Mutations and Molecular Techniques for Crop Improvement, pp. 275-285. Internat. Atomic Energy Agency publ. Vienna, Austria.

126. Pillai, S., R. P. Funke, and P. M. Gresshoff. 1996. Yeast and bacterial artificial chromosomes (YAC and BAC) clones of the model legume *Lotus japonicus*. Symbiosis 21:149-164.

127. Zhu, T., L. Shi, R. P. Funke, P. M. Gresshoff, and P. Keim. 1996. Characterization and application of soybean YACs to molecular cytogenetics. Mol. Gen. Genetics 252:483-488.

128. Rongwen, J., M. S. Akkaya, A. A. Bhagwat, U. Lavi, and P. B. Cregan. 1995. The use of microsatellite DNA markers for soybean genotype identification. Theor. Appl. Genet. 90:43-48.

129. Shoemaker, R. C., and J. E. Specht. 1995. Integration of the soybean molecular and classical genetic linkage maps. Crop Sci. 35:436-446.

130. Penmetsa, R. V. and D. R. Cook. 1997. A legume ethylene-insensitive mutant hyperinfected by its rhizobial symbiont. Science 275:527-530.

131. Sayavedra-Soto, L. A., S. S. Angermüller, R. R. Prabhu, and P. M. Gresshoff. 1995. Polypeptide patterns in leaves of *Glycine max* (L.) Merr. cv. Bragg and its supernodulating mutant during early nodulation. Physiol. Mol. Biol. Plants 1:27-36.

132. Gresshoff, P. M. and A. Filatov. 1998. Synteny mapping of the soybean supernodulation locus *nts*-1 using *Phaseolus vulgaris* probes. (in review).

133. Schwartz, D. C. and C. R. Cantor. 1984. Separation of yeast chromosome-sized DNAs by pulsed field gradient gel electrophoresis. Cell 37:67-75.

134. Gurley, W. P., A. G. Hepburn, and J. L. Key. 1979. Sequence organization of the soybean genome. Biochem. Biophys. Acta 561:167-183.

135. Yang, W-C., C. de Blank, I. Meskiene, H. Hirt, J. Bakker, A. van Kammen, H. Franssen, and T. Bisseling. 1994. *Rhizobium* Nod factors reactivate the cell cycle during infection and nodule primordium formation, but the cycle is only completed in primordium formation. Plant Cell 6:1415-1426.

136. Fortin, M. G., N. A. Morrison, and D. P. S. Verma. Nodulin 26, a peribacteroid membrane nodulin is expressed independently of the development of the peribacteroid compartment. Nucl. Acid Res. 15:813-824.

137. Weaver, D., N. H. Shomer, C. F. Louis, and D. M. Roberts. 1994. Nodulin 26, a nodule-specific symbiosome membrane protein from soybean, is an ion channel. J. Biol. Chem. 269:17858-17862.

138. Moore, G., M. D. Gale, N. Kurata, and R. B. Flavell. 1993. Molecular analysis of small grain legume genomes: status and prospects. Bio/Technol. 11:584-585.

139. Keim, P., B. W. Diers, T. C. Olson, and R.C. Shoemaker. 1990. RFLP mapping in soybean: association between marker loci and variation in quantitative traits. Genetics 126:735-742.

140. Webb, D. M., B. M. Baltazar, A. P. Rao-Arelli, J. Schupp, K. Clayton, P. Keim, and W. D. Beavis. 1995. Genetic mapping of soybean cyst nematode race-3 resistance loci in the soybean PI 437.654. Theor. Appl. Genet. 91:574-581.

141. Vodkin, L. O. 1994. Molecular exploitation of soybean genetic resources. In Plant Genome Analysis. ed. P. M. Gresshoff. pp 83-95. CRC Press, Boca Raton.

142. Keim, P., B. W. Diers, and R. C. Shoemaker. 1990. Genetic analysis of soybean hard seededness with molecular markers. Theor. Appl. Genet. 79:465-469.

143. Taranenko,N, K. Senoo, and P. M. Gresshoff. 1998. Isolation of several *cdc2* cDNA clones from a shoot cDNA library of soybean suggests a complex multigene family. J. Plant Physiol. (in press).

144. Handberg, K. and J. Stougaard. 1992. *Lotus japonicus*, an autogamous, diploid legume species for classical and molecular genetics. Plant J. 2:487-496.
145. Jiang, Q. and P. M. Gresshoff. 1993. *Lotus japonicus*: a model plant for structure-function analysis in nodulation and nitrogen fixation. Current Topics of Plant Mol. Biol. 2:97-110.
146. Huguet, Th., L. Tirichine, M. Ghérardi, M. Sagan, G. Duc, and J.-M. Prospéri. 1997. Molecular genetics of a model plant: *Medicago truncatula*. In Biological Fixation of Nitrogen for Ecology and Sustainable Agriculture. eds. A. Legocki, H. Bothe, and A. Pühler. pp 259-262. Springer Verlag Berlin Heidelberg.
147. Chaubaud, M., C. Larsonnaud, C. Marmouget, and Th. Huguet. 1996. Transformation of barrel medic (*Medicago truncatula* Gaertn.) by *Agrobacterium tumefaciens* and regeneration via somatic embryogenesis of transgenic plants with the MtENOD12 nodulin promoter fused to the GUS gene. Plant Cell Reports 15:305-311.
148. Bhatt, A. M., T. Page, E. J. R. Lawson, C. Lister, and C. Dean. 1996. The use of Ac as an insertional mutagen in *Arabidopsis*. Plant J. 9:935-945.
149. Koncz, C., K. Némenth, G. P. Rédei, and J. Schell. 1992. T-DNA insertional mutagenesis in *Arabidopsis*. Plant Mol. Biol. 20:963-976.
150. Walden, R., H. Haryashi, and J. Schell. 1991. T-DNA as a gene tag. Plant J. 1:281-288.
151. Stiller, J., L. Martirani, S. Tuppale, R.-J. Chian, M. Chiurazzi, and P. M. Gresshoff. 1997. High frequency transformation and regeneration of transgenic plants in the model legume *Lotus japonicus*. J. Exp. Bot. 48:1357-1365.
152. Stougaard, J., D. Abildsten, and K. A. Marcker. 1987. *Agrobacterium rhizogenes* pRi TL-DNA segment as a gene vector system for transformation of plants. Mol. Gen. Genet. 207:251-255.
153. Grill, E. and C. Somerville. 1991. Construction and characterization of a yeast artificial chromosome library of *Arabidopsis* which is suitable for chromosome walking. Mol. Gen. Genet. 226:484-490.
154. Shizuya, H., B. Birren, U.-J. Kim, V. Mancino, T. Slepak, Y. Tachiiri, and M. Simon. 1992. Cloning and stable maintenance of 300-kilobase-pair fragments of human DNA in *Escherichia coli* using an F-factor-based vector. Proc. Nat. Acad. Sci. USA 89:8794-8797.
155. Woo, S. S., V. K. Rastogi, H.-B. Zhang, A. H. Paterson, K. F. Schertz, and R. A. Wing. 1995. Isolation of megabase-size DNA from sorghum and applications for physical mapping and bacterial and yeast artificial chromosome library construction. Plant Molecular Biology Reporter 13:82-94.
156. Gresshoff, P. M. 1995. The interface between RFLP techniques, DNA amplification and plant breeding. In New Diagnostics in Crop Sciences, eds. J. H. Skerritt and R. Appels. pp. 101-125. CAB International, England.
157. Gresshoff, P. M. 1994. Genome analysis of soybean for the positional cloning of nodulation controlling genes. In Proc. First European Nitrogen Fixation Conference. eds. G. Kiss and G. Endre, pp. 174-179. Officiana Press, Szeged, Hungary.
158. Stacey, G. and N. Keen. 1994. Molecular Plant Microbe Interactions. Vol. 1. Chapman and Hall Publ., New York.
159. Franssen, H. J., P.-J. Nap, T. Gloudemans, W. Stiekema, H. V. Dam, F. Govers, J. Louwerse, A. van Kammen, and T. Bisseling. 1987. Characterization of cDNA for nodulin-75 of soybean: A gene product involved in early stages of root nodule development. Proc. Nat. Acad. Sci. USA 84:4495-4490.

160. Chen, R. and F. J. de Bruijn. 1995. Regulation and function of the *Sesbania rostrata* early nodulin gene *enod2*. In Nitrogen Fixation: Horizons and Application, eds. I. Tikhonovich, N. Provorov, V. Romanov, and W. E. Newton, p. 502. Kluwer Academic Publishers, Dortrecht, Netherlands.
161. Kolchinsky A. and P. M. Gresshoff. 1995. A major satellite DNA of soybean is 92 base pairs tandem repeat. Theor. Appl. Genet. 90:621-626.
162. Shi, L., T. Zhu, M. Morgante, J. A. Rafalski, and P. Keim. 1996. Soybean chromosome painting: a strategy for somatic cytogenetics. J. Hered. 87:308-313.
163. Monod, J. 1971. Chance and Necessity:an essay on the natural philosophy of modern biology. Knopf Publ., New York, NY.
164. Liang, P., and A. B. Pardee. 1992. Differential display of eukaryotic messenger RNA by means of the polymerase chain reaction. Science 257:967-971.
165. Scott, K. F., J. Hughes, P. M. Gresshoff, J. Beringer, B. G. Rolfe, and J. Shine. 1982. Molecular cloning of symbiotic genes from *Rhizobium trifolii*. J. Molec. Appl. Genet., 1:315-326.
166. McClintock, B. 1948. Mutable loci in maize. Carnegie Inst. Wash. Yearbook 47:155-169.
167. Federoff, N. 1989. Maize transposable elements. In Mobile DNA, eds. D. E. Berg and M. M. Howe. pp. 375-412 (American Society for Microbiology, Washington DC).
168. Johal, G. S. and S. P. Briggs. 1992. Reductase activity encoded by the *HM1* disease resistance gene in maize. Science 258:985-987.
169. Flor, H. H. 1947. Inheritance reaction to rust in flax. J. Agric. Res. 74:241-262.
170. Flor, H. H. 1956. The complementary genic system in flax and flax rust. Adv. Genet. 8:29-54.
171. Vodkin, L. O. 1996. Plant transposable elements: potential applications for gene tagging in soybean. In: Soybean: Genetics, Molecular Biology and Biotechnology. eds. D. P. S. Verma and R. C. Shoemaker. pp. 69-89. publ. CAB International, Wallingford, UK.
172. Baker, B., J. Schell, H. Lörz, and N. Federoff. 1986. Transposition of the maize controlling element "Activator" in tobacco. Proc. Nat. Acad. Sci. USA 83:4844-4848.
173. Chuck, G., T. Robbins, C. Nijjar, E. Ralston, N. Courney-Gutterson, and H. K. Dooner. 1993. Tagging and cloning of petunia flower color gene with the maize transposable element activator. Plant Cell 5:371-378.
174. Kanazin, V., L. Frederick Marek and R. C. Shoemaker. 1996. Resistance gene analogs are conserved and clustered in soybean. Proc. Nat. Acad. Sci. USA. 93:11746-11750.
175. Chasan, R. 1994. Plant-pathogen encounters in Edinburgh. Plant Cell 6:1332-1341.
176. Hammond-Kosack, K.E., D. A. Jones, and J. D. G. Jones. 1994. Identification of two genes required in tomato for full Cf-9 dependent resistance to *Cladosporium fulvum*. Plant Cell 6:361-374.
177. Saraste, M., P. R. Sibbald, and A. Wittinghofer. 1990. The P-loop: a common motif in ATP- and GTP-binding proteins. Trends Biochem. Sci. 15: 430-434.
178. Kolchinsky, A. M., R. P. Funke, and P. M. Gresshoff. 1994. Dissecting molecular mechanisms of nodulation: taking a leaf from *Arabidopsis*. Plant Molec. Biol. 26:549-552.
179. Feldmann, A. K. 1991. T-DNA insertion mutagenesis in *Arabidopsis*: mutational spectrum. Plant J. 1:71-82.
180. Meyerowitz, E. M. 1989. *Arabidopsis*, a useful weed. Cell 56:263-269.

181. Gresshoff, P.M. 1996. Molecular genetic analysis of soybean nodulation mutants. In Soybean: Genetics, Molecular Biology and Biotechnology. eds. D. P. S. Verma and R. C. Shoemaker. pp 189-217. CAB International; Wallingford, UK.

182. Walden, R. and J. Schell. 1994. Activation T-DNA tagging-a silver bullet approach to isolating plant genes. Agro-Food Ind. Hi-Tech 5:9-12.

183. Geffroy, V., F. Creusot, M. Sevignac, C. Riou, J. Falquet,J . V. Fabre, D. Sicard, C. Neema, A. F. Adam-Blondon, and M. Dron. 1996. Anthracnose resistance in Phaseolus vulgaris : how many genes and what is the molecular basis for resistance? Abstract W24, Plant Genome IV Conference, San Diego, CA. p24.

184. Szczyglowski, K., D. Hamburger, P. Kapranov, and F. de Bruijn. 1997. Construction of a *Lotus japonicus* late nodulin EST library and identification of novel nodule-specific genes. Plant Physiol. 114:1335-1346.

185. Welsh, J., K. Chanda, S. S. Dalal, R. Cheng, D. Ralph, and M. McClelland. 1992 Arbitrarily primed PCR fingerprinting of RNA. Nucl. Acid Res. 20:4965-4970.

186. Goormachtig, S., M. Valerio-Lepiniec, K. Szczyglowski, M. van Montagu, M. Holsters, and F. J. de Bruijn. 1995. Use of differential display to identify novel *Sesbania rostrata* genes enhanced by *Azorhizobium caulinodans* infection. Mol. Plant-Microbe Interact. 8:816-824.

187. Udvardi, M. K. and M. L. Kahn. 1991. Isolation and analysis of a cDNA clone that encodes an alfalfa (*Medicago sativa*) aspartate amino transferase. Mol. Gen. Genet. 231:97-105.

188. Hata, S., H. Kouchi, I. Suzuka, and T. Ishi. 1991. Isolation and characterization of cDNA clones for plant cyclins. EMBO J. 10:2681-2688.

189. Miao, G-H., Z. Hong, and D. P. S. Verma. 1993. Two functional soybean genes encoding p34^{cdc2} protein kinases are regulated by different plant developmental pathways. Proc. Nat. Acad. Sci. USA 90:943-947.

190. Steiner, H.-Y., W. Song, L. Zhang, F. Naider, J. M. Becker, and G. Stacey. 1994. An *Arabidopsis peptide* transporter is a member of a new class of membrane transport proteins. Plant Cell 6:1289-1299.

191. Cheon, C-I., N-G. Lee, A-B-M. Siddique, A. K. Bal, and D. P. S. Verma. 1993. Roles of plant homologs of *Rap1p* and *Rap7p* in the biogenesis of the peribacteroid membrane, a subcellular compartment formed (de novo) during root nodule symbiosis. EMBO J. 12:4125-4135.

192. Bond, J. E., R.-J. Chian, P. Joshi, R. E. McDonnell, S. Farrand, and P. M. Gresshoff. 1998. Chimeric soybean plants may develop abnormal and meristematic nodules after transformation with *Agrobacterium rhizogenes* strain K599. Plant Cell Physiol. (in review)

193. Staskawicz, B. J., F. Ausubel, B. J. Baker, J. Ellis, and J. D. G. Jones. 1995. Molecular genetics of plant disease resistance. Science 268:661-667.

CHAPTER 7

Iron and Biocontrol

Peter J. Weisbeek and Han Gerrits

The growth of plants is influenced by a multitude of biotic and abiotic factors. Separately or in combination, these factors form major selective forces that can drive the evolutionary development and adaptation of plant species. Such selective forces can strongly affect the agricultural yield of crop plants. For example, repression of growth and development by pathogenic microorganisms is a major cause of economic losses in agriculture. In the rhizosphere, the soil layer that is influenced by plant roots[81], many soil-borne microorganisms threaten the progress of the regular life cycle of the plant. Pathogens respond to signals emanating from the growing plant, leading to microbial migration towards the root and subsequent growth and pathogenesis.[86] Similarly, the plant responds to the presence of the pathogen by activating one or more of its defense mechanisms (see Chapters 4 and 5 in this volume).

One major target of agricultural research is to enhance the defensive capacity of crop plants. A second line of research aims at modification of the rhizosphere environment, such that the presence and growth of the pathogen *ex planta* is limited.[74] This can be done chemically, with fungicides and bactericides, but pesticide use is met with increasing public concern. It also can be done with biological agents, usually microorganisms.[80] The latter approach has been less effective than chemical protection, due to inconsistent performance and limitations of the pathogen-repressing mechanisms, but mostly due to lack of knowledge about resistance mechanisms and soil ecology.

Enhancement of microbial biocontrol can occur when, in addition to lessening the pathogen pressure via microbial antibiosis or competition for nutrients, resistance in the plant is induced. The microbial capacity to induce systemic resistance has been recognized for a long time, but evidence about its molecular basis has emerged only recently.[1] Biocontrol disease reducing mechanisms are active towards a large array of microbial pathogens and, as such, can form an important part of our arsenal for crop protection.

The growth of microorganisms in soil and in the rhizosphere is largely controlled by the availability of nutrients, together with a number of additional chemical and physical parameters.[81] In combination, they determine the potential of a specific environment to promote microbial growth. The soil and rhizosphere microbial population is usually complex and physiologically diverse. When a soil prevents the growth and proliferation of particular pathogenic microorganisms, but not of the microbial community as a whole, we call the soil 'suppressive'. When the major factor in this pathogen suppression is of biological origin, we speak of biocon-trol.[73,105] Although it is often suggested that such control is directed specifically at the pathogen, it is very probable that, in most cases, a group of organisms or species is affected that includes the pathogen. A similar situation is found for most synthetic pesticides.

Due to the complexity of the soil and the rhizosphere, the number of different microorganisms found in soil is large. The effect of specific treatments on the composition of the microbial flora is therefore difficult to measure. For that reason, research generally focuses on (sets of) specific strains that, due to some property, are easy to track. In this way, a number of mechanisms have been identified for the suppression or enhancement of specific microorganisms. This includes enhanced efficiency for nutrient utilization, such as carbon, nitrogen, phosphate and iron, the acquisition of the capacity to metabolize special soil or rhizosphere constituents, the production of toxins and antibiotics, or the capacity to survive harsh conditions like drought and extreme temperatures (reviewed in 1a).

This chapter describes how competition for iron, a limiting nutrient in most environments, can be used to influence the microbial population, and thereby function as a mechanism for microbial control. In addition, iron is of importance in the interaction between microorganisms and plants and can affect plant defense.

Iron

Virtually all organisms, with the exception of certain lactobacilli, require iron to support vital cellular processes. Iron is employed as a cofactor by many metabolic enzymes and regulatory proteins because of its two stable oxidation states, Fe^{2+}, the ferrous state and Fe^{3+}, the ferric state. Its redox capacity can also be used *in vivo* to activate regulatory proteins or choose between alternative enzymatic states.[2]

Approximately 5% of the solid state mineral component of natural soils is comprised of iron. Only free iron, either present as ferrous or ferric iron ions, is available to plants and microorganisms; the amount of iron available is limited by a number of factors. Iron is readily oxidized by atmospheric O_2 and exists predominantly as the Fe^{3+} mineral $Fe(OH)_3$ but also as hemarite (Fe_2O_3), magnetite (Fe_3O_4), and limonite $[FeO(OH)]$.[3] The insolubility of $Fe(OH)_3$ $(K_{sp} = 10^{-38})$[4] limits free iron at pH 7 to a concentration of approximately 10^{-18} M.[5] This does not meet plant growth requirements which are between 10^{-4} to 10^{-9} M, depending on other nutritional factors.[6,8] Similarly, optimal iron concentrations to support microbial growth is approximately 10^{-5}-10^{-7} M.[7] Because iron availability in the soil is influenced by several additional factors, plants usually are not subject to iron limitation. Soil pH is, however, a major determinant, as Fe(III) concentrations decrease 1000-fold for every unit increase in pH.[8,9] About one-third of the world's land surface exists of calcareous soils, with a pH of 8 or higher.[10] In these soils, iron acquisition by plants and microorganisms is a problem. Attempts have been made to improve iron availability in calcareous soils by adding natural materials and chemicals that possess chelating activity (e.g. poultry blood powder, vivianite $(Fe_3(PO_4)_2.8H_2O)$ and Fe-EDDHA).[11,12]

Other factors influencing iron availability in soil are moisture and the presence of microorganisms.[10,13,14] Under flooded conditions, metals tend to be retained more strongly to soil particles whereas microorganisms can enhance the iron availability at microsites in soil and rhizosphere.

Bacterial Uptake Mechanisms

Most bacteria can produce and secrete molecules designated siderophores (Greek for iron bearers) to solve their iron needs.[15] These are water-soluble, low molecular weight (500-1500 daltons) molecules which bind ferric iron with high affinity. Many bacteria have additionally evolved systems to exploit iron complexed to siderophores produced by other bacterial and fungal species. The ability to utilize siderophores is associated with the presence of transport systems, which can recognize and mediate the uptake of the ferric siderophore complexes into the cell. The structure of pseudobactin 358, which is produced by *Pseudomonas putida* WCS358, is given in Figure 7-1.[16] Like other characterized pseudobactins, it possesses a fluorescent quinoline group linked to an oligopeptide chain. The length and composition of the oligopeptide varies among pseudobactins. The alternating D- and L-amino acid residues probably provide protection against proteolytic cleavage. Binding of iron is mediated by the catecholate group of the chromophore and by either two hydroxamate groups or one hydroxamate group and one hydroxyaspartic acid group.[17]

The intracellular concentration of iron is carefully monitored in the bacterial cell. A shortage of iron will reduce the growth of bacteria, whereas high concentrations of the metal are toxic. Therefore, the expression of the iron acquisition system is regulated in response to the iron concentration, being increased under iron limitation. Iron-dependent regulation has been most extensively studied in *Escherichia coli*. The transcription of iron responsive genes is under direct control of a repressor protein called Fur, a cytoplasmic 17 kDa iron binding protein.[18,19] When complexed to divalent ferrous ions, it binds to a specific DNA sequence (the Fur-box), located in gene promoters, where it inhibits transcription. At lower iron concentrations, the metal disassociates from the protein and repression is abolished.[20]

Recent data on iron regulation in other bacteria show that genetically diverse bacteria rely on similar mechanisms. All bacteria studied thus far, such as *Yersinia*,[21] *Vibrio*,[22] *Neisseria*[23] and *Pseudomonas*,[24,25] possess a *fur*-like gene. In several of these bacterial species, the Fur protein is part of a global control system regulating not only iron assimilation, but also several other factors involved in pathogenicity.[26] In *Vibrio anguillarum*, for example, the production and transport of the siderophore anguibactin is negatively regulated via the Fur repressor but also controlled by anti-sense RNA[27] and positively regulated by activator proteins.[28]

The limited availability of iron in the soil, together with the presence of iron scavenging systems in most organisms, results in competition. One

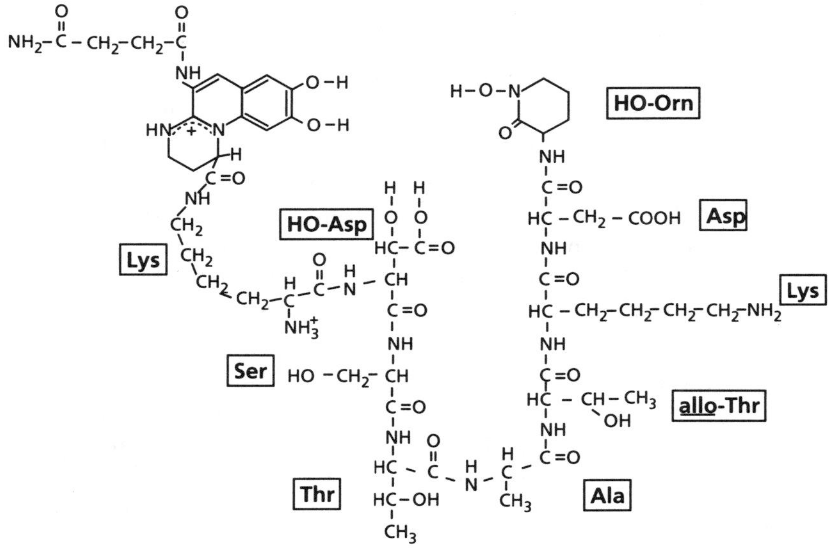

Fig. 7-1. Chemical structure of the yellow-green fluorescent siderophore pseudobactin 358, produced by *P. putida* WCS358.

would expect this to constitute a strong stimulus for the development of Fe acquisition systems with improved efficiency, thereby enhancing the competitiveness of that particular organism. There is indeed enormous variability in the capacity of microbial strains and species to sequester iron. This ranges from the production of siderophores with higher affinity for iron, via more efficient transport systems, to the presence of systems to use siderophores produced by other organisms. This has been described in greatest detail for *E. coli* and *Pseudomonas* but is likely true for most other microorganisms.[15] An excellent organism for research on iron uptake and regulation is *Pseudomonas putida* WCS358.[25] It produces a high affinity siderophore, pseudobactin 358, and it has many receptors for the uptake of iron via heterologous siderophores.[16] Among one thousand rhizosphere *Pseudomonas* strains tested, 99% produced a siderophore that can be utilized by strain WCS358. In contrast, the siderophore of strain WCS358 can be used by only 2% of the strains. Strain WCS358 benefits via its uptake of siderophores produced by other microorganisms whereas the other strains are denied access to the iron complexed to the WCS358 siderophore. This gives WCS358 a definite competitive advantage *in vitro*. This extreme specialization in iron competition is rather unique among soil pseudomonads. Therefore, in this review, we will focus on the *P. putida* WCS358 system, but it should be noted that other systems have also been studied and have their own unique aspects. A well-studied pseudomonad is *P. aeruginosa*, an important opportunistic pathogen of humans, that is well adapted to conditions of poor iron availability imposed by the host.[26] *P. aeruginosa* produces two siderophores, the fluorescent pseudobactin-type pyoverdin and pyochelin, a purple-blue hydroxamate siderophore. This species can also utilize a restricted number of heterologous ferricsiderophores. The regulation of the iron uptake systems is closely linked to the expression of pathogenic virulence factors.[26,29] *P. fluorescens* M114, a well studied plant growth-promoting strain, has very similar uptake and regulatory mechanisms as *P. putida* WCS358.[30]

Most bacterial strains are dependent on endogenously produced siderophores or can use only a limited set of heterologous siderophores. If one realizes that the production of secondary metabolites, including many antagonistic compounds, is to a large extent dependent on the intracellular concentration of iron, investment in an effective iron uptake system seems a particular good strategy.

Regulation of gene expression. The biosynthesis of pseudobactin 358, the fluorescent siderophore produced by *P. putida* WCS358, requires at least 15 genes distributed over five gene clusters.[31] This large number of genes is not surprising considering the chemical complexity of the siderophore. One major gene cluster of strain WCS358 has been well charac-terized and consists of at least five transcriptional units which are regulated at the transcriptional level by iron concentration.[31] Three genes, *pfrI, pfrA* and *fur* were isolated that are essential for transcriptional regulation.[25,32]

The *pfrI* gene was identified through analysis of a transposon insertion mutant defective in siderophore production that had lost the ability to activate transcription of a pseudobactin biosynthetic gene.[25] The *pfrI* gene was found to encode a protein of 176 amino acids (19.5-kDa). This protein contains a potential helix-turn-helix DNA-binding domain near the C-terminus, suggesting a role as a transcriptional activator. PfrI shows homol-ogy to several recently identified positive regulators that are involved in the regulation of siderophore mediated iron acquisition genes. The PfrI protein sequence is homologous (80% identical) to PbrA and PvdS, which are activators regulating pseudobactin production in *P. fluorescens* M114[33] and *P. aeruginosa*,[34] respectively. The PbrA regulator has been shown to activate a WCS358 pseudobactin biosynthetic promoter,[33] suggesting that PbrA and PfrI may be functionally interchangeable. The expression of the genes for both activators is regulated by iron.

The *pfrA* gene was identified by its capacity to increase the activity of a WCS358 pseudobactin biosynthetic gene promoter in a heterologous *Pseudomonas* strain.[32] PfrA is an 18 kDa protein with a primary structure that does not contain known DNA-binding motifs. The *pfrA* gene is expressed regardless of the iron concentration and its putative promoter contains a sequence very similar to the *E. coli* σ70 consensus. PfrA is homologous (58% identity) and functionally interchangeable with AlgQ (also known as AlgR2), a regulator of the production of the exopolysac-charide alginate in *P. aeruginosa*.[32]

It is not known whether the PfrA and PfrI regulators of strain WCS358 interact or whether they function in different parts of the pathway. Mutations in both genes abolish the production of pseudobactin 358 but the mutants are still able to utilize exogenously added pseudobactin 358, indicating that the transport systems are still expressed and functional. Transcription of the *pupA* gene, coding for the pseudobactin 358 receptor, however is reduced in the two mutants (V. Venturi, *unpublished results*). This shows that the PfrI and PfrA proteins influence the expression of the receptor gene for the homologous siderophore.

Promoters of several *Pseudomonas* iron regulatory genes contain regions with high similarity to the consensus Fur binding nucleotide sequence, suggesting that they are negatively regulated by Fur.[26] A *fur*-like gene was identified in strain WCS358 and found to be highly similar (58%) and functionally interchangeable with the *E. coli fur* gene.[25] Construction of a *fur* null mutant was unsuccessful in strain WCS358 and this hampered investigations on the possible role of this protein in the regulation of *pfrI* and *pupIR* gene expression (see below). However, the *pfrI* and *pupIR* promoters were found to bind the *E. coli* Fur protein, indicating that Fur is likely regulating their expression.[25] The promoter of the *P. aeruginosa pvdS* gene, a *PfrI* homolog, also contains a putative Fur-box consensus and is iron regulated via Fur.[34] In addition, a *P. aeruginosa fur* mutant is constitutive for *pvdS* gene expression and siderophore (pyoverdin and pyochelin) production.[24] The important role of negative regulation by Fur, as described in *E. coli* , has apparently been retained in *Pseudomonas*. However, *Pseudomonas* appears to possess added positive regulation elements. A tentative model for the Fur, PfrA and PfrI dependent expression of the *pupA* gene is presented in Figure 7-2.

The three very closely related regulators of pseudobactin biosynthesis (PfrI, PvdS and PbrA) display significant similarity (approximately 25%) to PupI and FecI, the two activators which regulate expression of ferric-pseudobactin BN8 transport in *P. putida* WCS358 and ferric-dicitrate transport in *E. coli*, respectively (see below). They all have domains with homology to the conserved domains of the *E. coli* σ70 protein family.[33,35] Members of this family contain four conserved regions, each responsible for a specific function, namely, RNA polymerase induction, promoter

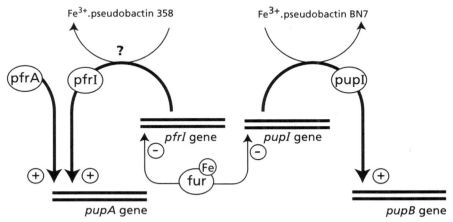

Fig. 7-2. Schematic representation of the two step transcriptional control of the *pupB* gene and a putative model for the expression of the *pupA* gene. The *pupI* and *pfrI* genes are negatively controlled by the fur repressor. Under iron limitation both genes are derepressed. The PupI protein is activated by the presence of ferric pseudobactin BN7 through the combined action of the PupR sensor in the inner membrane and the PupB receptor in the outer membrane. Activation of the *pupA* gene requires the action of PfrA and PfrI and is stimulated by the presence of pseudobactin 358.

recognition, DNA binding and open complex formation. The PfrI family of proteins shows particular conservation in regions 2 and 4.

Transport. A major component of the transport system is the outer membrane receptor, PupA.[36] This integral membrane protein spans the outer membrane several times and has a periplasmic N-terminal domain that is involved in siderophore-specific signal transduction.[37,41] Transport is highly specific. Only pseudobactin 358 is recognized and transport requires the inner membrane protein TonB and its associated proteins, ExbB and ExbD.[38] The latter three components are believed to transfer energy to the outer membrane transport process by direct interaction with the receptor protein.[39] Expression of the *pupA* gene in other fluorescent *Pseudomonas* strains allows the recipient to use pseudobactin 358. Hence the specificity for pseudobactin utilization resides largely in the outer membrane receptor. Transport across the inner membrane into the cytoplasm and the release of the iron from the ferric siderophore complex apparently do not depend on a specific structure of the siderophore.

Heterologous siderophore uptake. In addition to the PupA receptor, strain WCS358 possesses at least five and probably more than twenty additional outer membrane receptors that recognize heterologous pseudobactins.[37] Each receptor interacts with a single siderophore or with a small set of siderophores with very similar structure. The induction of these receptors is regulated independently of siderophore biosynthesis and expression is not affected in the *pfrI* and *pfrA* mutants. Iron concentration and the presence of the cognate pseudobactin control receptor synthesis. The transcriptional regulation of one of the heterologous receptor genes, *pupB*, has been studied in detail.[40,41] The PupB receptor mediates the uptake of iron complexed to pseudobactin BN8 and pseudobactin BN7, produced by *Pseudomonas* BN8 and BN7, respectively. Three genes, *pupI*, *pupR* and *pupB* collaborate in transcriptional activation of the *pupB* gene. Transcription of *pupI* and *pupR* is induced by iron-limitation, whereas expression of the *pupB* gene additionally requires the siderophore BN7 or BN8.

The cytoplasmic 19 kDa PupI protein has a DNA binding motif (helix-turn-helix domain) in the C-terminus, which implies a direct role of PupI in the transcriptional activation of the *pupB* gene by binding to the promoter region. The PupR protein is thought to modify the activity of PupI in response to siderophore availability.[41] The 36 kDa PupR protein has two potential membrane spanning domains, suggesting an inner membrane location. In a *pupR* mutant, the *pupB* gene is transcribed independently of pseudobactin BN8, indicating that PupR inhibits the activity of the PupI protein in the absence of the siderophore. The activity of the *pupB* promoter in this mutant is reduced relative to the wild-type strain, suggesting that PupR is required for maximal transcription of the *pupB* gene. How the PupR protein interacts with PupI is still unknown, but the stoichiometry of the two regulators appears to be very important.[41] Overproduction of PupI leads to induction of the *pupB* gene even in the absence of pseudo-bactin BN8 or BN7, suggesting that PupR is titrated out by excess PupI protein. The PupI and PupR proteins share sequence and functional similarity with the *E. coli* FecI and FecR proteins which control the expression of the ferric-dicitrate transport system.[42] The *fec* genes encoding this system are only transcribed under low iron conditions and ferric-dicitrate availability. The siderophore dependent regulation of the *fec* genes is exceptional since the components of the other siderophore transport systems in *E. coli* are controlled by iron concentration only, independently of the presence of the corresponding siderophores. The regulation of the *pupC* gene, that codes for the receptor for the heterologous siderophore pseudo-bactin B10, is controlled by this siderophore and uses a *pupI-pupR* related system (A. Fofana and M. Koster, *unpublished results*).

The inducible expression of the *PfeA* enterobactin receptor gene of *P. aeruginosa* requires iron-limitation and the presence of ferric-enterobac-tin.[43] Interestingly, synthesis of the enterobactin receptor is under the control of two proteins that show homology to histidine kinases and response regulators of the classical two-component systems, but not to the PupI and PupR sequences. Hence, different types of regulatory systems have evolved in *Pseudomonas* to allow substrate-dependent expression of ferric-sidero-phore transport systems.

Regulation of the *pupB* receptor gene is dependent on iron-limitation and the presence of the appropriate siderophore. These two conditions are fulfilled in a stepwise order. First, the expression of *pupI* and *pupR* is controlled by iron via the Fur repressor. Second, the PupI protein is activated through the presence of the BN7 and BN8 siderophores and acts upon the *pupB* gene promoter, leading to its expression.

The product of the third gene, the PupB receptor, transports the ferric-siderophore complex across the outer membrane. However, in addition, it was found to have a very specific role in the siderophore-responsive activation of transcription of the *pupB* gene itself.[41] This phenomenon was studied using chimeric receptor proteins. Hybrid receptor proteins, in which the N-terminal 86 amino acid residues of the 759 amino acids containing PupB protein were replaced by the corresponding domain of PupA, could still transport pseudobactin BN8, but had lost the ability to activate the *pupB* gene. The reverse chimera, the PupA protein with the first 86 amino acid residues of PupB, could still import ferric-pseudobactin 358 but had gained the capacity to activate the *pupB* promoter. The PupI-PupR system and the *pupB* gene now responded to the PupA-related siderophore pseudobactin 358, instead of the cognate pseudobactin BN8. These experiments demonstrated that the stimulus for the PupI-PupR system is not the ferric-pseudobactin itself, but rather a signal transduced by the receptor upon transport of its substrate. A model for the Fur, PupI and ferric-peudobactin BN7 dependent transcription of the *pupB* gene is presented in Figure 7-2.

Signal transduction across two membranes. The fact that *pupB* gene regulation can be reconstituted in heterologous *Pseudomonas* strains by introducing the *pupIRB* genes suggests that the signal transduction pathway is composed of only these three elements. The simplest model is that the PupB receptor transmits a signal to the PupR protein during transport of its substrate. As a result, PupR no longer represses but activates PupI resulting in transcriptional activation of the *pupB* gene (Figure 7-3). The ferric-siderophore receptor acts as a gated channel which is opened by the combined action of the ferric-pseudobactin complex and the TonB-energy coupling system, to allow passage through the outer membrane.[44,45] It is attractive to speculate that opening of the channel and the concomitant conformational change of the receptor is the signal which is sensed by the PupR protein. Consistent with this hypothesis is the observation that the TonB protein is required for transcriptional activation of the *pupB* gene.[41]

The experiments described above using the hybrid receptor proteins defines the first 86 amino acid residues as the region of the PupB protein involved in signal transduction. This N-terminus is located in the periplasm.[46] Comparison of siderophore receptors of different bacterial species shows that the pseudobactin-type receptors are all characterized by the presence of an extra N-terminal region of approximately 50 amino acids. Therefore, the PupA receptor of strain WCS358, the pyoverdine receptor of *P. aeruginosa*[29] and the pseudobactin M114 receptor PbuA of *Pseudomonas* sp. M114,[47] are also expected to function in signal transduction. It should be noted that the outer membrane receptor FecA, involved in the uptake of ferric-dicitrate in *E. coli*, lacks this N-terminal region.[42]

The proposed model for the regulation of heterologous receptor genes involves an inner membrane sensor protein (PupR) interacting with a cytoplasmic activator protein (PupI) and an outer membrane receptor protein (PupB) (Figure 7-3). The direct interaction of outer and inner membrane proteins to activate intracellular transcription in response to extracellular signals is new in prokaryotic signal transduction. The PupI, PupR, PupB system may become a paradigm for a new signaling mechanism that combines sensors, regulators and outer membrane transporters.

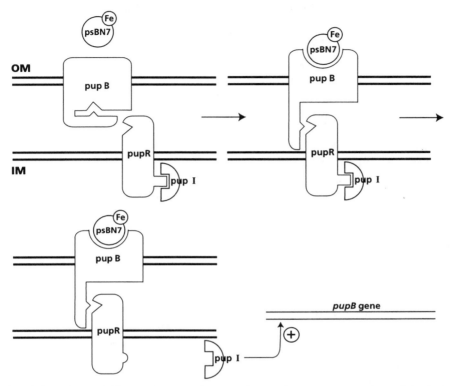

Fig. 7-3. Tentative model for signal transduction across inner and outer membrane. Extracellular ferric pseudobactin BN7, upon binding by and transport through the outer membrane PupB receptor, induces or modifies the interaction between PupB and the inner membrane PupR protein. This abolishes the binding and inactivation of the cytoplasmic PupI protein. PupI initiates the transcription of the *pupB* gene.

Iron and Plants

Iron is an essential element for plants, just as it is for microorganisms. Its redox properties makes it an ideal cofactor in oxidation/reduction reactions in processes such as photosynthesis and respiration. An important additional role of iron involves the generation of reactive oxygen species. In the chloroplast, when the absorbed light energy exceeds the capacity of the photosystems or when $NADP^+$ becomes limiting, electrons are translocated to molecular oxygen. The superoxide radicals (O_2^-) formed can react with hydrogen peroxide, with iron as a cofactor, resulting in the formation of highly reactive hydroxyl radicals ($OH\cdot$) which can react with any constituent of the cell.[48,49] In the chloroplast, a Fe-superoxide dismutase (Fe-SOD)[50] is present that converts O_2^- to hydrogen peroxide, thereby preventing the formation of hydroxyl radicals. In plants, the hypersensitive response (HR) utilizes the generation of reactive oxygens species to kill infected cells (see Chapter 4 in this volume).

Gene expression. Next to its role in electron chemistry, iron can act as signal for gene expression. In plants, only a few genes have been characterized in which expression is regulated by iron. In soybean cell suspension cultures, transcription of the iron storage protein, ferritin, was increased by the addition of Fe-citrate to the medium.[51] Stimulation of transcription correlated with the increase in cellular iron content. This contrasts with animal systems, where ferritin is regulated at the translational level.[52-56] In animals, ferritin mRNA contains an iron-responsive element (IRE) in its 5' untranslated region which, when iron is limiting, can be bound by the iron regulatory protein (IRP). IRP can also bind to a domain in the mRNA of the transferrin receptor (TfR), involved in cellular iron uptake. In this way, IRP blocks the translation of ferritin mRNA and prevents degradation of TfR mRNA. When iron concentrations increase, IRP acquires a 4Fe-4S cluster, no longer binds to the IRE and functions as a cytoplasmic aconitase. This results in the activation of the ferritin mRNA and degradation of the TfR mRNA.[52-56] It was recently shown that H_2O_2 can inhibit the expression of ferritin, linking oxidative stress to cellular iron metabolism.[57] The regulatory mechanisms and elements have yet to be elucidated in a plant system. However, a system involving iron-sulfur cluster mediated control over iron transport and storage is expected.

It is becoming increasingly evident that iron-sulfur clusters are used as biosensors of oxidants and iron. Oxidation of the iron leads to destruction of the cluster,[58] thereby changing the characteristics of the protein to which it is bound. In this way, iron-sulfur clusters can function as regulatory switches. In *E. coli*, function of the transcriptional factor, fumarate nitrate reduction protein (FNR), is dependent on the oxidation state of its [4Fe-4S] cluster. FNR is important in the cellular switch between the use of oxygen or alternative compounds as terminal electron acceptors.[59]

Other potential targets for iron regulation are the genes involved in iron stress responses (see below). In our lab, the pathogenesis-related (PR) β-1,3-glucanase gene *BGL2* of *Arabidopsis* was highly expressed when the synthetic iron chelator EDDHA was added to the medium. BGL2 expression was abolished when an equimolar concentration of $FeCl_3$ was added. Other *PR* genes, such as *PR1*, are not responsive to iron (H. Gerrits, in preparation). The functional significance of the *BGL2* gene induction therefore remains to be elucidated.

The genetic analysis of iron metabolism in plants has received little attention. For example, only three mutants have been isolated that are disturbed in iron metabolism. The *bronze* mutant of *Pisum sativum* l., cv.

sparkle is one such mutant. This mutant is defective in feedback inhibition of iron uptake. As a result, iron accumulates to extremely high concentrations in older leaves and bronze-colored spots appear.[60,61] Two tomato mutants have been described that are defective in iron metabolism.[62,63] The root system of the *fer* mutant cannot induce the characteristic responses to iron-deficiency and iron uptake is completely blocked. The *chloronerva* (*chln*) mutant lacks a gene that controls the synthesis of nicotianamine, a key component in iron physiology.[63] The *FER* gene is epistatic over the *CHLN* gene.

Iron deficiency. Plants have developed systems to control the cellular iron concentration through selective uptake and storage. Synthesis of the iron storage protein, ferritin, serves two purposes. First, iron is stored for times of limitation. Second, excess iron is separated from cellular metabolism, thereby preventing its toxic effects.

Symptoms of iron deficiency include a reduction in the chlorophyll content of younger leaves, because of inhibition of the biosynthesis of chlorophyll.[64] Haem iron and iron-sulfur clusters are present in the chloroplast photosynthetic machinery and function in electron transfer. In the prolonged absence of iron, electrons that are generated in photosystem I and II can no longer be translocated over the thylakoid membrane and can exert toxic effects on the cell. However, these toxic effects are likely prevented by the fact that chlorophyll decrease precedes photosystem II inactivation.[65] This likely occurs through a mechanism that involves iron sensing and subsequent down regulation of the chlorophyll biosynthetic genes.

How does the plant manage iron uptake in an environment where iron solubility is severely limited and where microorganisms scavenge the rhizosphere for iron with powerful siderophores? Upon iron limitation, an iron stress response is initiated.[66] This is accompanied by morphological changes in the root tips. In the apical root zone, roots thicken and the root hair number and length is increased. When ethylene synthesis or action is inhibited, typical symptoms of iron stress do not appear upon iron limitation.[66] This suggests that ethylene is part of the signaling pathway between iron limitation and iron stress responses.

In dicots and non-grass monocots, large amounts of protons are secreted by the roots acidifying the rhizosphere and increasing the solubility of iron.[67] Organic acids and phenolics are also released that lower the pH and chelate solubilized iron.[68] A new reductase[69,70] is induced in the cell wall that reduces ferric ions to ferrous ions, probably releasing the complexed iron. Whether all these responses are induced in combination or separately is unknown. The mechanism of iron uptake by the plant cell is unknown.

A second strategy has only been found in graminaceous plants. These plants are capable of synthesizing phytosiderophores that are secreted into the rhizosphere and, when complexed with iron, re-enter the cell.[70] Based upon competition experiments with FeEDDHA, it was concluded that a specific uptake mechanism for these phytosiderophores (perhaps homologous to bacterial siderophore uptake) must be present.[70] However, no such uptake system has been found. Examples of phytosiderophores are mugineic acid, avenic acid and 3-epihydroxymugineic acid that are synthesized from L-methionine through 2'-deoxymugineic acid.[71]

Rhizosphere Interactions

The rhizosphere, the volume of soil that surrounds the root and is influenced biologically and physically by the living root, is an important interface for microbial and microbe-plant interactions.[81] The complexity of the rhizosphere microbial community and soil changes frequently in response to various parameters (e.g., metabolism of the plant, seasonal conditions and agricultural activities). Depending on the conditions, 20% or more of the plant dry weight is released from the roots.[72] Rhizosphere bacteria can influence and enhance the release of exudates through the production of compounds (e.g., plant growth factors, toxins and lytic enzymes) that affect plant metabolism.[73]

The microbial world in the rhizosphere has been the subject of analysis for several years. Due to its heterogeneity, chemical and microbial diversity, analysis has been largely restricted to the behavior and fate of separate microbial strains. Fluorescent pseudomonads, as prominent rhizosphere colonizers, have received most attention. However, new knowledge and techniques, should allow for better analysis in the future. The results obtained will enhance our understanding of rhizosphere processes and could form the basis for the creation of a biased rhizosphere supporting pathogen control and plant protection.[74]

The rhizosphere is a place of strong competition and selection, as can be judged from the fate of introduced bacteria. Although inoculation usually occurs at high density, only a small fraction of the introduced cells can be recovered after a few months. After one growth season, it is often difficult to reisolate the introduced strain.[75,76] Introduced strains colonize roots efficiently, and may temporarily dominate the microbial population, but are replaced slowly by endogenous strains. A single advantageous trait does not give sufficient dominance for an extended period of time. Therefore, ways must be found to either prolong survival (e.g., as in suppressive soils), or inoculation will have to be repeated during the growing season.

More information about rhizosphere conditions and interactions is needed to better control the growth of pathogenic microorganisms. In the absence of such data, we must rely on naturally occurring pathogen-control ecosystems (e.g., suppressive soils) or use massive introductions of microbial strains with powerful antibiotic mechanisms.

The rhizosphere is also an environment where major beneficial interactions occur within consortia of microorganisms. This is indicated by the estimated large number of viable but non-culturable microorganisms.[77] These organisms may survive through the mutual nutritional exchange of nutrients and metabolites.

An analogy has been suggested between plant protection mediated by antibiotic production by the rhizosphere microbial population and *in planta* defense mediated by phytoalexin production.[78] This is based on the structural similarities between antibiotics and phytoalexins and on homologies between plant and microbial genes involved in their production.[126] This latter observation even raises the possibility of gene transfer between plant and rhizosphere associated microorganisms.

Microbial Growth

Plant root colonization by microorganisms is essential for symbiotic and pathogenic interactions and for the successful application of microbial inoculants for biocontrol, biofertilization and phytoremediation. Microorganisms need to colonize and concentrate at the root surface to be effective as agents for rhizosphere biocontrol. Microbial chemotaxis to specific plant exudates may be involved in concentrating microorganisms in the rhizosphere. Colonization requires the potential to establish a more permanent association with the root surface. Many factors (e.g., microbial metabolic capacity, motility, composition of the plant exudates, etc.) have been described that are relevant to a discussion of microbial growth in the rhizosphere.[79-81] Much less is known about plant or microbial components important for root colonization. Various publications have implicated adhesins, lectins and cellulose fibrils in mediating root colonization but specific mechanisms have yet to be described.[82-85]

Direct visualization of bacterial colonization on roots shows most of the root surface to be free of bacteria but with micro-colonies appearing at the root base and in junctions between epidermal cells.[86,87] This localization may indicate that these sites are places where exudate release is maximal but this remains to be determined. One interesting report indicated the presence of a semitransparent film, probably of plant origin, covering the microcolonies.[87]

In bacteria, a network of cellular transduction mechanisms integrates signals from the environment into control of gene expression leading to an appropriate bacterial response (e.g., chemotaxis, mating, cell division etc.). One such signaling mechanism, related to cell density, involves acylated homoserine lactones.[88] Many gram-negative bacteria produce N-acyl-L-homoserine lactones (A-HSL). They are secreted and function as signals to the producing cell (autoinducers) or to other bacteria. These molecules function in the regulation of processes that depend on cell density (quorum sensing), such as pathogenesis, nodulation, conjugation and bioluminescence. They control the biosynthesis of a variety of factors (e.g., luciferase, antibiotics, surfactants and exoenzymes).[88] The structure of the signaling molecule, particularly in the fatty acyl chain, and the particular response associated with it can differ considerably. Cross talk between different A-HSL type signals and transduction pathways is commonly observed.[89,90]

Siderophores, whose production is not controlled by cell density, can also serve as important signals in microbial communities. They can function as autoinducers, to maximize siderophore production and uptake[90,91], but can also be recognized by a variety of other rhizosphere bacteria and possibly fungi. Recognition of heterologous siderophores by specific receptor proteins can lead to the activation of specific genes and responses (see above). Thus far, such research has largely focused on iron-uptake related processes but it is conceivable that other cellular responses are regulated through siderophore signals within microbial communities.

Microbial Competition

Most fluorescent pseudomonads have siderophores with high affinity for iron. Therefore, these siderophores are potential antagonists for any microorganism that produces a less potent siderophore. The dominance of fluorescent *Pseudomonas* spp. in suppressive soils may be due, in part, to the high iron affinity of their siderophores. Likewise, siderophore production probably contributes to the efficacy of biocontrol bacteria. However, there is no current evidence demonstrating that siderophore production alone is sufficient for biocontrol success.

Antagonism between microorganisms based on competition for iron via siderophores is readily observed on plates.[73,127] However, the interpretation of these observations is not as straightforward as generally assumed. The usual conclusion, that organisms are outcompeted on the basis of the strength of the cognate siderophore, is often incorrect. This is explained by a close examination of the actual experimental assay conditions. Plates are usually inoculated with the first strain and then incubated for several hours or even days, after which they are sprayed with a culture of the second strain. Under such conditions, before the second strain has the opportunity to produce a siderophore, the first strain has produced large quantities of siderophores that have likely sequestered most or all of the available ferric ions. Therefore, the inability of the second strain to grow is actually a measure of the relative amounts of the different siderophores at the start of competition and not a true indication of the siderophores' relative binding constants for iron. Only in the case of siderophores with large differences in affinity for iron, can release of iron from the iron-siderophore complex influence the outcome of the competition. Most microbial siderophores have high and comparable affinities for Fe(III). Ligand exchange therefore is extremely slow (220 hr for two hydroxamate siderophores)[92] and usually does not relate to the length of the experiment. Absence of competition under these conditions is not an indication of a higher affinity for Fe(III) but in general indicates the capacity to recognize and utilize the sidero-phore present on the plate.[93]

Under the dynamic conditions of the rhizosphere, the outcome of the competition for iron is determined by the balance of the relative Fe(III) affinities of the various siderophores, the quantities of the siderophores produced and the microbial capacity to utilize homologous and heterolo-gous siderophores. The competitive outcome can be affected temporarily through the input of high amounts of siderophore or microbial cells. In the absence of information to predict the microbial balance in the rhizosphere, such short term manipulations are all that can be accomplished at present.

Although little is known about the microenvironment of rhizosphere bacteria, the general consensus is that growth is likely iron limited and, therefore, the genes for iron acquisition and adaptation to low iron are expressed.[73] The special character of the rhizosphere, with the plant actively involved in iron acquisition, may create situations where the bacteria can rely on root metabolism and not experience iron limitation. This may be of particular importance in maintaining the microbial microcolonies observed on root surfaces.

The production of siderophores and their relevance for competition in soil and the rhizosphere was studied using potato and radish roots colonized with different *Pseudomonas* strains and siderophore negative (*sid*) mutants. In cross-feeding experiments,[94] the *sid* mutant JM218[31] of *P. putida* WCS358 was applied to soil that was subsequently planted with rooted potato seedlings that had been dipped in a suspension of siderophore producing *P. putida* WCS358. Growth of the *sid* mutant on the root-surface was monitored. It was found that colonization by the mutant was enhanced by more than fifty-fold on roots that had been treated with the wild-type strain, as compared to roots that had not been treated or had received only the *sid* mutant. Reisolation of the *sid* bacteria showed that they had retained their siderophore negative phenotype. Therefore, genetic recombination between the mutant and wild-type cells does not appear to be an explanation for the results. The data obtained in this experiment support the notion that rhizosphere microbial strains are subjected to iron limitation.

In situ expression studies of Loper and Lindow[125] with *Pseudomonas syringae* and *Pseudomonas fluorescens* strains containing a sensitive reporter system fused to an iron dependent siderophore promoter, indicated that iron concentrations in the rhizosphere were sufficiently low to allow for the activation of the siderophore promoter.

Similar experiments were performed by Raaymakers *et al.* with *P. fluorescens* WCS374.[95] The results of these experiments indicated that the wild-type strain can cross-feed *in situ* a *sid* mutant of *P. putida* WCS374. Moreover, the growth of *P. putida* WCS358 *sid* mutants in the rhizosphere was stimulated by the addition of the *P. fluorescens* WCS374 wild-type strain. The latter result confirms the observation that *P. putida* WCS358 can uptake Fe(III) via the siderophore produced by *P. fluorescens* WCS374. The two conclusions from these experiments were that siderophore is produced by the parent, indicating iron limitation, and that the mutant and parent strains were located on the root surface such that the siderophore of the wild -type strain was accessible to the mutant.

The root surface is likely not saturated with siderophore over its complete length, suggesting that there may be favored locations where bacteria grow and interact. This idea is supported by scanning electron microscopic observation of roots showing localized areas with high bacterial cell density.[87]

Competition experiments where mixtures of the wild-type strains *P. putida* WCS358 and *P. fluorescens* WCS374 were applied to radish roots showed that the number of WCS374 cells was reduced three-fold after one radish growth cycle and thirty-fold after a second period of radish growth when compared to roots grown in the absence of strain WCS358.[76] Since WCS374 cannot utilize the siderophore produced by WCS358 while WCS358 can use the WCS374 siderophore, these results suggest that WCS358 profits from its capacity to obtain iron not only from its own siderophore but also that produced by WCS374. To test this hypothesis, the *pupA* siderophore receptor gene of strain WCS358 was introduced into strain WCS374 and the competition experiment was repeated. It was found that expression of the heterologous siderophore receptor in strain *P. fluorescens* WCS374 resulted in similar densities of this strain in the radish rhizosphere when grown in the presence of strain WCS358. Under these conditions there was no apparent antagonism between the two strains. In this case, apparently only expression of the correct receptor was necessary to overcome antagonism between the strains. The other components of the iron uptake machinery were apparently no involved. These results clearly demonstrate the occurrence *in situ* of competition for iron and the ecological advantage of being able to use heterologous siderophores. These results provide an experimental rational for using siderophores to restrict the growth of rhizosphere microbial pathogens.

Plant-Microbe Interaction

The close interaction between plants and microbes in the rhizosphere suggests the possibility that plants have evolved mechanisms to use microbial siderophores for their own iron requirements. A number of studies have addressed this possibility.[96-101] For example, it was shown that several plant species can bind iron via microbial siderophores. However, these studies were performed under non-axenic conditions and used excised roots. Therefore, it could not be excluded that the measured iron uptake through Fe-siderophore complexes reflects iron uptake by colonizing microorganisms and not by the root itself. Bar-Ness and co-workers[102] conducted experiments in maize and oat in which they measured the translocation rate from the rhizosphere to the shoot of ^{55}Fe complexed to either pseudobactin, EDDHA or epi-3-hydroxy-mugineic acid. It was found that ^{55}Fe iron, as a complex with pseudobactin, efficiently associates with plant roots only when applied together with the pseudobactin-producing *Pseudomonas* cells. However, this iron was not translocated to shoots. Under axenic conditions, iron uptake from ferric-pseudobactin complexes was 10- to 30-fold lower than when ferric EDTA was applied. These studies provide good evidence that the higher uptake rates measured in previous studies were caused by colonizing *Pseudomonas cells* and were not due to direct uptake by the plant.[102] Indeed, Becker *et al.*[103] showed that addition of *Pseudomonas* siderophores to plant growth medium resulted in a strong reduction of iron uptake and accumulation in the shoots of maize and pea. They interpreted this as competitive iron binding by the siderophores and an inability of the roots to dissociate or take up the Fe-siderophore complex.

We have performed experiments on iron uptake in *Arabidopsis thaliana* growing on MS agar under axenic conditions (H. Gerrits, in preparation). Iron was supplied as a complex with EDTA, at a concentration of 100 mM. Under these conditions, EDTA has no iron-limiting effects and Fe-EDTA is an excellent iron source for plants. Addition of the strong iron chelator EDDHA to these plates resulted in the manifestation of an iron stress response. When *P. fluorescens* WCS417 was added at high concentrations (10^9-10^{10} cells/ml) instead of EDDHA, a similar iron stress response was induced (as visualized by an increased lateral branching of the root, root tips that are expanded radially, and an increased number of root hairs with increased length). The shoots of these plants remained small. Similar results were obtained with *P. fluorescens* WCS374. We conclude that the production of high amounts of siderophore in the rhizosphere leads to a strong reduction of iron-availability for the plant. This agrees with the results of Becker *et al.*[103]

The results obtained from the above studies strongly indicate that plant roots do not acquire iron via microbial siderophores and that high siderophore concentrations can interfere with a plant's capacity to obtain sufficient iron resulting in iron stress. However, one cannot unequivocally exclude the possibility that there are particular situations or plant-microbe combinations where plants make use of microbial siderophores.

Biocontrol

Microbial Antagonism

As discussed above, siderophore production in the soil and in the rhizosphere by fluorescent pseudomonads can result in antagonism towards other microorganisms. In combination with the observation that pseudomonads form a significant portion of the microflora of suppressive soils[80], this provides a rationale for a biocontrol strategy where competition for iron is used to limit the growth and development of pathogenic microorganisms.

Inoculation of carnation seeds with the *Alcaligenes* strain MFA1 lowered colonization of the root by *Fusarium oxysporum* f. sp. *dianthi*.[104] Experiments by Duijff[105] showed that application of *P. putida* WCS358 to carnation roots could reduce disease-symptoms caused by subsequent *Fusarium* inoculation, but only under conditions of iron limitation. Mutants defective in siderophore production had no effect on disease development. Raaymakers.[76] has described similar siderophore-dependent protection of radish from Fusarium pathogenesis

Whereas the involvement of siderophore production in biocontrol protection by *P. putida* WCS358 has been clearly demonstrated (see above), the biocontrol activity of other *Pseudomonas* strains appears to be independent of siderophore action. Although the siderophores of *P. fluorescens* strains WCS417 and WCS374 are just as antagonistic *in vitro* as the siderophore of *P. putida* WCS358, siderophore negative mutants of strains WCS417 and WCS374 are still effective in disease suppression when tested on *Arabidopsis* (Gerrits, *unpublished results*) and radish[106], respectively. No other antagonistic compounds have been isolated from cultures of strains WCS417 and WCS374. Clearly, siderophore biosynthesis is an important prerequisite for microbial growth in the iron-limited and competitive rhizosphere. Mutants defective in siderophore production have lower growth rates.[95] Therefore, irrespective of the mechanism of biocontrol action, pathogen control by a microbial biocontrol agent will be enhanced by the ability to uptake iron within the rhizosphere.

In some cases, iron sufficiency has been shown to be a prerequisite for efficient biocontrol. For example, the plant growth-promoting strain, *P. fluorescens* CHA0, suppresses the soil-borne tobacco pathogen *Thielaviopsis basicola*. Strain CHA0 produces several secondary metabolites with antifungal properties, such as pyoverdine, 2,4-diacetyl phloroglucinol, pyoluteorin and hydrogen cyanide (HCN),[107] of which the last three are under control of cell density regulation and the global regulator gene *gacA*.[90,108] HCN is a major factor in the antagonism against *T. basicola* and has been proposed to act directly on the pathogen without damaging the plant. Synthesis of HCN is induced by high ferric iron concentrations whereas conditions of low iron are inhibitory. These findings appear to contradict the general consensus that iron is limiting in the rhizosphere. However, there may be specific rhizosphere niches where sufficient iron exits to support HCN production by *P. fluorescens* strain CHA0.

Induced Systemic Resistance

Induced systemic disease resistance is a defense response to the local presence and activity of a variety of biotic and abiotic agents. Induction can occur through biotic factors (e.g., infecting or feeding pathogens) or through chemical agents (e.g., salicylate). Systemic acquired resistance (SAR) is induced during compatible and incompatible plant-pathogen interactions. In incompatible interactions, the local hypersensitive response (HR) is an effective and major defense reaction of the plant. In a compatible interaction, induced systemic responses appear to function to limit progress of pathogenesis and to avoid additional infections (for reviews see references 109 and 110).

Plant growth-promoting rhizobacteria (PGPR) are able to protect the plant against microbial pathogen pressure via antibiosis and nutrient competition. Their potential for the induction of systemic defense systems in plants has been acknowledged for some time, but it is only recently that supporting evidence has become available. As described below, the distinction between direct pathogen antibiosis and plant induced pathogen suppression was elucidated using bioassays where pathogen and PGPR strains were inoculated on separate parts of carnation and radish plants. Application to the root of the bacterial inoculant or cell wall fractions obtained from particular strains was found to protect leaves against subsequent infection with pathogenic microorganisms.

Van Peer et al.[111] showed that bacterization of carnation roots with *P. fluorescens* WCS417 reduced significantly the wilting caused by *Fusarium oxysporum* f. sp. *dianthi*. Upon pathogen infection, higher amounts of the anthranilate derived phytoalexins dianthalexin (2-phenyl-7-hydroxy-1,3-benzoxazin-4H-one) and a group of dianthramides[112] accumulated in plants that had received a root-pretreatment with strain WCS417 than in untreated control plants.

Similarly, Leeman et al.[106] showed that treatment of radish roots with *P. fluorescens* WCS374 two days prior to pathogen inoculation resulted in a significant decrease in the growth of *Fusarium oxysporum* f. sp. *raphani* and disease development. In these experiments, the plant root and aerial portions were enclosed by separate polythene bags, excluding migration or cross-contamination. Mauerhofen et al.[113] also showed that root inoculation of *Nicotiana tabacum* with the biocontrol strain *P. fluorescens* CHA0 induced leaf resistance against tobacco necrosis virus (TNV). Similar experiments, in which cucumber seeds were inoculated with several plant growth-promoting strains,[114] resulted in protection of emerging leaves against infection by the fungal pathogen, *Colletotrichum orbiculare*. The observation that, in the absence of direct contact between the biocontrol strain and the pathogen, disease progression is still reduced, excludes direct antagonism and competition as mechanism of the protective response. Therefore, a more likely explanation is that root colonization with biocontrol microorganisms induces a systemic resistance response that protects against pathogen attack. Induction of plant defense may also explain why strains *P. fluorescens* WCS417 and WCS374 can have biocontrol activity without the production of antimicrobial compounds (see above). These strains apparently produce signals that are not antagonistic to microorganisms but may be perceived by plants and activate one or more defense systems. Identification of the signal and the pathway leading to plant defense is the topic of present research.

Rhizosphere induced systemic resistance (ISR) has been shown to occur in radish,[106] tobacco,[108] carnation,[111] common bean,[115] cucumber[116] and the model plant *Arabidopsis thaliana*.[1] It is active against pathogens of viral, fungal and bacterial origin. The *Arabidopsis* system, with *P. fluorescens* strain WCS417 as the biocontrol agent and *P. syringae* pv. *tomato* strain DC3000 as bacterial pathogen, is currently the best studied model. The elegant experiments of Pieterse *et al.*[1] showed that rhizosphere induced systemic resistance is distinct from pathogen induced systemic resistance. A different signal-transduction route is used during the biocontrol response when compared to pathogen-induced resistance (for review see reference 117). SAR involves an increase in endogenously synthesized salicylate that mediates the expression of SAR genes, including pathogenesis-related (PR) proteins. The importance of salicylate in SAR was demonstrated by the inability of pathogens to induce SAR in plants that do not accumulate salicylic acid due to the expression of the bacterial salicylate hydroxylase *nahG* gene.[117] Strain WCS417 is able to induce ISR in NahG plants[1] and, therefore, ISR does not require salicylate accumulation. Similarly, induction of the transcription of the *Arabidopsis PR1, 2* and *5* genes could not be detected using competitive PCR or RNA gel blot analyses.[1] These results were confirmed at the protein level using transgenic *Arabidopsis* lines expressing *PR1* promoter-LUC and *BGL2* promoter-GUS fusions (H. Gerrits, in preparation). An antimicrobial activity has been identified in *Arabidopsis* leaves (Figure 7-4) that is specifically induced by application of *P. fluorescens* WCS417 to the roots (H. Gerrits, in preparation). Several bacterial products have been implicated as signals for the induction of ISR, such as outer membrane lipopolysaccharide (LPS), pseudobactin, salicylate and ethylene.[118,119] However, a description of the functional role of each of these compounds awaits further research.

Root inoculation of *Nicotiana tabacum* with the biocontrol strain *P. fluorescens* CHA0 induced leaf resistance against tobacco necrosis virus (TNV).[113] Strain CHA0, in contrast to *P. fluorescens* WCS417 on *Arabid-*

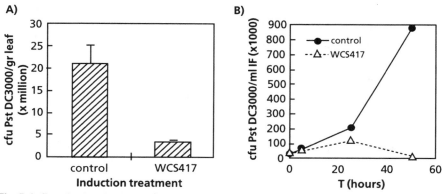

Fig. 7-4. Growth of pathogenic *Pseudomonas syringae* pv. *tomato* strain DC3000 in leaves of intact *Arabidopsis thaliana* plants (**A**) and in intracellular fluid (IF) obtained from *Arabidopsis* leaves (**B**). WCS417 indicates inoculation of roots prior to leaf infection with the pathogen or to isolation of leaf fluid.

opsis, caused the accumulation of several PR proteins (i.e., PR1, β-1,3-glucanases and endochitinases) in the intercellular fluid of leaves. Mutant CHA0 strains, defective in the GacA regulator and, therefore, unable to synthesize the antagonistic compounds hydrogen cyanide and 2,4-diacetyl-phloroglucinol, induced resistance to TNV to similar levels as that of the wild-type strain. Likewise, bacterial siderophore negative mutants were still effective but with lower efficiencies.

Iron Induced Resistance

It was found recently that one of the responses associated with iron stress is induced resistance, the activation of specific plant defense pathways. Under sterile conditions *Arabidopsis* plants, growing on medium supplemented with 100 mM of the synthetic iron chelator EDDHA, showed a high degree of resistance against infection with *P. syringae* pv. *tomato* strain DC3000 (H. Gerrits, in preparation). Growth of radish on medium with pseudobactin 374 (the siderophore of *P. fluorescens* WCS374) resulted in reduction of Fusarium wilt.[119] At the molecular level, the pathogen related β-1,3-beta-glucanase gene *BGL2* of *Arabidopsis* was found to be highly induced in plants growing on medium containing the synthetic iron chelator EDDHA. Induction was abolished when the chelator was complexed with iron. However, other *PR* genes, such as *PR1*, were not responsive to iron. It is interesting to note that ethylene has also been shown to induce expression of β-1,3-glucanase but not PR1 in bean.[117,120] Ethylene has often been proposed to play a role in defense responses.

An important question is why plants activate defense mechanisms in response to iron stress. In pathogenesis, the iron status of the host is of extreme importance to the pathogen as this determines its capacity to grow (e.g., in animal systems excess iron is a major cause of infections[121]). In plant and animal pathogenesis, the siderophores of the pathogen are important virulence determinants.[122] The *in planta* regulated production of microbial siderophore was found to contribute significantly to invasive growth of *Erwinia chrysanthemi*.[123,124] Therefore, pathogen infection and growth can lead to (local) iron limitation and plant cells may use iron stress as an indication of pathogen infection.

The specific induction in *Arabidopsis* of the *BGL2* gene separates iron-induced resistance from induced resistance caused by the biocontrol strain *P. fluorescens* WCS417 (ISR), as in the latter process no *PR* genes are activated (H. Gerrits, in preparation). In ISR, iron limitation apparently plays little or no role. Only at much higher cell concentrations can root inoculation cause iron limitation and add to plant defense. The absence of induction of the *PR1* gene in both iron- and WCS417-induced resistance sets them both apart from the systemic resistance induced by pathogenic microorganisms (SAR).

Whereas strain WCS417 is unable to induce expression of the *Arabidopsis PR* gene, the biocontrol strain *P. fluorescens* CHA0 induces resistance in tobacco with the concomitant expression of several *PR* genes.[113] CHA0-induced resistance seems to be linked, in part, to the production of siderophores, as *sid*-negative mutants are less effective in resistance induction. Clearly, further research is needed to understand the different induced resistance defense mechanisms and to decide if or how they converge into an integrated program for pathogen control.

Conclusions and Prospects

Microbial competition for iron as a mechanism for the repression of pathogens has yet to be demonstrated as a functional biocontrol mechanism. However, several experimental results show that iron-acquisition systems can be effective tools for microbial competition in the rhizosphere. Two aspects of the iron uptake apparatus are of importance, the first is the affinity of the siderophore for its ligand and the other is the efficiency by which the siderophore-iron complex is taken up.

The affinity for iron can vary considerably between different siderophores allowing for direct competition in the binding and exchange of iron. Whether this indeed occurs in the heterogeneous environment of the rhizosphere depends on the strength of the siderophore of the pathogen and the amounts produced. Competition based on siderophore affinity remains to be demonstrated. A complicating factor is that microbes can have several competitive mechanisms, several of which are sensitive to iron stress, thereby obscuring the impact of each separate system.

Efficient uptake will enhance the competitive effects of strong siderophores. Uptake is influenced by the number of receptors and their affinity for the siderophore-iron complex but little attention has been paid to these factors. Most interest has been given to the activities of receptors for heterologous siderophores. It has been demonstrated that certain rhizosphere strains have specialized in iron uptake through heterologous siderophores and that the presence of extra receptors can result in a significant competitive advantage. The effect of this on biocontrol is unclear since pathogens may also utilize this strategy.

Microbial growth in the rhizosphere can have significant effects on the host plant itself, through the production of compounds that interfere with specific processes of the plant. It was demonstrated recently that rhizosphere bacteria can activate plant defense mechanisms. Although the molecular nature of the bacterial signals and the plant responses have not been defined, it is clear that several different signals exist (one being iron limitation) and that the plant response can be diverse. Induced resistance exists in several forms, each with a particular and probably overlapping set of protective compounds, and, therefore, this system holds good promise for resistance to a variety of pathogens.

The ability of biocontrol strains to induce plant defense opens up new avenues to study the complexity of plant protection (Figure 7-5). However,

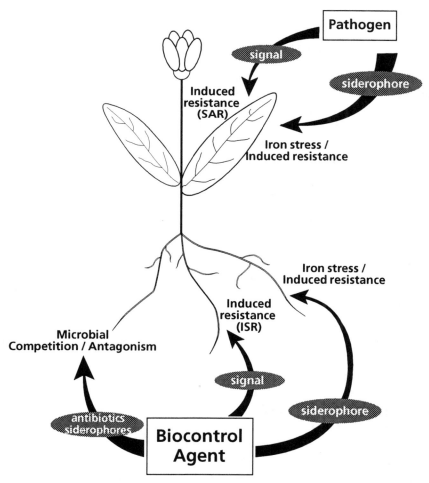

Fig. 7-5. Plant protection by microbial action. A schematic representation of the various mechanisms whereby microorganisms can limit pathogen growth, inside and outside the plant.

much research remains to be done. Once the microbial signals and plant receptors have been identified, it may be possible to induce an adequate defense response through manipulation of the microbial rhizosphere population. Biocontrol strains used in this way could benefit by the introduction of additional receptors for heterologous siderophores. Rhizosphere microorganisms can function as allies of the plant, by direct control of soil pathogens and induction of the plant defense response.

References

1. Pieterse, C. M. J., S. C. M. van Wees, E. Hoffland, J. A. van Pelt and L. C. van Loon. 1996. Systemic resistance in *Arabidopsis* induced by biocontrol bacteria is independent of salicylic acid accumulation and pathogenesis-related gene expression. Plant Cell 8:1255-1237.

1a. Thomashow, L. S. and D. M. Weller. 1996. Current concepts in the use of introduced bacteria for biological disease control: Mechanisms and antifungal metabolites. In: Plant-Microbe Interactions, Volume 1, eds. G. Stacey and N. Keen, pp. 187-235, Chapman and Hall Publ., N. Y.

2. Zellkovich, L., J. Libman and A. Shanzer. 1995. Molecular redox switches based on chemical triggering of iron translocation in triple-stranded helical complexes. Nature 374:790-792.

3. Schwertmann,U. and R. M. Taylor. 1989. Iron oxides. In: Minerals in Soil Environments, eds. J. B. Dixon and S. B. Weed, pp. 379-438. Soil Science Society of America, Madison, WI.

4. Latimer, W. M. 1952. Oxidation Potentials. Prentice-Hall, Inc., New York.

5. Raymond, K. N. and C. J. Corrano. 1979. Coordination chemistry and microbial iron transport. Acc. Chem. Res. 12:183-190.

6. Römheld, V. and H. Marschner. 1981. Rhythmic iron stress reactions in sunflower at suboptimal iron supply. Physiol. Plant 53:347-353.

7. Lankford, C. E. 1973. Bacterial assimilation of iron. Crit. Rev. Microbiol. 2:273-331.

8. Lindsay, W. L. and A. P. Schwab. 1982. The chemistry of iron in soils and its availability to plants. J. Plant Nutr. 5:821-840.

9. Schwab, A. P. and W. L. Lindsay. 1983. Effect of redox on solubility and availability of iron. Soil. Sci. Soc. Am. J. 47: 201-205.

10. Crowley, D. E., C. P. P. Reid and P. J. Szaniszlo. 1987. Microbial siderophores as iron sources for plants. In: Iron Transport in Microbes, Plants and Animals, eds. G. Winkelman, D. van der Helm and J. B. Neilands, pp. 375-386. VCG Verslagsgesellschaft, Weinheim.

11. Kalbasi, M. and H. Shariatmadari. 1993. Blood powder, a source of iron for plants. J. Plant Nutr. 16:2213-2223.

12. Eynard, A., M. C. del Campillo, V. Barron and J. Torrent. 1992. Use of vivianite to prevent iron chlorosis in calcareous soils. Fert. Res. Int. J. Fert. Use. Technol. 31:61-67.

13. Gambrell, R. P. 1994. Trace and toxic metals in wetlands. J. Environ. Qual. 23:883-891.

14. Sah, R. N., D. S. Mikkelsen and A. A. Hafez. 1989. Phosphorus behavior in flooded-drained soils. II. Iron transformation and phosphorus sorption. Soil Sci. Soc. Am. J. 53:1723-1729.

15. Neilands, J. B. 1981. Microbial iron compounds Annu. Rev. Biochem. 50:715-731.
16. van der Hofstad, G. A. M. J., J. D. Marugg, H. B. Nielander, I. van Megen, A. M. M. van Pelt, K. Recourt and P. J. Weisbeek. 1986. The iron-uptake system of the plant growth-stimulating *Pseudomonas putida* WCS358: genetic analysis and properties and structure analysis of its siderophore. In: Recognition in Microbe-Plant Symbiotic and Pathogenic Interactions, ed. B. Lugtenberg, pp. 405-408. Springer-Verlag, Berlin.
17. Abdallah, M. A. 1991. Pyoverdins and pseudobactins. In: CRC Handbook of Microbial Iron Chelates, ed. G. Winkelmann, pp. 139-153. CRC Press, Boca Raton, FL.
18. Hennecke, H. 1990. Regulation of bacterial gene expression by metal-protein complexes. Mol. Microbiol. 4:1621-1628.
19. Silver, S. and M. Walderhaug. 1992. Gene regulation of plasmid- and chromosome-determined inorganic ion transport in bacteria. Microbiol. Rev. 56: 195-228.
20. de Lorenzo, V., S. Wee, M. Herrore and J. B. Neilands. 1987. Operator sequences of the aerobactin operon of plasmid ColV-K30 binding the ferric uptake regulation (*fur*) repressor. J. Bacteriol. 169:2624-2630.
21. Staggs, T. M. and R. D. Perry. 1992. Fur regulation in *Yersinia* species. Mol. Microbiol. 6: 2507-2516.
22. Litwin, C. M. and S. B. Calderwood. 1993. Cloning and genetic analysis of the *Vibrio vulnificus fur* gene and construction of a *fur* mutant by *in vivo* marker exchange. J. Bacteriol. 175:706-715.
23. Thomas, C. E. and P. F. Sparling. 1994. Identification and cloning of a *fur* homologue from *Neisseria meningitidis*. Mol. Microbiol. 11:725-737.
24. Prince, R. W., C. D. Cox and M. L. Vasil. 1993. Coordinate regulation of siderophore and exotoxin A production: molecular cloning and sequencing of the *Pseudomonas aeruginosa fur* gene. J. Bacteriol. 175:2589-2598.
25. Venturi, V., C. Ottevanger, M. Bracke and P. J. Weisbeek. 1995. Iron regulation of siderophore biosynthesis and transport in *Pseudomonas putida* WCS358: involvement of a transcription activator and of the Fur protein. Mol. Microbiol. 15:1081-1093.
26. Litwin, C. M. and S. B. Calderwood. 1993. Role of iron in regulation of virulence genes. Clin. Microbiol. Rev. 6:137-149.
27. Waldbeser, L. S., M. E. Tolmalsky, L. A. Actis and J. H. Crosa. 1993. Mechanisms for negative regulation by iron of the *fatA* outer-membrane protein gene expression in *Vibrio anguillarum* 755. J. Bacteriol. 268:10433-10439.
28. Salinas, P. C., M. E. Tolmasky and J. H. Crosa. 1989. Regulation of the iron uptake system in *Vibrio anguillarum*. Evidence for a cooperative effect between two transcriptional activators. Proc. Natl. Acad. Sci. USA 86:3529-3533.
29. Poole, K., L. Young and S. Neshat. 1990. Enterobactin-mediated iron transport in *Pseudomonas aeruginosa*. J. Bacteriol. 172:6991-6996.
30. O'Sullivan, D. J., J. Morris and F. O'Gara. 1990. Identification of an additional ferric-siderophore uptake gene clustered with receptor, biosynthesis, and fur-like regulatory genes in fluorescent *Pseudomonas* sp. strain M114. Appl. Environ. Microbiol. 56:2056-2064.
31. Marugg, J. D., H. B. Nielander, A. JG. Horrevoets, I. van Megen, I. van Genderen and P. J. Weisbeek. 1988. Genetic organization and transcriptional analysis of a major gene cluster involved in siderophore biosynthesis in *Pseudomonas putida* WCS358. J. Bacteriol. 170:1812-1819.

32. Venturi,V., C. Ottenvanger, J. Leong and P. Weisbeek. 1993. Identification and characterization of a siderophore regulatory gene (*pfrA*) of *Pseudomonas putida* WCS358: homology to the alginate regulatory gene algQ of *Pseudomonas aeruginosa*. Mol. Microbiol. 10:63-73.

33. Sexton, R., P. R. Gill Jr., M. J. Callanan, D. J. O'Sullivan, D. N. Dowling and F. O'Gara. 1995. Iron-responsive gene expression in *Pseudomonas fluorescens* M114: cloning and characterization of a transcription-activating factor PbrA. Mol. Microbiol. 15:297-306.

34. Cunliffe, H. E., T. R. Merriman and I. L. Lamont. 1995. Cloning and characterization of *pvdS*, a gene required for pyoverdine synthesis in *Pseudomonas aeruginosa*: PvdS is probably an alternative sigma factor. J. Bacteriol. 177:2744-2750.

35. Lonetto, M. A., K. L. Brown, K. E. Rudd and M. J. Buttner. 1994. Analysis of the *Streptomyces coelicolor sigE* gene reveals existence of a subfamily of eubacterial RNA polymerase sigma factors involved in the regulation of extracytoplasmic function. Proc. Natl. Acad. Sci. USA 91:7573-7577.

36. Bitter, W., J. D. Marugg, L. A. de Weger, J. Tommassen and P. J. Weisbeek. 1991. The ferric-pseudobactin receptor PupA of *Pseudomonas putida* WCS358: Homology to TonB-dependent *E.coli* receptors and specificity of the protein. Mol. Microbiology 5:647-655.

37. Koster, M. 1994. Molecular genetics of siderophore mediated iron acquisition in *Pseudomonas putida*. Ph. D. Diss., Utrecht University, Utrecht, The Netherlands.

38. Bitter, W., J. Tommassen and P. J. Weisbeek. 1993. Identification and characterization of the *exbB, exbD* and *tonB* genes of *Pseudomonas putida* WCS358: their involvement in ferric-pseudobactin transport. Mol. Microbiol. 7:117-131

39. Postle, K. 1990. TonB and the gram-negative dilemma. Mol. Microbiol. 4:2019-2025.

40. Koster, M., J. van de Vossenberg, J. Leong and P. Weisbeek. 1993. Identification and characterization of the *pupB* gene encoding an inducible ferric-pseudobactin receptor of *Pseudomonas putida* WCS358. Mol. Microbiol. 8:591-601.

41. Koster, M., K. van Klompenburg, W. Bitter, J. Leong and P. Weisbeek. 1994. Role of the outer membrane ferric siderophore receptor pupB in signal transduction across the bacterial cell envelope. EMBO J. 13:2805-2813.

42. van Hove, B., H. Staudenmaier and V. Braun. 1990. Novel two component transmembrane transcription control: regulation of iron dicitrate transport in *Escherichia coli* K-12. J. Bacteriol. 172:6749-6758.

43. Dean, C. R. and K. Poole. 1993. Expression of the ferric enterobactin receptor (PfeA) of *Pseudomonas aeruginosa*: involvement of a two-component regulatory system. Mol. Microbiol. 8:1095-1103.

44. Killman, H., R. Benz and V. Braun. 1993. Conversion of the FhuA transport protein into a diffusion channel through the outer membrane of *Escherichia coli*. EMBO J. 12:3007-3016.

45. Rutz, J. M., J. Liu, J. Lyons, J. Goranson, S. K. Armstrong, M. A. McIntosh, J. B. Feix and D. E. Klebba. 1992. Formation of a gated channel by a ligand specific transport protein in the bacterial outer membrane. Science 258:471-475.

46. Bitter, W. 1992. Ferric-pseudobaction transport across the outer membrane of *Pseudomonas putida* WCS358. Ph. D. Diss., Utrecht University, Utrecht, The Netherlands.

47. Morris, J., D. F. Donnely, E. O'Neil, F. McConnell and F. O'Gara. 1994. Nucleotide sequence analysis and potential environmental distribution of a ferric

pseudobactin receptor gene of *Pseudomonas* sp. strain M114. Mol. Gen. Genet. 242:9-16.

48. Cadenas, E. 1989. Biochemistry of oxygen toxicity. Ann. Rev. Biochem. 58:79-110.

49. Fridovich, I. 1989. Superoxide dismutases. An adaptation to a paramagnetic gas. J. Biol. Chem. 264:1-11.

50. Van Camp, W., C. Bowler, R. Villaroel, E. W. T. Tsang, M. van Montagu and D. Inz. 1990. Characterization of iron superoxide dismutase cDNAs from plants obtained by genetic complimentation in *Escherichia coli.* Proc. Natl. Acad. Sci. USA 87:9903-9907.

51. Proudhon, D., J. Briat and A. Lescure. 1989. Iron induction of ferritin synthesis in soybean cell suspensions. Plant Physiology 90:586-590.

52. Gray, N. K. and M. W. Hentze. 1994. Iron regulatory protein prevents binding of the 43S translation pre-initiation complex to ferritin and eALAS mRNAs. EMBO J. 12:3882-3891.

53. Klausner, R. D., T. Rouault and J. B. Harford. 1993. Regulating the fate of mRNA: the control of cellular iron metabolism. Cell 72:19-28.

54. Binder, R., J. A. Horowitz, J. P. Basilion, D. M. Koeller, R. D. Klausner and J. B. Harford. 1994. Evidence that the pathway of transferrin receptor mRNA degradation involves an endonucleolytic cleavage within the 3'UTR and does not involve poly(A) tail shortening. EMBO J. 13:1969-1980.

55. Haile, D. J., T. A. Rouault, J. B. Harford, M. C. Kennedy, G. A. Blondin, H. Beinert and R. D. Klausner. 1992. Cellular regulation of the iron-responsive element binding protein: disassembly of the cubane iron-sulfur cluster results in high-affinity RNA binding. Proc. Natl. Acad. Sci. USA 89:11735-11739.

56. Kennedy, M. C., L. Mende-Mueller, G. A. Blondin and H. Beinert. 1992. Purification and characterization of cytosolic aconitase from beef liver and its relationship to the iron-responsive element binding protein. Proc. Natl. Acad. Sci. USA 89:11730-11734.

57. Pantopoulos, K. and M. W. Hentze. 1995. Rapid responses to oxidative stress mediated by iron regulatory protein. EMBO J. 14:2917-2924.

58. Ciurli, S., M. Carrié, J. A. Weigel, M. J. Carney, T. D. P. Stack, G. C. Papaefthymiou and R. H. Holm. 1990. Subsite-differentiated analogues of native [4Fe-4S]$^{2+}$ clusters: preparation of clusters with five and six-coordinate subsites and modulation of redox potentials and charge distributions. J. Am. Chem. Soc. 112:2654-2664.

59. Khoroshilova, N., H. Beinert and P. J. Kiley. 1995. Association of a polynuclear iron-sulfur center with a mutant FNR protein enhances DNA binding. Proc. Natl. Acad. USA 92:2499-2503.

60. Welch, R. M. and T. A. LaRue. 1990. Physiological characteristics of Fe accumulation in the bronze mutant of *Pisum sativum* L., cv sparkle. Plant Physiology 93:723-729.

61. Grusak, M. A., R. M. Welch and L. V. Kochian. 1990. Physiological characterization of a single-gene mutant of *Pisum sativum* exhibiting excess iron accumulation. Plant Physiology 93:976-981.

62. Bienfait, H. F. 1988. Proteins under the control of the gene for Fe efficiency in tomato. Plant Physiology 88:785-787.

63. Ling, H. Q., A. Pich, G. Scholz and M. W. Ganal. 1996. Genetic analysis of two tomato mutants affected in the regulation of iron metabolism. Mol. Gen. Genet. 252:87-92.

64. Terry, N. and J., Abadia. 1986. Function of iron in chloroplasts. J. Plant Nutr. 9:609-646.
65. Morales, F., A. Abadia and F. Aadia. 1991. Chlorophyll fluorescence and photon yield of oxygen evolution in iron-deficient sugar beet (*Beta vulgaris* L.) leaves. Plant Physiology 97:886-893.
66. Romera, F. J and E. Alcantara. 1994. Iron-deficiency stress responses in cucumber (*Cucumis sativus* l.) roots. Plant Physiology 105:1133-1138.
67. de Vos, C. R., J. Lubberding and H. F Bienfait. 1986. Rhizosphere acidification as a response to iron deficiency in bean plants. Plant Physiology 81:842-846.
68. Boyer, R. F., H. M. Clark and S. Sanchez. 1989. Solubilization of ferrihydrite iron by plant phenolics: a model for rhizosphere processes. J. Plant Nutr. 12:581-592.
69. Bienfait, H. F. 1988. The turboreductase in plant plasma membranes. NATO ASI Ser. Ser. A. Life Sci. 157:89-93.
70. Bienfait, F. 1987. Biochemical basis of iron efficiency reactions in plants. In: Iron Transporting Microbes, Plants and Animals, eds. G. Winkelman, D. van der Helm and J. B. Neilands, pp. 339-349. VCG Verlagsgesellschaft Weinheim.
71. Ma, J. F. and K. Nomoto. 1993. Two related biosynthetic pathways of mugineic acids in gramineous plants. Plant Physiology 102:373-378.
72. Whipps, J. M. 1987. Carbon loss from the roots of tomato and pea seedlings grown in soil. Plant Soil 102:95-100.
73. Schippers, B. A., A. Bakker and P. A. H. M. Bakker. 1987. Interactions of deleterious and beneficial rhizosphere microorganisms and the effect of cropping practices. Ann. Rev. Phytopathol. 25:339-358.
74. O'Connell, K. P., R. M. Goodman and J. Handelsman. 1996. Engineering the rhizosphere: expressing a bias. Tibtech 14:83-88.
75. Drahos, D. J., G. F. Barry, B. C. Hemmings, E. J., Brandt and E. L. Kline. 1992. Spread and survival of genetically marked bacteria in soil. In: Release of Genetically Engineered and Other Microorganisms, eds. J. C. Fry and M. J. Day, pp. 147-159. Cambridge Univ. Press, Great Britain.
76. Raaymakers, J. 1994. Microbial interactions in the rhizosphere: root colonization by *Pseudomonas* spp. and suppression of fusarium wilt. Ph. D. Diss., Utrecht University, Utrecht, The Netherlands.
77. Roszak, D. B. and R. R. Colwell. 1987. Survival strategies of bacteria in the natural environment. Microbiol. Rev. 51:365-79.
78. Cook, R. J., L. S. Thomashow, D. M. Weller, D. Fujimoto, M. Mazzola, G. Bangera and Dal-soo Kim. 1995. Molecular mechanisms of defense by rhizobacteria against root disease. Proc. Natl. Acad. Aci. USA 92:4197-4201.
79. de Weger, L. A. de, C. I. M. van der Vlugt, A. H. M. Wijfjes, P. A. H. M. Bakker, B. Schippers and B. Lugtenberg. 1987. Flagella of a plant growth-stimulating *Pseudomonas fluorescens* strain are required for colonization of potato roots. J. Bacteriol. 169:2769-73.
80. Weller, D. M. 1988. Biological control of soilborne plant-pathogens in the rhizosphere with bacteria. Ann. Rev. Phytopathol. 26:379-407.
81. Curl, E. A. and B. Truelove. 1986. The Rhizosphere. Springer-Verlag, Berlin.
82. Smit, G., T. J. J. Logman, M. E. T. I. Boerrigter, J. W. Kijne and B. J. J. Lugtenberg. 1989. Purification and partial characterization of the *Rhizobium leguminosarum* biovar *viciae* Ca^{2+}-dependent adhesin, which mediates the first step in attachment of cells of the family *Rhizobiaceae* to plant root hair tips. J. Bacteriol. 171:4054-4062.

83. Dienema, M. H. and L. P. T. M. Zevenhuizen. 1971. Formation of cellulose fibrils by gram-negative bacteria and their role in bacterial flocculation. Arch. Mikrobiol. 78:42-57.
84. Matthysse, A. G., K. V. Holmes and R. H. G. Gurlitz. 1981. Elaboration of cellulose fibrils by *Agrobacterium tumefaciens* during attachment to carrot cells. J. Bacteriol. 145:583-595.
85. Kijne, J. W., G. Smit, C. L. Diaz and B. J. J. Lugtenberg. 1988. Lectin-enhanced accumulation of manganese-limited *Rhizobium leguminosarum* cells on pea root hair tips. J. Bacteriol. 170:2994-3000.
86. Bowen, G. D. and A. D. Rovira. 1976. Microbial colonization of plant roots. Ann. Rev. Phytopathol. 14:121-44.
87. Chin-A-Woeng, Th. F. C., W. de Priester, A. J. van der Bij and B. J. J. Lugtenberg. 1997. Description of the colonization of a gnotobiotic tomato rhizosphere by *Pseudomonas fluorescens* biocontrol strain WCS365, using scanning electron microscopy. Mol. Plant-Microbe Int. 10, 79-86.
88. Salmond, G. P. C., B. W. Bycroft, G. S. Steward and P. Williams. 1995. The bacterial 'enigma': cracking the code of cell-cell communication. Mol. Microbiol. 16:615-624.
89. Swift, S., J. P. Throup, P. Williams, G. P. C. Salmond and G. S. A. B. Stewart. 1996. Quorum sensing: a population-density component in the determination of bacterial phenotype. TIBS 21:214-219.
90. Dowling, D. N. and F. O'Gara. 1994. Metabolites of *Pseudomonas* involved in the biocontrol of plant disease. Tibtech 12:133-141.
91. Venturi, V., P. J. Weisbeek and M. Koster. 1995. Gene regulation of siderophore-mediated iron acquisition in *Pseudomonas*: not only the Fur repressor. Mol. Microbiol. 17:603-610.
92. Tufano, T. P. and K. N. Raymond. 1981. Coordination chemistry of microbial iron transport compounds. 21. Kinetics and mechanisms of iron exchange in hydroxamate siderophore complexes. J. Am. Chem. Soc. 103:6617-6624.
93. Buyer, J. S. and J. Leong. 1986. Iron transport-mediated antagonism between plant growth-promoting and plant deleterious *Pseudomonas* strains. J. Biol. Chem. 261:791-794.
94. Bakker, P. A. H. M. 1989. Siderophore-mediated plant growth promotion and colonization of roots by strains of *Pseudomonas* spp. Ph. D. Diss. Utrecht University, Utrecht, The Netherlands.
95. Raaymakers, J. M., I. van der Sluis, M. Koster, P. A. Bakker and B. Schippers. 1995. Utilization of heterologous siderophores and rhizosphere competence of fluorescent *Pseudomonas* spp. Can. J. Microbiol. 41:126-135.
96. Stutz, E. 1964. Aufname von ferrioxamine B durch tomatenpflanzen. Experientia 20:430-431.
97. Orlando, J. A. and J. B. Neilands. 1982. In: Chemistry and Biology of Hydroxamic Acids, eds. H. Kehl, pp. 123-129. Karger, Basel.
98. Powell, P. E., P. J. Szaniszlo, G. R. Cline and C. P. P. Reid. 1982. Hydroxamate siderophores in the iron nutrition of plants. J. Plant Nutr. 5:653-673.
99. Cline, G., C. P. P. Reid, P. E. Powell and P. J. Szaniszlo. 1984. Effects of a hydroxamate siderophore on iron absorption by sunflower and sorghum. Plant Physiology 76:36-40.
100. Reid, C. P. P., D. E. Crowley, P. E. Powell, H. J. Kim and P. J. Szaniszlo. 1984. Utilization of rion by oat when supplied as ferrated synthetic chelate or as ferrated hydroxamate siderophore. J. Plant Nutr. 7:437-447.

101. Römheld, V. and H. Marschner. 1986. Evidence for a specific uptake system for iron phytosiderophores on roots of grasses. Plant Physiology 80:175-180.
102. Bar-Ness, E., Hadar, Y., Chen, Y. Römheld, V. and H. Marschner. 1992. Short-term effects of rhizosphere microorganisms on Fe uptake from microbial sidero-phores by maize and oat. Plant Physiology 100:451-456.
103. Becker, J. O., R. Hedges and E. Messens. 1985. Inhibitory effects of pseudobactin on the uptake of iron by higher plants. Appl. Environ. Microbiol. 49:1090-1093.
104. Yuen, G. Y. and M. N. Schroth. 1986. Inhibition of *Fusarium oxysporum* f. sp. *dianthi* by iron competition with an *Alcaligenes* sp. Phytopathology 76:171-176.
105. Duijff, B. J. 1994. Suppression of fusarium wilt by fluoresent *Pseudomonas* spp.: mechanisms, influence of environmental factors and effects on plant iron nutrition. Ph. D. Diss., Utrecht University, Utrecht, The Netherlands.
106. Leeman, M., J. A. van Pelt, F. M. den Ouden, M. Heinsbroek, P. A. H. M. Bakker and Schippers, B. 1995. Induction of systemic resistance by *Pseudomonas fluorescens* in radish cultivars differing in susceptibility to fusarium wilt, using a novel bioassay. Eur. J. Plant. Pathol. 101:655-664.
107. Voisard, C., C. Keel, D. Haas and G. Défago. 1989. Cyanide production by *Pseudomonas fluorescens* helps suppress black rot of tobacco under gnotobiotic conditions. EMBO J. 8:351-358.
108. Laville, J., C. Voisard, C. Keel, M. Maurhofer, G. Défago and D. Haas. 1992. Global control in *Pseudomonas fluorescens* mediating antibiotic synthesis and suppression of black root rot of tobacco. Proc. Natl. Acad. Sci. USA 89:1562-1566.
109. Hammerschmidt, R. and J. Kuc. 1995. Induced Resistance to Disease in Plants. Kluwer Academic Publishers, Dordrecht, The Netherlands.
110. Cameron, R. K., R. A. Dixon and C. J. Lamb. 1994. Biologically induced systemic acquired resistance in *Arabidopsis thaliana*. Plant J. 5:715-725.
111. van Peer, R., G. J. Nieman and B. Schippers. 1991. Induced resistance and phytoalexin accumulation in biological control of Fusarium wilt of carnation by *Pseudomonas* spp. strain WCS417r. Phytopathology 81:728-734.
112. Niemann, G. J., J. Leim, A. van der Kerk van Hoof and W. M. A. Niessen. 1992. Phytoalexins, benzoxazinones, N-asoylanthranilates and N-aroylanilines from *Fusarium* infected carnation stems. Phytochemistry 31:3761-3767.
113. Mauerhofen, M., C. Hase, P. Meuwly, J. P. Métraux and G. Défago. 1994. Induction of systemic resistance of tobacco to tobacco necrosis virus by the root-colonizing *Pseudomonas fluorescens* strain CHA0: Influence of the *gacA* gene and of pyoverdine production. Phytopathology 84:139-146.
114. Wei, G., J. W. Kloepper and S. Tuzun. 1991. Induction of systemic resistance of cucumber to *Colletotrichum orbiculare* by selected strains of plant growth-promoting rhizobacteria. Phytopathology 81:1508-1512.
115. Alström, S. 1991. Induction of disease resistance in common bean susceptible to halo blight bacterial pathogen after seed bacterisation with rhizosphere pseudo-monads. J. Gen. Appl. Microbiol. 37:495-501.
116. Zhou, T. and T. C. Paulitz. 1994. Induced resistance in the biocontrol *Phythium aphanidermatum* by *Pseudomonas* spp. on cucumber. J. Phytophatol. 142:51-63.
117. Neuenschawander, U., K. Lawton and J. Ryals. 1995. Systemic Acquired Resistance. In: Plant-Microbe Interactions I, eds. G. Stacey and N. T. Keen, pp 81-106. Chapman and Hall, New York.
118. van Peer, R. and B. Schippers. 1992. Lipopolysccharides of plant growth-promoting *Pseudomonas* sp. WCS417r induce resistance in carnation to Fusarium wilt. Neth. J. Pl. Path. 98:129-139.

119. Leeman, M., J. A. van Pelt, F. M. den Ouden, M. Heinsbroek, P. A. H. M. Bakker and B. Schippers. 1995. Induction of systemic resistance against Fusarium wilt of radish by lipopolysaccharides of *Pseudomonas fluorescens*. Phytopathology 85: 1021-1027.
120. Mauch, F. and L. A. Staehelin. 1989. Functional implications of the subcellular localization of ethylene-induced chitinase and β-1,3-glucanase in bean leaves. Plant Cell 1:447-457.
121. Weinberg, E. D. 1978. Iron and infection. Microbiol. Rev. 42:45-60.
122. Enard, C., A. Diolez and D. Expert. 1988. Systemic virulence of *Erwinia chrysanthemi* 3937 requires a functional iron assimilation system. J. Bacteriol. 170:2419-2426.
123. Masclaux, C. and D. Expert. 1995. Signalling potential of iron in plant-microbe interactions: the pathogenic switch of iron transport in *Erwinia chrysantheuri*. Plant J. 7:121-128
124. Expert, D., C. Enard and C. Masclaux. 1996. The role of iron in plant host-pathogen interactions. Trends Microbiol. 4:232-237.
125. Loper, J. E., and S. E. Lindow. 1994. A biological sensor for iron available to bacteria in their habitats on plant surfaces. Appl. Environ. Microbiol. 60:1934-1941.
126. Bangera, M. G., D. M. Weller, and L. S. Thomashow. 1995. Genetic analysis of the 2,4-diacetylphloroglucinol locus from *Pseudomonas fluorescens* Q2-87. In: Advances in Molecular Genetics of Plant-Microbe Interactions, vol. 3, eds. M. J. Daniels, J. A. Downie, and A. E. Osbourn, pp. 383-386. Kluwer Academic Publishers, Dordrecht.
127. Kloepper, J. W., J. Leong, M. Teintze, and M. N. Schroth. 1980. *Pseudomonas* siderophores: a mechanism explaining disease-suppressive soils. Curr. Microbiol. 4:317-320.

CHAPTER 8

Transcriptional Regulation of Plant Genes Responsive to Pathogens and Elicitors

Paul J. Rushton and Imre E. Somssich

Plants are continuously confronted with diverse potential pathogens within their environment. Nevertheless, disease development is the exception rather than the rule, revealing that plants must have evolved highly efficient defense systems. Indeed, over the past two decades biochemical, molecular and genetic studies have uncovered many components of the plant defense arsenal and have helped us to gain a more detailed, though still fragmentary, picture of the complex mechanisms involved in establishing resistance (for recent reviews see references 1–4). One can distinguish between two types of plant resistance, race-specific resistance and non-host resistance. Race-specific resistance is genetically defined by the direct or indirect interaction between the product of a dominant or semi-dominant major plant resistance gene (R) and the product of the corresponding dominant pathogen avirulence (avr) gene[5,6]. Thus, the presence or absence of race/cultivar-specific resistance in a given host plant species is dependent on the genotypes of the interacting partners. By contrast, non-host resistance (basic incompatibility) is exhibited by all plant species responding to potential pathogens without apparent R/avr gene combinations. Apart from this distinction, however, many of the subsequent biochemical reactions triggered in the plant are similar to both resistance types.

Perception of the pathogen by the plant immediately triggers various early defense responses via a number of signal transduction pathways[3,7]. A major target of signal transduction is the cell nucleus where the terminal biochemical signals lead to the transcriptional activation of numerous genes often arbitrarily referred to as 'pathogen-responsive' or 'defense-related' genes. We will use the term 'pathogen-responsive' throughout this review, rather than 'defense-related', since many of the genes which are upregulated during the defense response also have important functions in various other processes during growth and development. Here we will focus primarily on what is known about the transcriptional regulation of these plant genes. In particular we will try to outline what information is available concerning the *cis*-acting elements and the *trans*-acting factors involved in mediating pathogen-induced expression of such genes.

Pathogen-Responsive Plant Genes

Studies using a number of plant pathosystems have revealed that *de novo* synthesis of plant proteins is required to mount an effective defense response[8,9]. Based on transcription inhibitor experiments and nuclear run-on transcription assays it appears that in most cases this is the consequence of transcriptional activation of genes. For this reason, considerable efforts have been made to devise molecular screening approaches to distinguish and to isolate plant genes whose expression is strongly and rapidly influenced by signals generated during pathogenesis. In this respect, the use of homogeneously responding cultured plant cells in combination with defined pathogen-derived molecules or chemical compounds (elicitors[10]) capable of mimicking various facets of the infection process proved to be another valuable tool in the isolation of such genes. These screenings have led to the cloning of a large variety of pathogen-responsive genes[1,3,8,11-13]. Table 8-1 lists genes identified from various plant species whose expression is transcriptionally upregulated upon treatment with diverse pathogens and elicitors. Based on the function of their encoded products, such studies substantiated earlier biochemical work showing that the active response of plants to attempted pathogen ingress is associated with dramatic reprogramming of metabolism. Many of these genes code for enzymes involved in the flow of carbon from primary into secondary metabolism, e.g., enzymes of the shikimate and the general phenylpropanoid pathways, along with various enzymes from subsequent branch pathways leading to the production of low molecular weight antimicrobial phenolic compounds (phytoalexins), lignin and other wall-bound phenolics[14-17]. One arbitrarily defined subset of genes, termed pathogenesis-related (PR) genes, have often been found to be associated with the defense response in numerous plant species[3]. The function of many of these PR proteins still remains unknown. However, the various functions encoded by other genes demonstrate that a plethora of biochemical activities including hydrolases, kinases, transcription factors and enzymes of diverse metabolic pathways are also modulated by the pathogen/elicitor-triggered incoming stimuli. Despite the fact that a causal relationship between the encoded gene products and the establishment of resistance is still entirely lacking, many of these genes are proving to be extremely valuable tools for the investigation of the regulatory components comprising the penultimate steps of defense signal transduction cascades.

Table 8-1. Plant genes showing induced expression upon pathogen attack or by elicitor treatment

Gene	Function	Plant species	Induced by[a,b]
G6PDH	Glucose 6-phosphate dehydrogenase	Parsley	Fungal elicitor[17]
GAPDH	Glyceraldehyde 3-phosphate dehydrogenase	Parsley	Fungi[17]
SHH	S-adenosyl-L-homocysteine hydrolase	Parsley, periwinkle, alfalfa	Fungi[17,84,85]
SMS	S-adenosyl-L-methionin synthetase	Parsley, alfalfa	Fungal elicitor[17,84]
ACC-S	1-aminocyclopropane-1-carboxylate synthase	Tomato, parsley	Fungal elicitor, fungi[86]
ACO-1	1-aminocycloprpane-1-carboxylate oxidase	Melon	Bacteria[87]
SOD	Superoxide dismutase	Tobacco, tomato	Bacteria, SA, nematode[88,89]
PLD	Phospholipase D	Rice	Bacteria[90]
LOX	Lipoxygenase	Tomato, rice, tobacco, bean, Arabidopsis, wheat, cucumber, potato, oats, soybean, coffee	Fungi, viruses, bacteria, MJ, elicitor, herbivore, viroid[91]
cw-3FAD	Fatty acid desaturase (cytosolic)	Parsley	Fungal elicitor, fungi[17]
pw-3FAD	Fatty acid desaturase (plastidic)	Parsley	Fungal elicitor[17]
cw-6FAD	Fatty acid desaturase (cytosolic)	Parsley, tomato	Fungal elicitor, viroid[17,92]
SFR2	Receptor-like kinase	B. oleracea	Bacteria, SA[93]
DS22	Protein kinase	Tobacco	TMV[94]
GST	Glutathione S-transferase	Potato, bean, wheat, tobacco, Arabidopsis	Fungal elicitor, fungi, JA, SA, bacteria[125,95]
GP	Glutathione peroxidase	Soybean	Elicitor[95]
VSP	Acidic phosphotase	Soybean, Arabidopsis	Elicitors[96]
LAP	Leucine aminopeptidase	Tomato, potato	Elicitor, Mj[97]
SerCP	Carboxypeptidase	Tomato	Elicitor[98]
CDI	Aspartic proteinase inhibitor	Tomato, potato	Elicitor, MJ[97]
CYS	Cysteine proteinase inhibitor	Tomato, potato	Elicitor, MJ[97]
TD	Threonine deaminase	Tomato, potato	Elicitor, MJ[97]
NDPK	Nucleotide diphosphate kinase	Tomato	Elicitor[98]
AspP	Aspartic proteinase	Tomato	Elicitor[98]
CysP	Cysteine proteinase	Tomato	Elicitor[98]
DAHP	3-Deoxy-D-arabino-heptulosonate-7-phosphate synthase	Parsley, tomato, Arabidopsis	Fungal elicitor, bacteria[17,99]
CS1	Chorismate synthase	Parsley, tomato	Fungal elicitor[17,99]
SK	Shikimate kinase	Tomato	Fungal elicitor[99]
EPSPS	5-Enolpyruvyl-shikimate 3-phosphate	Tomato	Fungal elicitor[99]
CM	Chorismate mutase	Parsley, tomato	Fungal elicitor[17,99]
PAL	Phenylalanine ammonia-lyase	Parsley, alfalfa, bean, potato, tobacco, pea, Arabidopsis, soybean, tomato, poplar, pearl millet, wheat, pine, carrot	Fungal elicitors, bacteria, fungi, virus, elicitors[44,100]
C4H	Cinnamate-4-hydroxylase	Parsley, alfalfa, bean, wheat	Fungal elicitor[17,101]

(continued on next page)

[a] Reference numbers are shown as superscripts.
[b] SA = salicylic acid; MJ = methyl jasmonate; JA = jasmonic acid; TMV = tobacco mosaic virus; and CEV = citrus exocortis viroid.

Table 8-1. (*continued*)

Gene	Function	Plant species	Induced by[a,b]
4CL	4-Coumaroyl-CoA ligase	Parsley, alfalfa, pine, potato, soybean, poplar, tobacco, *Arabidopsis*	Fungal elicitor, fungi, bacteria[3]
CPR1	NADPH:cytochrome P450 oxidoreductase	Parsley	Fungal elicitor, fungi[102]
BAH	Benzoic acid 2-hydroxylase	Tobacco	TMV[103]
TDC	Tryptophan decarboxylase	Periwinkle	Fungal elicitor[104]
Tyr-DC	Tyrosine decarboxylase	Parsley, *Arabidopsis*, tobacco, *E. californica, T. rugosum*	Fungal elicitor, fungi, TMV[105,106]
OXO	Oxalate oxidase	Barley	Fungi[107]
OxOLP	Oxalate oxidase-like	Barley	Fungi[108]
ODC	Ornithine decarboxylase	Tobacco	TMV[105]
ELI3	Benzyl alcohol dehydrogenase	*Arabidopsis*	Fungal elicitor, bacteria[109]
IFR	Isoflavone reductase	Alfalfa	Fungal elicitor[101]
IOMT	Isoflavone 7-*O*-methyltransferase	Alfalfa	Fungi[110]
DOMT	Daidzein-*O*-methyltransferase	Alfalfa	Fungi[111]
COMT	Caffeic acid *O*-methyltransferase	Alfalfa, pine, *Arabidopsis*	Fungal elicitor, bacteria[112]
CCOMT	Caffeoyl-CoA-*O*-methyltransferase	Parsley, alfalfa, carnation, carrot, safflower, grapevine	Fungal elicitor[112,113]
OMT1	Orthodiphenol-*O*-methyltransferase	Tobacco	TMV, fungal elicitor[114]
BH72-F1	Flavonoid-*O*-methyltransferase	Barley	Fungi[107]
CAD	Cinnamyl alcohol dehydrogenase	Parsley, bean, alfalfa, wheat, pine, spruce	Fungal elicitor, fungi[115]
CCR	Cinnamoyl coA reductase	Parsley	Fungal elicitor (*unpublished*)
ACCase	Acetyl coA carboxylase	Alfalfa	Fungal elicitor[101]
BH72-08	Homoeriodictyl chalcone synthase	Barley	Fungi[107]
CHS	Chalcone synthase	Bean, alfalfa, tobacco, soybean, sorghum	Fungal elicitor, fungi, bacteria[101]
CHR	Chalcone reductase	Alfalfa	Fungal elicitor[101]
CHI	Chalcone isomerase	Bean, alfalfa	Fungal elicitor, fungi[101]
PPO	Polyphenol oxidase	Tomato, cucumber	Viruses, fungi, elicitor[97]
HMGR	Hydroxymethyl glutaryl CoA reductase	Potato, tobacco	Fungal elicitor, virus, bacteria, elicitor[116]
BBE	Berberine bridge enzyme	*E. californica*	Fungal elicitor[117]
EAS	Sesquiterpene cyclase	Tobacco	Fungal elicitor, bacteria[62]
BMT	Bergaptol *O*-methyltransferase	Parsley	Fungal elicitor[17]
CaS	Casbene synthetase	Castor bean	Fungal elicitor, fungi[16]

(*continued on next page*)

Table 8-1. (*continued*)

Gene	Function	Plant species	Induced by[a,b]
ASA	Anthranilate synthase	*Arabidopsis*	Bacteria[118]
Vst	Resveratrol synthase	Grapevine	Fungal elicitor[16]
SSS	Strictosidine synthase	Periwinkle	Fungal elicitor[104]
HRGP	Hydroxyproline-rich glycoproteins	Parsley, bean, melon, carrot, cucumber, tobacco	Fungal elicitor, fungi, bacteria, virus[3]
PR1	Unknown	Tobacco, *Arabidopsis*, tomato, maize, barley, radish, cowpea	Viruses, fungi, bacteria, SA[119]
PR2	β-1,3 Glucanase	Tobacco, *Arabidopsis*, potato, tomato, oats, barley, sugar beet, alfalfa, bean, pea, wheat, rice, cucumber	Fungal elicitor, viruses, SA, fungi, bacteria[119]
PR3	Chitinase	Parsley, tobacco, pea, *Arabidopsis*, potato, tomato, bean, sugar beet, rape, maize, rice, turnip, barley, cucumber, soybean, melon, onion, spruce, black pine, carrot, peanut, wheat	Fungal elicitor, viruses, SA, fungi, bacteria[119]
PR4	Unknown	Tobacco, potato, barley, tomato, *Arabidopsis*	Viruses, fungi, bacteria, SA[119]
PR5	Unknown	Tobacco, *Arabidopsis*, potato, soybean, sorghum, barley, maize, wheat	Viruses, fungi, bacteria, SA[119]
PR6	Proteinase inhibitor	Potato, tomato, tobacco, melon, barley, soybean, maize, squash	Herbivores, viruses, fungi, viroid, JA bacteria, fungal elicitor[119]
PR7	Proteinase	Tomato	CEV, fungi[119]
PR8	Lysozyme/chitinase	Tobacco, *Arabidopsis*, cucumber, bean, chick pea	TMV, SA[119]
PR9 /POD	Peroxidase	Parsley, tobacco, cucumber, tomato, *S. humilis*, barley, pine wheat, *Arabidopsis*	Fungal elicitor, fungi, viruses, viroid[119]
PR10	Unknown	Parsley, bean, potato, pea, asparagus, apple, soybean, alfalfa, rice, lupine, sorghum, clover	Fungal elicitor, fungi, bacteria, virus, elicitors[119]
Thi	Thionin	Barley, *Arabidopsis*	Fungi[119]
Hsr515	P450 monooxygenase	Tobacco	Bacteria[120]
ELI8	P450 monooxygenase	Parsley	Fungal elicitor[121]
ELI6	Unknown	Parsley	Fungal elicitor[121]
ELI10	Unknown	Parsley	Fungal elicitor[121]
ELI7	Unknown	Parsley	Fungal elicitor, fungi[121]
ELI12	Unknown	Parsley	Fungal elicitor[121]

(*continued on next page*)

Table 8-1. (*continued*)

Gene	Function	Plant species	Induced by[a,b]
ELI15	Unknown	Parsley	Fungal elicitor[121]
ELI16	Unknown	Parsley	Fungal elicitor[121]
ELI17	Unknown	Parsley	Fungal elicitor[121]
WRKY 1	Transcription factor (zinc-finger class)	Parsley	Fungal elicitor, fungi[21]
WRKY3	Transcription factor (zinc-finger class)	Parsley	Fungal elicitor[21]
BPF-1	Transcription factor (myb-like)	Parsley	Fungal elicitor, fungi[11]
myb1	Transcription factor (myb-class)	Tobacco	TMV, bacteria, SA[68]
LTP	Lipid-transfer protein	Barley	Fungi[122]
pRP1 to *6*	Unknown	Barley, wheat	Fungi[123]
WIR1	Unknown	Barley, wheat	Fungi[124]
GRP	Unknown	Barley, tobacco	TMV, SA[125]
GRP94	Endoplasmin	Barley	Fungi[107]
14-3-3	Unknown	Barley	Fungi[107]
BH72-Q3	Unknown	Barley	Fungi[107]
BH16-12	Unknown	Barley	Fungi[107]
NDR1	Unknown	*Arabidopsis*	Bacteria, fungi[126]
NPR1/ NIM1	Unknown	*Arabidopsis*	Fungi, SA[127]
PIG2	Unknown	*Arabidopsis*	Bacteria[128]
PIG18	Unknown	*Arabidopsis*	Bacteria[128]
AIG1	Unknown	*Arabidopsis*	Bacteria[129]
AIG2	Unknown	*Arabidopsis*	Bacteria[129]
PDF1.2/ AFP	Unknown	*Arabidopsis*, radish	Fungi[130]
hsr203J	Unknown	Tobacco	Bacteria[120]
str246C	Unknown	Tobacco	Bacteria, fungi[120]
Hsr201	Unknown	Tobacco	Bacteria[120]
hin	Unknown	Tobacco	Elicitor, bacteria[131]
pl39	Unknown	Pea	Fungi[132]
SLRR	Unknown	Sorghum	Fungi[133]
pl230	Unknown	Pea	Fungi[132]

Regulatory Components of Pathogen-Responsive Genes

The expression of pathogen-responsive genes shows both a temporal and spatial hierarchy. Some genes undergo a rapid localized activation at the site of infection, whereas others are more slowly activated either locally and/or systemically throughout the plant[18]. The differences in these expression patterns appear to be a result of the architecture of the promoters; pathogen-responsive promoters being dissimilar in the number, order and type of *cis*-acting elements present. This therefore leads to a different array of *trans*-acting factors that are able to bind and places the promoters at the end of one or more signal transduction pathways set in motion as the plant responds to potential pathogens.

A number of plant hormones, including salicylic acid (SA), ethylene and jasmonic acid (JA), have been implicated in acting as secondary signals following pathogen attack[4] and enhanced expression of many pathogen-responsive genes can be induced by these plant hormones. In addition the expression of most, if not all, pathogen-responsive genes is also developmentally and spatially regulated. It is becoming clear, not only that a number of signal transduction pathways meet at the promoters of pathogen-responsive genes but also that many meet earlier with cross talk occurring between signaling pathways set in motion by different stimuli[19].

Pathogen-Responsive *cis*-Acting Elements

We now have evidence that certain *cis*-acting elements are found in the promoter regions of many pathogen-responsive genes. We have grouped these elements below based on their core sequences.

W Boxes. There is a considerable body of evidence suggesting that W Boxes are important pathogen-responsive *cis*-acting elements. This comes from *in vivo* footprint analysis, gel shifts, methylation interference experiments and functional studies in transient expression systems as well as in transgenic plants[20-24]. There appears to be two types of W Boxes (Table 8-2)[21], the first contains the hexamer sequence TTGACC whilst the second consists of the two tetramer half sites TGAC-N_x-GTCA. Initial evidence for a role for these *cis*-acting elements during pathogenesis came from *in vivo* DNA footprint experiments. A W Box in the parsley PR-10 class gene, *PR1-1*, was shown to be the site of a fungal elicitor-inducible DNA-binding activity[20]. A W Box from the tobacco class I chitinase gene *CHN50* was

Table 8-2. Pathogen-responsive *cis*-acting elements and their cognate *trans*-acting factors

Element	Sequence	Isolated factors	Factor type
W Boxes	1. TTGACC	WRKY 1/2/3	WRKY
	2. TGAC-N_x-GTCA		
GCC Boxes	AGCCGCC	EREBP-1/2/3/4	EREBP
		Pti 4/5/6	
		AtEBP	
H/E/L/P Boxes	CC(A/T)(A/T)CC	BPF-1	MYB-like
		KAP-1/2	
G Boxes	CACGTG	G/HBF-1	bZIP
as-1-like	CTGACGTAAGGGATGACGCAC	TGA1a	bZIP
		SARP	
		OCSTF	
		ASF-1	
PR2-*d* SARE	TTCGACCTCC	Unknown	Unknown

also shown by gel shift experiments to be the binding site for a similar activity[22]. The binding site was pinpointed by methylation interference analysis where the five G residues identified were also the five G residues found within the W Boxes[22]. Subsequent experiments have shown not only that W Boxes are functional elicitor response elements (EREs) in the parsley *PR1* genes[21], the maize PR-1 class gene *PRms*[24] and the tobacco class I chitinase gene *CHN50*[23] but also that they alone are sufficient to direct elicitor responsive expression in transient expression systems[21,24], thus confirming them as functional EREs.

Inspection of many other pathogen-responsive promoters reveals the presence of W Box sequences within functional areas of the promoters. These include the potato glutathione S-transferase *gst1* (*prp1*)[25], PR10 genes from asparagus (*AoPR1*)[26] and potato (*PR-10a*)[27,28] and the stilbene synthase gene, *Vst1*, from grapevine[29]. These data suggest that W Boxes may be a general feature of the promoters of a large subset of pathogen-responsive genes. It is also interesting that a methyl jasmonate-responsive region from the promoter of the barley lipoxygenase 1 gene, *Lox1*, contains the TGAC core sequence of the W Boxes[30] and it is possible that this represents a link between W Boxes and the signaling molecule jasmonate that has been implicated in the plant defense response.

GCC Boxes. The GCC Box (AGCCGCC; Table 8-2) is found in the promoter regions of many pathogen-responsive genes[31-36] and possibly in all basic PR-protein encoding genes[31,32]. This *cis*-acting element (also called the PR-Box or AGC Box) has been shown to function as an ethylene response element[35]. Ethylene is involved in the plant response to pathogen attack as well as other stresses and senescence and the rate of biosynthesis of ethylene increases rapidly during plant-pathogen interactions[37]. Interestingly, although there are a large number of reports suggesting that ethylene acts through the GCC Box during pathogenesis to induce the expression of pathogen-responsive genes[19,31-36,38], the GCC Box has yet to be found in the promoters of ethylene-regulated genes involved in some of the other ethylene responses[35], such as fruit ripening[39]. It seems therefore that ethylene perception may involve distinct regulatory mechanisms.

Similar elements also act as low temperature response elements and water deficit response elements[40] suggesting that a subset of ethylene-responsive genes have GCC Boxes in their promoters and that these genes may play roles in the plant response to a number of stresses including pathogenesis.

P and L Boxes. The parsley phenylalanine ammonia-lyase (PAL) gene *PAL1* has been studied by *in vivo* DNA footprint analysis. This led to the identification of Boxes P and L as the sites of elicitor-inducible DNA-protein interactions[41]. Similar motifs are also present in the parsley 4-coumarate:CoA ligase (4CL) promoters, the enzymes of which are coordinately regulated with PAL, suggesting that Boxes P and L are *cis*-acting elements involved in the response to various stimuli such as fungal elicitor, UV light or wounding[41,42]. The P and L Boxes appear similar, many of the elements containing a CC(A/T)(A/T)CC core sequence that fits the type II MYB consensus sequence (MBSII) A(A/C)C(A/T)A (A/C)C, suggesting that they are MYB recognition elements (MREs). In addition, binding studies using the MYB factors bMYB1 and bMYB305 clearly showed that Boxes P and L from the parsley *PAL1* and *4CL1/2* promoters can be specifically bound these two MYB proteins and therefore that they are probably functional MYB recognition elements[43]. The exact sequence requirements for function, however, have not yet been established as neither Box P nor Box L was able to direct elicitor-inducible expression alone in a transient expression system[44]. This may be of little surprise as MYB proteins are known to activate expression in combination with other transcription factors such as bZIP proteins, bHLH proteins (MYCs) and homeodomain (HD) proteins[45].

H-Boxes. The H-Box (CCTACC(N)$_7$CT) was originally defined in the promoter of the bean chalcone synthase gene *Chs15* and is a *cis*-acting element required for pathogen-responsive expression[46]. Like Boxes P and L, the H-Box contains a CC(A/T)(A/T)CC core, suggesting that these three elements are synonymous. The similarity of the P, L and H Boxes is apparent from the observation that Box L in the parsley *PAL-3* gene[44] exactly fits the H-Box consensus[46] and the Box L sequences from parsley *PAL4, 4CL1* and *4CL2* fit the Box P consensus[44]. Related motifs appear in other elicitor-responsive promoters such as Box E in the parsley caffeoyl-CoA *O*-methyltransferase gene[47]. Although there is no evidence to show that Box E is a functional *cis*-acting element, it is the site of a specific DNA-protein interaction and contains the core sequence CCATCA[47], suggesting that it is similar to Boxes H, L and P . As this group of *cis*-acting elements are implicated in playing roles during the plant defense response it is tempting to call them HELP Boxes(!) and it is highly likely that all are MREs.

In a similar manner to Boxes P and L the H-Box does not appear to function to a high level alone. However, gain of function experiments show that it is active in combination with a G Box element in transgenic tobacco plants in establishing the characteristic tissue-specific pattern of expres⁻sion[48] and mutations in either the H-Box or G Box resulted in a reduction in the response to tobacco mosaic virus (TMV) infection[49].

G Boxes. G Boxes (CACGTG) have been shown to function during the regulation of diverse genes by environmental cues, such as abscisic acid

(ABA), light, UV radiation and wounding, as well as pathogen signals[50]. G Boxes are members of the family of ACGT-containing *cis*-acting elements and have been implicated in the expression of a number of genes as a result of pathogen attack[46,51]. They often seem to function in concert with other cis-acting elements. As mentioned above, gain-of-function experiments using transgenic tobacco plants with the bean *Chs15* promoter show that neither the G Box nor the H-Box alone can direct high levels of expression, but that a synthetic tetramer containing both leads to tissue-specific expression and expression in response to TMV infection[48,49].

Additional pathogen-responsive cis-acting elements. The elements listed above are the best-defined pathogen-responsive *cis*-acting elements and have been shown to be present in the promoters of a number of pathogen-responsive genes. It is not surprising, however, that other, as yet less well characterized, *cis*-acting elements have also been implicated as being pathogen-responsive.

Salicylic acid has been shown to be an important signal in plant defense[52] and a SA-response element (SARE) has been identified in the promoter region of the tobacco *PR2-d* gene. Functional studies showed that a fragment from -364 to -288 conferred a 20-fold induction by SA in transgenic tobacco plants[53]. Mutational and protein binding studies indicated that the core sequence of this SARE is TTCGACCTCC[53] and it has been suggested that this may be a similar element to the W Boxes[53], although the conserved TGAC core is not present. Further studies are required to establish similarity between these two elements.

Inspection of a number of stress-inducible genes led to the observation that a 10 bp sequence (TCATCTTCTT) , called the TCA element, is often present within their promoters[54]. This sequence is, however, neither sufficient nor required for the SA-mediated induction of the tobacco *PR-2d* promoter *in vivo*[37,53]. Although a 40-kD tobacco nuclear protein (TCA1) can bind to this sequence, its DNA-binding specificity has not been investigated[54]. Clearly, more data is required before the TCA element can be considered a functional *cis*-acting element.

Another class of pathogen-responsive *cis*-acting elements consists of *as-1*-like elements. The *as-1* (or *ocs*) element (CTGACGTAAGGGATGA-CGCAC) was initially identified in the *35S* promoter of cauliflower mosaic virus (CaMV) and the *nos* and *ocs* promoters of *Agrobacterium tumefaciens*[55,56]. The *as-1* element confers responsiveness to a number of signals, including SA, auxin, jasmonates and H_2O_2.[4] Although it is well documented that the *as-1* element is responsible for the rapid activation of a number of pathogen genes that are transferred to the plant cell, it is not clear whether functional *as-1*-like elements exist in plant genes. A search of plant gene databases identified a number of promoters that contain *as-1*-like sequences, however, only in the case of the soybean heat-shock gene *Gmhsp26-A* was this shown to be functional[57]. How widespread functional *as-1*-like elements are is therefore not known[58,59].

The tobacco *hsr203J* gene is of unknown function and has been shown to be specifically activated during the early steps of incompatible plant/pathogen interactions and to be a molecular marker for rapid localized cell death (the hypersensitive response)[60]. A 28 bp element located between -106 and -79 is necessary and sufficient for activation in response to an avirulent pathogen. It appears that this functions as a hypersensitive response element in this system. The sequence itself shows no homology with known regulatory sequences and is therefore most probably a novel type of *cis*-acting element (D. Roby, *personal communication*).

In a number of cases, promoter areas have been delineated that contain pathogen-responsive *cis*-acting elements that have yet to be clearly defined[29,61-65]. Often these promoter regions contain one or more copies of sequences similar to already characterized *cis*-acting elements. The above list of pathogen-responsive *cis*-acting elements is by no means complete. For example, the parsley *PR2* promoter (another *PR10* class gene) contains none of the above *cis*-acting elements. Instead, in addition to a homeodomain (HD) protein binding site, it has two novel elements, both of which alone are capable of directing elicitor-responsive expression[66] (Rushton, Carrasco, and Somssich, *unpublished*).

Factors Implicated in the Transcriptional Activation of Pathogen-Responsive Genes

Considerably less information is available concerning *trans*-acting factors than the pathogen-responsive *cis*-acting elements, largely because characterization of the *cis*-acting elements is normally a prerequisite for the study of the factors themselves. The understanding of these transcription factors is however an area that has seen significant advances in the last few years. There is now evidence for a role for WRKY proteins[21], EREBP/AP2-like proteins[35], bZIP proteins[67] and MYB proteins[68] during the plant defense response. It is of interest that some of these factors (bZIPs and MYBs) have counterparts in animals, but that others (WRKYs and EREBPs) appear to be present only in plants, perhaps reflecting the fact that plants have developed a defense system unlike that found in animals.

WRKY proteins. Evidence suggests that W Boxes are the binding sites for the WRKY family of DNA-binding proteins[21,69]. WRKY proteins are found widely in plants but are absent from animals[21,70]. It has been demonstrated that they bind specifically to W Boxes or W Box-like sequences[21,70] and random binding site selection using the Arabidopsis WRKY protein ZAP1 shows that the optimal binding site for this protein consists of two W Boxes[69]. In parsley, three WRKY proteins, WRKY1, -2 and -3, have been shown to bind specifically to functional W Boxes in the *PR1-1* and *PR1-2* promoters[21]. Fungal elicitor induces rapid and transient changes in the mRNA levels of WRKY1, -2 and -3, strongly implicating them in a role in plant defense.

WRKY proteins are defined by the presence of the WRKY domain. This domain contains the amino acid sequence <u>WRKY</u>GQK at the N-terminal end that is conserved in all WRKY domains so far analyzed. C-terminal to this conserved amino acid motif is located what appears to be a novel zinc finger-like structure[21,69-71]. The WRKY proteins themselves can be divided into three groups based on the number of WRKY domains (either one or two) and the pattern of cysteine and histidine residues that are part of the putative zinc finger-like structure (Rushton and Somssich, *unpublished*). Outside of the WRKY domain most WRKY proteins are dissimilar but do contain features found in many transcription factors, including potential nuclear localization signals and glutamine-rich, acidic, serine/threonine-rich and proline-rich domains. In addition it has been shown that the WRKY protein ZAP1 can transactivate, both in plant cells and in yeast[69] and that this transactivation is dependent on the presence of the W Box binding site for ZAP1. Furthermore, the parsley WRKY1 protein has been shown to be targeted to the nucleus (Eulgem and Somssich, *unpublished*).

Taken together, the specificity of binding to the elicitor-responsive W Boxes, the ability to transactivate, nuclear localization and the rapid changes in mRNA levels following the addition of fungal elicitor strongly suggest that WRKY proteins are transcription factors that play a role during the plant defense response by binding to W Boxes in the promoters of pathogen-responsive genes thereby activating transcription.

WRKY proteins may also play roles during other plant processes such as germination[21] and carbohydrate-regulated gene expression[71]. In this context it is interesting to note that preliminary evidence demonstrates that there are at least 30-50 WRKY genes in the Arabidopsis genome (Robatzek, Rushton, and Somssich, *unpublished*).

EREBPs. A number of proteins that bind to GCC Boxes have been isolated and in each case they were found to be members of the EREBP (ethylene-responsive element binding protein) family of DNA-binding proteins[19,33,35,40]. EREBPs contain a DNA-binding domain that is also present in the APETALA2 family of proteins[72] and appear, like WRKY proteins, to only be present in plants, again showing that the plant defense response has a number of features that have no obvious counterparts in animals. The first EREBPs to be isolated were tobacco EREBP-1, -2, -3 and -4[35]. All four proteins bind specifically to GCC Boxes and are ethylene-inducible in leaves.

CBF1, a protein of the EREBP family, was isolated by a yeast one-hybrid screen using a dehydration response element (GCCGAC) as the 'bait' DNA[40]. This suggests that members of the EREBP family may play a wider role in the response to various stresses such as low temperature and water stress[40]. In addition, it was shown that CBF1 is capable of transactivation in yeast and that this transactivation requires a functional *cis*-acting element[40].

Further evidence for an important role for EREBPs in plant defense has come from yeast two-hybrid experiments using the tomato resistance gene *Pto* as a 'bait' protein[19]. Pto is a protein kinase[73] and was shown to interact with three proteins: Pti4, Pti5 and Pti 6. Pti4/5/6 are members of the EREBP family and bind specifically to GCC Boxes (here called the PR Box). In tobacco carrying a *Pto* transgene, the expression of an EREBP gene was specifically enhanced upon *Pto-avrPto* recognition. These observations suggest a direct link between a resistance gene and the specific activation of plant defense genes and imply that the Pto kinase confers resistance by activating a signaling pathway that leads through EREBPS to activation of PR genes containing GCC Boxes.

bZIP proteins. Unlike EREBP-like proteins and WRKY proteins, bZIP proteins are well-characterised transcription factors that are also found in animals. So far, two types of *cis*-acting elements that are bound by bZIP factors have been implicated in the plant defense response: the G Box and the *as-1* or *ocs* element. Although it is unclear what role the *as-1* element itself plays during plant defense[57] there are a number of reports that demonstrate a role for bZIP factors. Büttner and Singh[33] have illustrated that an *ocs* element-binding protein (OBF4) interacts with an EREBP suggesting that cross-coupling between EREBP and bZIP transcription factors may be involved in the regulation of gene expression during the plant defense response.

A bZIP protein from soybean has been isolated that binds to the G Box in the bean *Chs15* promoter[67]. The protein, called G/HBF-1, can also bind to the adjacent H-Box. The significance of this relaxed binding specificity is unclear[67]. Although the mRNA and protein levels of G/HBF-1 do not increase during the induction of its putative target genes, the protein itself is rapidly phosphorylated and *in vitro* phosphorylation enhances binding to the *Chs15* promoter. A serine kinase capable of phosphorylating G/HBF-1 was also identified and this kinase is rapidly and transiently stimulated in elicitor-treated cells[67].

It seems clear that bZIP proteins play important roles during the activation of a number of genes during the plant defense response but our knowledge of the factors responsible is still at a preliminary stage.

MYB proteins. Evidence that MYB proteins play a role during the plant defense response comes from a number of observations. The elicitor-response *cis*-acting elements Boxes H, E, L and P appear to be MREs and Boxes P and L from the parsley *PAL1* and *4CL1/2* promoters have been shown to be binding sites for MYB proteins[43]. It has also been shown that a flower-specific MYB can activate transcription from the *Arabidopsis gPAL2* promoter through similar elements[45]. BPF-1 from parsley was the first protein isolated that binds to elements of this type[11] and is a MYB-like protein[43]. BPF-1 mRNA accumulates rapidly in elicitor-treated parsley cells suggesting that it participates in the plant defense response. More evidence for a role for MYBs has come from the work of Yang and Klessig[68]. Tobacco *MYB1* is induced by TMV during both the hypersensitive response and the development of systemic acquired resistance and, in addition, during the hypersensitive response by an avirulent race of the bacteria *Pseudomonas syringae*. Interestingly, the addition of 1mM SA induced expression of *MYB1* within 15 minutes. The MYB1 protein was shown to bind to a MYB consensus binding site found in the tobacco *PR1-a* promoter although no data has been presented to show that this sequence is indeed a functional *cis*-acting element[68].

MYB proteins are known to activate gene expression in combination with other transcription factors[45]. In this respect, it has been shown that the H Box from the bean *Chs15* promoter requires a G Box for its function[46] and it has been suggested that tobacco MYB1 cannot function alone[68]. It is possible that not only a combination of MYB and bZIP factors play a role in the plant defense response but also that MYBs may function together with bHLH factors.

Additional factors. A number of other less well-characterized proteins have also been implicated in the transcriptional regulation of defense-related genes. The G Box (CACGTG), as well as being a binding site for bZIP proteins, is also a potential bHLH binding site (CACNTG)[74]. bHLH factors have been isolated from plants[75,76] but although MYB/bHLH interacting partners have been shown to control flower pigmentation[77] no role has yet been assigned to them during plant defense.

Matton *et al.*[28] showed that an upstream regulatory region was located between -135 and -52 within the potato *PR-10a* gene and that an elicitor and wound-inducible factor from tuber nuclear extracts bound to the -135 to -105 region[28]. The factors binding to this ERE were further characterized by Despres *et al.*[27]. The ERE was specifically recognized by two nuclear factors, PBF-1 and PBF-2, and evidence was presented showing that wounding and elicitor treatment induce the phosphorylation of PBF-1[27]. Data has also been provided suggesting a role for a protein kinase C (PKC) in this process, as inhibitors of PKC abolished the elicitor-induced binding of PBF-2, whereas TPA, an activator of PKC, was able to substitute for the elicitor in increasing the binding of PBF-2[78]. Neither PBF-1 nor PBF-2 has been cloned. However, the ERE contains the sequence TGAC-N_6-GTCA[28] that is very similar to the Box W3 sequence (TGAC-N_6-GTCA) from parsley[21] suggesting that the *PR10-a* ERE contains a W Box and by implication that PBF-1 and -2 may be WRKY proteins.

Two H-Box binding activities, KAP-1 and KAP-2, have been purified from bean[79]. Elicitation with glutathione did not affect their total cellular activities but instead caused an increase in their specific activities in the nuclear fraction. Due to their sizes, both KAP-1 and KAP-2 are different from the other characterized H-Box binding factor G/HBF[67].

G Boxes and G Box binding factors have been shown to be associated with 14-3-3 proteins[80,81]. 14-3-3 proteins are involved in many signaling pathways, interacting with protein kinases[80,82], transcription factors[81] and even ion channels in membranes[83]. It has been shown that the G Box from the tobacco osmotin gene (a protein of the PR-5 group) is associated with a 14-3-3 protein, suggesting that bZIP proteins involved in the regulation of this gene during pathogenesis may be regulated, at least in part, by forming a complex on the G Box with 14-3-3 proteins[38]. It has also been demonstrated that a 14-3-3 protein from barley is pathogen-induced and it was suggested that it might regulate protein kinase activity[82].

Concluding Remarks and Future Prospects

The last few years have marked a period of significant progress in our understanding of how pathogens trigger the plants defense system thereby resulting in counteractive measures. Transcriptional activation of genes constitutes one such vital constituent. We are now starting to define specific DNA elements within the regulatory regions of these genes and the protein factors with which they interact and thereby mediate induced expression . The modular nature of these promoters is becoming clear and many promoters contain combinations of a limited number of pathogen-responsive *cis*-acting elements that may represent endpoints of common signal transduction chains. One set of promoters contains multiple W Boxes. Others contain functional W Boxes and GCC-like Boxes or GCC-like Boxes alone. Another subset of promoters contains MREs and these are often associated with G Boxes or other elements. These *cis*-acting elements seem in turn to be the binding sites for an array of *trans*-acting factors that appear to include WRKY proteins, EREBPs, bZIP proteins, MYB proteins and possibly bHLH proteins. It is also clear that there are additional elements and factors whose places in the plant's defense response will be made clear in the near future.

The knowledge gained by these studies is not only of academic interest but could also be of biotechnological importance in the production of plants with increased disease resistance. The design of transgenic plants capable of rapidly expressing antifungal or antibacterial compounds highly restricted to infection sites requires precise delineation of the regulatory elements mediating this pathogen-induced response.

Acknowledgments

We would like to thank all colleagues who provided us with preprints and unpublished data and Klaus Hahlbrock, Robert Cormack, Thomas Eulgem, Silke Robatzek and Bernd Weisshaar (all Köln) for discussions and critical reading of the manuscript.

References

1. Alexander, D., K. Lawton, S. Uknes, E. Ward and J. Ryals. 1994. Defense-related gene induction in plants. In: Genetic Engineering, Vol. 16, eds., J. K. Setlow, pp. 195-212. Plenum Press, New York.
2. Hammond-Kosack, K. E. and J. D. G. Jones. 1996. Resistance gene-dependent plant defense responses. Plant Cell 8:1773-1791.
3. Kombrink, E. and I. E. Somssich. 1995. Defense responses of plants to pathogens. In: Advances in Botanical Research, Vol. 21, eds., J. H. Andrews and I. C. Tommerup, pp. 1-34. Academic Press Limited, London.
4. Yang, Y., J. Shah and D. F. Klessig. 1997. Signal perception and transduction in plant defense responses. Genes Dev. 11:1621-1639.
5. Hammond-Kosack, K. E. and J. D. G. Jones. 1997. Plant disease resistance genes. Annu. Rev. Plant Physiol. Plant Mol. Biol. 48:575-607.
6. Staskawicz, B. J., F. M. Ausubel, B. J. Baker, J. G. Ellis and J. D. G. Jones. 1995. Molecular genetics of plant disease resistance. Science 268:661-667.
7. Dixon, R. A., M. J. Harrison and C. J. Lamb. 1994. Early events in the activation of plant defense responses. Annual Reviews of Phytopathology 32:479-501.
8. Dixon, R. A. and C. J. Lamb. 1990. Molecular communication in interactions between plants and microbial pathogens. Annu. Rev. Plant Physiol. Plant Mol. Biol. 41:339-367.
9. He, S. Y., D. W. Bauer, A. Collmer and S. V. Beer. 1994. Hypersensitive response elicited by *Erwinia amylovora* Harpin requires active plant metabolism. Mol. Plant-Microbe Interact. 7:289-292.
10. Ebel, J. and D. Scheel. 1997. Signals in host-parasite interactions. In: The Mycota Part A, Plant Relationships, Vol. V, eds., G. Carroll and P. Tudzynski, pp. 85-105. Springer-Verlag, Berlin.

11. da Costa e Silva, O., L. Klein, E. Schmelzer, G. F. Trezzini and K. Hahlbrock. 1993. BPF-1, a pathogen-induced DNA-binding protein involved in the plant defense response. Plant J. 4:125-135.
12. Bowles, D. J. 1990. Defense-related proteins in higher plants. Annu. Rev. Biochem. 59:873-907.
13. Collinge, D. B., P. L. Gregersen and H. Thordal-Christensen. 1994. The induction of gene expression in response to pathogenic microbes. In: Mechanisms of Plant Growth and Improved Productivity: Modern Approaches and Perspectives, eds., A. S. Basra, pp. 391-433. Marcel Dekker, New York.
14. Herrmann, K. M. 1995. The shikimate pathway: early steps in the biosynthesis of aromatic compounds. Plant Cell 7:907-919.
15. Nicholson, R. L. and R. Hammerschmidt. 1992. Phenolic compounds and their role in disease resistance. Annu. Rev. Phytopathol. 30:369-389.
16. Smith, C. J. 1996. Accumulation of phytoalexins: defence mechanism and stimulus response system. New Phytol. 132:1-45.
17. Somssich, I. E. and K. Hahlbrock. 1998. Pathogen defence in plants - a paradigm of biological complexity. Trends Plant Sci. 3:86-90.
18. Hahlbrock, K., D. Scheel, E. Logemann, T. Nürnberger, M. Parniske, S. Reinold, W. R. Sacks and E. Schmelzer. 1995. Oligopeptide elicitor-mediated defense gene activation in cultured parsley cells. Proc. Natl. Acad. Sci. USA 92:4150-4157.
19. Zhou, J., X. Tang and G. B. Martin. 1997. The Pto kinase conferring resistance to tomato bacterial speck disease interacts with proteins that bind a *cis*-element of pathogenesis-related genes. EMBO J. 16:3207-3218.
20. Meier, I., K. Hahlbrock and I. E. Somssich. 1991. Elicitor-inducible and constitutive in vivo DNA footprints indicate novel *cis*-acting elements in the promoter of a parsley gene encoding pathogenesis-related protein 1. Plant Cell 3:309-315.
21. Rushton, P. J., J. T. Torres, M. Parniske, P. Wernert, K. Hahlbrock and I. E. Somssich. 1996. Interaction of elicitor-induced DNA binding proteins with elicitor response elements in the promoters of parsley PR1 genes. EMBO J. 15:5690-5700.
22. Fukuda, Y. and H. Shinshi. 1994. Characterization of a novel *cis*-acting element that is responsive to a fungal elicitor in the promoter of a tobacco class I chitinase gene. Plant Mol. Biol. 24:485-493.
23. Fukuda, Y. 1997. Interaction of tobacco nuclear proteins with an elicitor-responsive element in the promoter of a basic class I chitinase gene. Plant Mol. Biol. 34:81-87.
24. Raventós, D., A. B. Jensen, M.-B. Rask, J. M. Casacuberta, J. Mundy and B. San Segundo. 1995. A 20 by *cis*-acting element is both necessary and sufficient to mediate elicitor response of a maize PRms gene. Plant J. 7:147-156.
25. Hahn, K. and G. Strittmatter. 1994. Pathogen defence gene *prp1-1* from potato encodes an auxin-responsive glutathion *S*-transferease. Eur. J. Biochem. 226:619-626.
26. Warner, S. A. J., R. Scott and J. Draper. 1993. Isolation of an asparagus intracellular *PR* gene (*AoPR1*) wound-responsive promoter by the inverse polymerase chain reaction and its characterization in transgenic tobacco. Plant J. 3:191-201.
27. Després, C., R. Subramaniam, D. P. Matton and N. Brisson. 1995. The activation of the potato *PR-10a* gene requires the phophorylation of the nuclear factor PBF-1. Plant Cell 7:589-598.
28. Matton, D. P., G. Prescott, C. Bertrand, A. Camirand and N. Brisson. 1993. Identification of cis-acting elements involved in the regulation of the 17 kDa pathogenesis-related gene *STH-2* in potato. Plant Mol. Biol. 22:279-291.

29. Schubert, R., R. Fischer, R. Hain, P. H. Schreier, G. Bahnweg, D. Ernst and H. J. Sandermann. 1997. An ozone-responsive region of the grapevine resveratrol synthase promoter differs from the basal pathogen-responsive sequence. Plant Mol. Biol. 34:417-426.

30. Rouster, J., R. Leah, J. Mundy and V. Cameron-Mills. 1997. Identification of a methyl jasmonate-responsive region in the promoter of a lipoxygenase 1 gene expressed in barley grain. Plant J. 11:513-523.

31. Sato, F., S. Kitajima, T. Koyama and Y. Yamada. 1996. Ethylene-induced gene expression of osmotin-like protein, a neutral isoform of tobacco PR-5, is mediated by the AGCCGCC cis-sequence. Plant Cell Physiol. 37:249-255.

32. Alonso, E., F. d. C. Niebel, P. Obregón, G. Gheysen, D. Inzé, M. Van Montagu and C. Castresana. 1995. Differential in vitro DNA binding activity to a promoter element of the gn1 b-1,3-glucanase gene in hypersensitively reacting tobacco plants. Plant J. 7:309-320.

33. Büttner, M. and K. B. Singh. 1997. Arabidopsis thaliana ethylene-responsive element binding protein (AtEBP), an ethylene-inducible, GCC box DNA-binding protein interacts with an ocs element binding protein. Proc. Natl. Acad. Sci. USA 94:5961-5966.

34. Vögeli-Lange, R., C. Fründt, C. M. Hart, F. Nagy and F. J. Meins. 1994. Developmental, hormonal, and pathogenesis-related regulation of the tobacco class I b-1,3-glucanase B promoter. Plant Mol. Biol. 25:299-311.

35. Ohme-Takagi, M. and H. Shinshi. 1995. Ethylene-inducible DNA binding proteins that interact with an ethylene-responsive element. Plant Cell 7:173-182.

36. Hart, C. M., F. Nagy and F. J. Meins. 1993. A 61 bp enhancer element of the tobacco β-1,3-glucanase B gene interacts with a regulated nuclear protein(s). Plant Mol. Biol. 21:121-131.

37. Yang, S. F. and N. E. Hoffman. 1984. Ethylene biosnythesis and its regulation in higher plants. Annu. Rev. Plant Phys. 35:155-189.

38. Liu, D., M. I. Narasimhan, Y. Xu, K. G. Raghothama, P. M. Hasegawa and R. A. Bressan. 1995. Fine structure and function of the osmotin gene promoter. Plant Mol. Biol. 29:1015-1026.

39. Cordes, S., J. Deikman, L. J. Margossian and R. L. Fischer. 1989. Interaction of a developmentally regulated DNA-binding factor with sites flanking two different fruit-ripening genes from tomato. Plant Cell 1:1025-1034.

40. Stockinger, E. J., S. J. Gilmour and M. F. Thomashow. 1997. Arabidopsis thaliana CBF1 encodes an AP2 domain-containing transcriptional activator that binds to the C-repeat/DRE, a cis-acting DNA regulatory element that stimulates transcription in response to low temperature and water deficit. Proc. Natl. Acad. Sci. USA 94:1035-1040.

41. Lois, R., A. Dietrich, K. Hahlbrock and W. Schulz. 1989. A phenylalanine ammonia-lyase gene from parsley: structure, regulation and identification of elicitor and light responsive cis-acting elements. EMBO J. 8:1641-1648.

42. Logemann, E., S.-C. Wu, J. Schröder, E. Schmelzer, I. E. Somssich and K. Hahlbrock. 1995. Gene activation by UV light, fungal elicitor or fungal infection in Petroselinum crispum is correlated with repression of cell cycle-related genes. Plant J. 8:865-876.

43. Feldbrügge, M., M. Sprenger, K. Hahlbrock and B. Weisshaar. 1997. PcMYB1, a novel plant protein conraining a DNA-binding domain with one MYB repeat, interacts in vivo with a light-regulatory promoter unit. Plant J. 11:1079-1093.

44. Logemann, E., M. Parniske and K. Hahlbrock. 1995. Modes of expression and common structural features of the complete phenylalanine ammonia-lyase gene family in parsley. Proc. Natl. Acad. Sci. USA 92:5905-5909.
45. Sablowski, R. W. M., E. Moyano, F. A. Culianez-Macia, W. Schuch, C. Martin and M. Bevan. 1994. A flower-specific Myb protein activates transcription of phenylpropanoid biosynthetic genes. EMBO J. 13:128-137.
46. Loake, G. J., O. Faktor, C. J. Lamb and R. A. Dixon. 1992. Combination of H-box [CCTACC(N)$_7$CT] and G-box (CACGTG) *cis* elements is necessary for feed-forward stimulation of a chalcone synthase promoter by the phenyl-propanoid-pathway intermediate *p*-coumaric acid. Proc. Natl. Acad. Sci. USA 89:9230-9234.
47. Grimmig, B. and U. Matern. 1997. Structure of the parsley caffeoyl-CoA *O*-methyltransferase gene, harbouring a novel elicitor responsive *cis*-acting element. Plant Mol. Biol. 33:323-341.
48. Faktor, O., G. Loake, R. A. Dixon and C. J. Lamb. 1997. The G-box and H-box in a 39 bp region of a French bean chalcone synthase promoter constitutes a tissue-specific regulatory element. Plant J. 11:1105-1113.
49. Faktor, O., J. M. Kooter, G. J. Loake, R. A. Dixon and C. J. Lamb. 1997. Differential utilization of regulatory *cis*-elements for stress-induced and tissue-specific activity of a French bean chalcone synthase promoter. Plant Sci. 124: 175-182.
50. Menkens, A. E., U. Schindler and A. R. Cashmore. 1995. The G-box: a ubiquitous regulatory DNA element in plants bound by the GBF family of bZIP proteins. Trends Biochem. Sci. 20:506-510.
51. Kim, S.-R., J.-L. Choi, M. A. Costa and G. An. 1992. Identification of G-box sequence as an essential element for methyl jasmonate response of potato pro-teinase inhibitor II promoter. Plant Physiol. 99:627-631.
52. Durner, J., J. Shah and D. F. Klessig. 1997. Salicylic acid and disease resistance in plants. Trends Plant Sci. 2:266-274.
53. Shah, J. and D. F. Klessig. 1996. Identification of a salicylic acid-responsive element in the promoter of the tobacco pathogenesis-related β-1,3-glucanase gene, *PR-2d*. Plant J. 10:1089-1101.
54. Goldsbrough, A. P., H. Albrecht and R. Stratford. 1993. Salicyclic acid-inducible binding of a tobacco nuclear protein to a 10 bp sequence which is highly conserved amongst stress-inducible genes. Plant J. 3:563-571.
55. Lam, E., P. N. Benfey, P. M. Gilmartin, R.-X. Fang and N.-H. Chua. 1989. Site-specific mutations alter *in vitro* factor binding and change promoter expression pattern in transgenic plants. Proc. Natl. Acad. Sci. USA 86:7890-7894.
56. Ellis, J. G., D. J. Llewellyn, J. C. Walker, E. S. Dennis and W. J. Peacock. 1987. The ocs element: a 16 base pair palindrome essential for activity of the octopine synthase enhancer. EMBO J. 6:3203-3208.
57. Ellis, J. G., J. G. Tokuhisa, D. J. Llewellyn, D. Bouchez, K. Singh, E. S. Dennis and P. W.S. 1993. Does the *ocs*-element occur as a functional component of the promoters of plant genes? Plant J. 4:433-443.
58. van der Zaal, B. T., F. N. J. Droog, F. J. Pieterse and P. J. J. Hooykaas. 1996. Auxin-sensitive elements from promoters of tobacco *GST* genes and a consensus *as-1*-like element differ only in relative strength. Plant Physiol. 110:79-88.
59. Chen, W., G. Choa and K. B. Singh. 1996. The promoter of a H_2O_2-inducible, *Arabidopsis* glutathione S-transferase gene contains closely linked OBF- and OBP1-binding sites. Plant J. 10:955-966.

60. Pontier, D., L. Godiard, Y. Marco and D. Roby. 1994. *hsr203J*, a tobacco gene whose activation is rapid, highly localized and specific for incompatible plant/ pathogen interactions. Plant J. 5:507-521.
61. Warner, S. A. J., A. Gill and J. Draper. 1994. The developmental expression of the asparagus intracellular PR protein *(AoPR1)* gene correlates with sites of phenylpropanoid biosynthesis. Plant J. 6:31-34.
62. Yin, S., L. Mei, J. Newman, K. Back and Chappell. 1997. Regulation of sesquiterpene cyclase gene expression. Characterization of an elicitor- and pathogeninducible promoter. Plant Physiol 115:437-451.
63. Meller, Y., G. Sessa, Y. Eyal and R. Fluhr. 1993. DNA-protein interactions on a *cis*-DNA element essential for ethylene regulation. Plant Mol. Biol. 23:453-463.
64. Martini, N., M. Egen, I. Rüntz and G. Strittmatter. 1993. Promoter sequences of a potato pathogenesis-related gene mediate transcriptional activation selectively upon fungal infection. Mol. Gen. Genet. 236:179-186.
65. Hagiwara, H., M. Matsuoka, M. Ohshima, M. Watanabe, D. Hosokawa and Y. Ohashi. 1993. Sequence-specific binding of factors to two independent promoter regions of the acidic tobacco pathogenesis-related-1 protein (PR-1). Molecular and General Genetics 240:197-205.
66. van de Löcht, U., I. Meier, K. Hahlbrock and I. E. Somssich. 1990. A 125 bp promoter fragment is sufficient for strong elicitor-mediated gene activation in parsley. EMBO J. 9:2945-2950.
67. Dröge-Laser, W., A. Kaiser, W. P. Lindsay, B. A. Halkier, G. J. Loake, P. Doerner, R. A. Dixon and C. Lamb. 1997. Rapid stimulation of a soybean protein-serine kinase that phosphorylates a novel bZIP DNA-binding protein, G/HBF-1, during the induction of early transcription-dependent defenses. EMBO J. 16:726-738.
68. Yang, Y. and D. F. Klessig. 1996. Isolation and characterization of a tobacco mosaic virus-inducible *myb* oncogene homolog from tobacco. Proc. Natl. Acad. Sci. USA 93:14972-14977.
69. de Pater, S., V. Greco, K. Pham, J. Memelink and J. Kijne. 1996. Characterization of a zinc-dependent transcriptional activator from *Arabidopsis*. Nucl. Acid Res. 24:4624-4631.
70. Rushton, P. J., H. Macdonald, A. K. Huttly, C. M. Lazarus and R. Hooley. 1995. Members of a new family of DNA-binding proteins bind to a conserved *cis*-element in the promoters of *a-Amy2* genes. Plant Mol. Biol. 29:691-702.
71. Ishiguro, S. and K. Nakamura. 1994. Characterization of a cDNA encoding a novel DNA-binding protein, SPF1, that recognizes SP8 sequences in the 5' upstream regions of genes coding for sporamin and ß-amylase from sweet potato. Mol. Gen. Genet. 244:563-571.
72. Jofuku, K. D., B. G. W. den Boer, M. V. Montagu and J. K. Okamuro. 1994. Control of Arabidopsis flower and seed development by the homeotic gene *APETALA2*. Plant Cell 6:1211-1225.
73. Zhou, J., Y.-T. Loh, R. A. Bressan and G. Martin. 1995. The tomato gene *Pti1* encodes a serine/threonine kinase that is phorphorylated by Pto and is involved in the hypersensitive response. Cell 83:925-935.
74. Kawagoe, Y. and N. Murai. 1996. A novel basic region/helix-loop-helix protein binds to a G-box motif CACGTG of the bean storage protein beta-phaseolin gene. Plant Sci. 116:47-57.
75. Mol, J., G. I. Jenkins, E. Schaefer and D. Weiss. 1996. Signal perception, transduction and gene expression involved in anthocyanin biosynthesis. Crit. Rev. Plant Sci. 15:525-557.

76. de Pater, S., K. Pham, J. Memelink and J. Kijne. 1997. RAP-1 is an *Arabidopsis* MYC-like R protein homologue, that binds to G-box sequence motifs. Plant Mol. Biol. 34:169-174.
77. Martin, C. and J. Paz-Ares. 1997. MYB transcription factors in plants. Trends Genet. 13:67-73.
78. Subramaniam, R., C. Després and N. Brisson. 1997. A functional homolog of mammalian protein kinase C participates in the elicitor-induced defense response in potato. Plant Cell 9:653-664.
79. Yu, L. M., C. J. Lamb and R. A. Dixon. 1993. Purification and biochemical characterization of proteins which bind to the H-box *cis*-element implicated in transcriptional activation of plant defense genes. Plant J. 3:805-816.
80. Chen, Z., H. Fu, D. Liu, P.-F. L. Chang, M. Narasimhan, R. Ferl, P. M. Hasegawa and R. A. Bressan. 1994. A NaCl-regulated plant gene encoding a brain protein homolog that activates ADP ribosyltransferase and inhibits protein kinase C. Plant J. 6:729-740.
81. de Vetten, N. C., G. Lu and R. J. Ferl. 1992. A maize protein associated with the G-box binding complex has homology to brain regulatory proteins. Plant Cell 4:1295-1307.
82. Brandt, J., H. Thordal-Christensen, K. Vad, P. L. Gregersen and D. B. Collinge. 1992. A pathogen-induced gene of barley encodes a protein showing high similarity to a protein kinase regulator. Plant J. 2:815-820.
83. Jahn, T., A. T. Fuglsang, A. Olsson, I. M. Bruntrup, D. B. Collinge, D. Volkmann, M. Sommarin, M. G. Palmgren and C. Larsson. 1997. The 14-3-3 protein interacts directly with the C-terminal region of the plant plasma membrane H(+)-ATPase. Plant Cell 9:1805-1814.
84. Edwards, R. 1996. S-adenosyl-L-methionine metabolism in alfalfa cell cultures following treatment with fungal elicitors. Phytochemistry 43:1163-1169.
85. Schröder, G., A. Waitz, M. Hotze and J. Schröder. 1994. cDNA for S-adenosyl-L-homocysteine hydrolase from *Catharanthus roseus*. Plant Physiol. 104:1099-1100.
86. Morgan, P. W. and M. C. Drew. 1997. Ethylene and plant responses to stress. Physiol. Plant. 100:620-630.
87. Lasserre, E., F. Godard, T. Bouquin, J. A. Hernandez, J.-C. Pech, D. Roby and C. Balagué. 1997. Differential activation of two ACC oxidase gene promoters from melon during plant development and in response to pathogen attack. Mol. Gen. Genet. 256:211-222.
88. Bowler, C., T. Alliotte, M. De Loose, M. Van Montagu and D. Inzé. 1989. The induction of manganese superoxide dismutase in response to stress in *Nicotiana plumbaginifolia*. EMBO J. 8:31-38.
89. Zacheo, G. and T. Bleve-Zacheo. 1988. Involvement of superoxide dismutase and superoxide radicals in the susceptibility and resistance of tomato plants to *Meloidogyne incognita* attack. Physiol. Mol. Plant Pathol. 32:313-322.
90. Young, S., X. Wang and J. E. Leach. 1996. Changes in the plasma membrane distribution of rice phospholipase D during resistant interactions with *Xanthomonas oryzae* pv *oryzae*. Plant Cell 8:1079-1090.
91. Rosahl, S. 1996. Lipoxygenases in plants - Their role in development and stress response. Z. Naturforsch. 51 c:123-138.
92. Gadea, J., M. E. Mayda, V. Conejero and P. Vera. 1996. Characterization of defense-related genes ectopically expressed in viroid-infected tomato plants. Mol. Plant-Microbe Interact. 9:409-415.

93. Pastuglia, M., D. Roby, C. Dumas and J. M. Cock. 1997. Rapid induction by wounding and bacterial infection of an *S* gene family receptor-like kinase gene in *Brassica oleracea*. Plant Cell 9:49-60.
94. Seo, S., H. Sano and Y. Ohashi. 1997. Transgenic manipulation of signaling pathways of plants resistance to pathogen attack. In: Biotechnology Annual Review, Vol. 3, eds., M. R. El-Gewely, pp. 197-225. Elsevier, Amsterdam.
95. Lamb, C. and R. A. Dixon. 1997. The oxidative burst in plant disease resistance. Annu. Rev. Plant Physiol. Plant Mol. Biol. 48:251-275.
96. Creelman, R. A. and J. E. Mullet. 1995. Jasmonic acid distribution and action in plants: regulation during development and response to biotic and abiotic stress. Proc. Natl. Acad. Sci. USA 92:4114-4119.
97. Schaller, A., D. R. Bergey and C. A. Ryan. 1995. Induction of wound response genes in tomato leaves by bestatin, an inhibitor of aminopeptidases. Plant Cell 7:1893-1898.
98. Bergey, D. R., G. A. Howe and C. A. Ryan. 1996. Polypeptide signaling for plant defensive genes exhibits analogies to defense signaling in animals. Proc. Natl. Acad. Sci. USA 93:12053-12058.
99. Görlach, J., H.-R. Raesecke, D. Rentsch, M. Regenass, P. Roy, M. Zala, C. Keel, T. Boller, N. Amrhein and J. Schmid. 1995. Temporally distinct accumulation of transcripts encoding enzymes of the prechorismate pathway in elicitor-treated, cultured tomato cells. Proc. Natl. Acad. Sci. USA 92:3166-3170.
100. Dixon, R. A. and N. L. Paiva. 1995. Stress-induced phenylpropaniod metabolism. Plant Cell 7:1085-1097.
101. Dixon, R. A., M. J. Harrison and N. L. Paiva. 1995. The isoflavonoid phytoalexin pathway: from enzymes to genes to transcription factors. Physiol. Plant. 93:385-392.
102. Koopmann, E. and K. Hahlbrock. 1997. Differentially regulated NADPH:cytochrome P450 oxidoreductase in parsley. Proc. Natl. Acad. Sci. USA 94:14954-14959.
103. León, J., N. Yalpani, I. Raskin and M. A. Lawton. 1993. Induction of benzoic acid 2-hydroxylase in virus-infected tobacco. Plant Physiol. 103:323-328.
104. Pasquali, G., O. J. M. Goddijn, A. de Waal, R. Verpoorte, R. A. Schilperoort, J. H. C. Hoge and J. Memelink. 1992. Coordinated regulation of two indole alkaloid biosynthetic genes from *Catharanthus roseus* by auxin and elicitors. Plant Mol. Biol. 18:1121-1131.
105. Negrel, J., J.-C. Vallée and C. Martin. 1984. Ornithin decarboxylase activity and the hypersensitive reaction to tobacco mosaic virus in *Nicotiana tabacum*. Phytochemistry 23:2747-2751.
106. Kawalleck, P., H. Keller, K. Hahlbrock, D. Scheel and I. E. Somssich. 1993. A pathogen-responsive gene of parsley encodes tyrosine decarboxylase. J. Biol. Chem. 268:2189-2194.
107. Gregersen, P. L., H. Thordal-Christensen, H. Förster and D. E. Collinge. 1997. Differential gene transcript accumulation in barley leaf epidermis and mesophyll in response to attack by *Blumeria graminis* f. sp. *hordei* (syn. *Erysiphe graminis* f. sp. *hordei*). Physiol. Mol. Plant Pathol. 51:85-97.
108. Wei, Y., Z. Zhang, C. H. Andersen, E. Schmelzer, P. L. Gregersen, D. B. Collinge, V. Smedegaard-Petersen and H. Thordal-Christensen. 1998. An epidermis/papilla-specific oxalate oxidase-like protein in the defence response of barley attacked by powdery mildew fungus. Plant Mol. Biol. 36:101-112.

109. Somssich, I. E., P. Wernert, S. Kiedrowski and K. Hahlbrock. 1996. *Arabidopsis thaliana* defense-related protein ELI3 is an aromatic alcohol:NADP⁺ oxidoreductase. Proc. Natl. Acad. Sci. USA 93:14199-14203.
110. Xian-Zhi, H., J. T. Reddy and R. A. Dixon. 1998. Stress response in alfalfa (*Medicago sativa* L.). XXII. cDNA cloning and characterization of an elicitor-inducible isoflavone 7-*O*-methyltransferase. Plant Mol. Biol. 36:43-51.
111. Dalkin, K., R. Edwards, B. Edington and R. A. Dixon. 1990. Stress responses in alfalfa (*Medicago sativa* L.) I. Induction of phenylpropanoid biosynthesis and hydrolytic enzymes in elicitor-treated cell suspension cultures. Plant Physiol. 92:440-446.
112. Ni, W., V. J. H. Sewalt, K. L. Korth, J. W. Blount, G. M. Ballance and R. A. Dixon. 1996. Stress responses in alfalfa. Plant Physiol. 112:717-726.
113. Busam, G., K. T. Junghanns, R. E. Kneusel, H.-H. Kassemeyer and U. Matern. 1997. Characterization and expression of caffeoyl-coenzyme A 3-*O*-methyltransferase proposed for the induced resistance response of *Vitis vinifera* L. Plant Physiol. 115:1039-1048.
114. Jaeck, E., B. Dumas, P. Geoffroy, N. Favet, D. Inzé, M. Van Montagu, B. Fritig and M. Legrand. 1992. Regulation of enzymes involved in lignin biosynthesis: induction of *O*-methyltransferase mRNAs during the hypersensitive reaction of tobacco to tobacco mosaic virus. Mol. Plant-Microbe Interact. 5:294-300.
115. Logemann, E., S. Reinold, I. E. Somssich and K. Hahlbrock. 1997. A novel type of pathogen defense-related cinnamyl alcohol dehydrogenase. Biol. Chem. 378:909-913.
116. Yang, Z., H. Park, G. H. Lacy and C. L. Cramer. 1991. Differential activation of potato 3-hydroxy-3-methylglutaryl Coenzyme A reductase genes by wounding and pathogen challenge. Plant Cell 3:397-405.
117. Blechert, S., W. Brodschelm, S. Hölder, L. Kammerer, T. M. Kutchan, M. J. Mueller, Z.-Q. Xia and M. H. Zenk. 1995. The octadecanoic pathway: signal molecules for the regulation of secondary pathways. Proc. Natl. Acad. Sci. USA 92:4099-4105.
118. Niyogi, K. K. and G. R. Fink. 1992. Two anthranilate synthase genes in *Arabidopsis*: defense-related regulation of the tryptophan pathway. Plant Cell 4:721-733.
119. Kombrink, E. and I. E. Somssich. 1997. Pathogenesis-related proteins and plant defense. In: The Mycota Part A, Plant Relationships, Vol. V, eds., G. Carroll and P. Tudzynski, pp. 107-128. Springer-Verlag, Berlin.
120. Marco, I., F. Ragueh, L. Godiard and D. Froissard. 1990. Transcriptional activation of 2 classes of genes during hypersensitive redaciton of tobacco leaves infiltrated with an incompatible isolate of the phytopathogenic bacterium *Pseudomonas solanacearum*. Plant Mol. Biol. 15:145-154.
121. Somssich, I. E., J. Bollmann, K. Hahlbrock, E. Kombrink and W. Schulz. 1989. Differential early activation of defense-related genes in elicitor-treated parsley cells. Plant Mol. Biol. 12:227-234.
122. Kader, J.-C. 1996. Lipid-transfer proteins in plants. Annu. Rev. Plant Physiol. Plant Mol. Biol. 47:627-654.
123. Davidson, A. D., J. M. Manners, R. S. Simpson and K. J. Scott. 1987. cDNA cloning of mRNAs induced in resistant barley during infection by *Erysiphe graminis* f. sp. *Hordei*. Plant Mol. Biol. 8:77-85.
124. Schweizer, P., W. Hunziker and E. Mösinger. 1989. cDNA cloning, in vitro transcription and partial sequence analysis of mRNAs from winter wheat (*Triticum*

aestivum L.) with induced resistance to *Erysiphe graminis* f. sp. *tritici*. Plant Mol. Biol. 12:643-645.

125. Molina, A., M. Mena, P. Carbonero and F. García-Olmedo. 1997. Differential expression of pathogen-responsive genes encoding two types of glycine-rich proteins in barley. Plant Mol. Biol. 33:803-810.

126. Century, K. S., A. D. Shapiro, P. P. Repetti, D. Dahlbeck, E. Holub and B. J. Staskawicz. 1997. *NDR1*, a pathogen-induced component required for *Arabidopsis* disease resistance. Science 278:1963-1965.

127. Ryals, J., K. Weymann, K. Lawton, L. Friedrich, D. Ellis, H.-Y. Steiner, J. Johnson, T. P. Delaney, T. Jesse, P. Vos and S. Uknes. 1997. The Arabidopsis *NIM1* protein shows homology to the mammalian transcription factor IkB. Plant Cell 9:425-439.

128. Rogers, E. E. and F. M. Ausubel. 1997. Arabidopsis enhanced disease susceptibility mutants exhibit enhanced susceptibility to several bacterial pathogens and alterations in *PR-1* gene expression. Plant Cell 9:305-316.

129. Reuber, T. L. and F. M. Ausubel. 1996. Isolation of Arabidopsis genes that differentiate between resistance responses mediated by the *RPS2* and *RPM1* disease resistance genes. Plant Cell 8:241-249.

130. Broekaert, W., F. R. G. Terras, B. P. A. Cammue and R. W. Osborn. 1995. Plant defensins: novel antimicrobial peptides as components of the host defense system. Plant Physiol. 108:1353-1358.

131. Gopalan, S., W. Wei and S. Y. He. 1996. hrp gene-dependent induction of *hin1*: a plant gene activated rapidly by both harpins and the *avrPto* gene-mediated signal. Plant J. 10:591-600.

132. Daniels, C. H., B. Fristenky, W. Wagoner and L. A. Hadwiger. 1988. Pea genes associated with non-host disease resistance to *Fusarium* are also active in race-specific disease resistance. Plant Mol. Biol. 8:309-316.

133. Hipskind, J. D., R. L. Nicholson and P. B. Goldsbrough. 1996. Isolation of a cDNA encoding a novel leucine-rich repeat motif from *Sorghum bicolor* inoculated with fungi. Mol. Plant-Microbe Interact. 9:819-825.

Index

ABC (ATP binding cassette) transporter, 17
abscisic acid, 201, 259
Acacia, 168
  albida, 16
Acer pseudoplatanus, 102
acetosyringone, 5, 164
acetyl-CoA, 14
O-acetyltransferase, 16
aconitase, 228
actinomycetes, 163, 168
active oxygen, (AO), 81, 121, 145, 227
  burst, 81, 91, 145
  definition, 82
  direct effects, 97
    antimicrobial effects, 98
    cell wall changes, 98
  measurement, 90
  production, 91
acyl carrier protein, 13
acylated homoserine lactone, 232
adhesins, 231
AFLP (*see* selective restriction fragment
  amplification)
agamous, 198
Agrobacterium, 164, 172, 184, 189, 197
  rhizogenes, 169, 190, 204
  tumefaciens, 169, 190, 198, 260
air pollution, 81
alfalfa (*see also Medicago sativa*), 167, 168,
  189, 255
Alkaligenes, 236
alkaline phosphatase, 133
Allium, 127
allopurinol, 107
α-tocopherol, 85
Alternaria, 123, 127, 128
  alternata f. sp. *lycopersici*, 106
alternative oxidase, 71
anguibactin, 220
angular leaf spot disease, 50
anthocyanin, 196
anthracnose, 201
antibiotics, 121, 232
antioxidant, 81, 93, 99
antisense, 204

Antirrhinum, 196
AP2-like protein, 261
apetela, 198, 262
apigenin, 5
apoplast, 42
apoptosis, 106
APS-kinase, 15
Arabidopsis thaliana, 42, 92, 99, 133, 169,
  170, 172, 178, 181, 185, 189, 190, 194,
  196, 197, 198, 199, 228, 235, 237, 238,
  240, 255, 262, 264
arabinose, 14
arabinosyl transferase, 16
arachidonic acid, 103
arbitrarily primed-PCR (AP-PCR), 177, 179
Ascochyta rabiei, 126
ascomycete, 59
ascorbate/glutathione cycle, 88, 99
ascorbate peroxidase, 88, 99
ascospores, 59, 67
as-1-like elements, 260, 263
asparagus, 258
ATPase, 134, 141, 144
ATP-sulfurylase, 15
avenic acid, 230
avr/pth family
  avrB6 alleles, 45
  cloned members, 44
  host species-specific, 47
  involved in dispersal, 47
  molecular analysis, 48
  name, 43
  not required by all Xanthomonads, 46
  pthA alleles, 46
  required for some diseases, 43
avr genes, 39, 41, 51, 96, 164
  avrA, 96
  avrB, 42
  avrB4, 48, 50
  avrB6, 42, 45, 48, 49, 50, 51
  avrB7, 48
  avrB101, 48
  avrB102, 48
  avrBIn, 48
  avrBs2, 42

RNA, double-stranded (*continued*)
  CHV2-NB58, 68
  CHV3-GH2, 68
RNA helicase, 67
RNA polymerase, 223
RNA profiling (*see* differential display)
Robinia nigra, 168
rolB gene, 169
root exudates, 5
root hair, 1
  curling (hac), 19, 173
  deformation (had), 19, 166

S-adenosylmethionine, 13
salicylate hydrolase (*nahG*), 239
salicylhydroxamic acid, 147
salicylic acid, 100, 237, 239, 257, 260
salinity, 81
saponins, 121
SAR (*see* systemic acquired resistance)
SA-response element (SARE), 260
scallion bulb rot, 127
scanning electron microscope, 135, 234
selective restriction fragment amplification
  (SRFA), 178, 182, 189
SEM (*see* scanning transmission electron
  microscope)
sequence characterized amplified regions
  (SCARS), 179
Sesbania rostrata, 1, 167, 202
sesquiterpenoid, 132
shikimate, 252, 253
siderophore, 219, 224, 229, 232, 234
  negative mutants, 233, 241
  phytosiderophore, 230
  regulation of biosynthesis, 221
signal transduction, 91, 164, 201, 224, 226
simple sequence repeats (SSR), 178, 189
singlet oxygen, 83, 86, 103, 108
siratro, 6, 7, 18, 20, 189
SOD (*see* superoxide dismutase)
soybean (*see also Glycine max*) , 92, 94, 98, 99,
  131, 135, 136, 138, 143, 168, 172, 175,
  178, 181, 183, 186, 189, 193, 195, 204,
  253, 261
sporangia, 125
Stemphylium sarcinaeforme, 127
stilbene synthase, 258
strawberry, 127
Streptomyces hygroscopicus, 16
sulphation, 17
sulphotransferase, 15, 16
supernodulation, 170, 172, 173, 181, 182, 186
superoxide, 83, 84, 101, 108, 125, 128, 146,
  227
superoxide dismutase, 84, 88, 99, 101, 147,
  227, 253
supprescins, 126, 136
suppressive soil, 218, 232, 236

suppressor, 121, 123
  from fungi, 124
surfactant, 232
sym genes, 181
symbiosis, 163, 182
symbiosome, 185, 204
symbiotic plasmid (pSym), 2
synteny, 186
syrM (*see nod* genes)
systemic acquired resistance, 100, 169, 237,
  264

T-DNA, 167, 169, 170, 184, 189, 198
telomere, 195
Tephrosia vogeli, 8
thermocycler, 176, 202
Thielaviopsis basicola, 237
thylakoid membranes, 107
tobacco, 94, 97, 98, 100, 105, 129, 194, 196,
  197, 237, 238, 253, 260, 263, 265
tobacco mosaic virus (TMV), 197, 259, 264
tobacco necrosis virus (TNV), 238
tomato, 46, 94, 96, 105, 106, 125, 127, 129,
  181, 185, 189, 194, 197, 229
TonB, 224, 226
touch-down PCR, 186
toxin, 86, 106, 123, 164, 196, 230
toxin reductase, 196
trans-acting nuclear factors, 132, 257
transferrin, 228
transformation, 189, 195
transmembrane signaling, 137
transport, 219, 224, 226
transposon, 167, 170, 184, 189, 190, 195, 196,
  222
Trifolium
  pratense, 128
  repens (*see also* clover), 6, 18, 168
Trigonella, 1, 6
triplet state, 82
trisomic, 193
two-component regulator, 225
tyrosine phosphatase, 169

ultraviolet light, 81, 260
umbelliferone, 5
undecaprenyl phosphate, 17
urate oxidase, 107
uricase, 184
Uromyces phaseoli, 107

vanadate, 147
vanillin, 6
vegetative incompatibility genes (*see*
  vic genes)
Vibrio, 220
  anguillarum, 220
vic genes, 73
Vicia sativa, 5, 18